Discrete Mathematics and Theoretical Computer Science

Springer

London
Berlin
Heidelberg
New York
Hong Kong
Milan
Paris
Tokyo

Jürg Kohlas

Information Algebras

Generic Structures For Inference

Springer

Jürg Kohlas, MSc, PhD
Department of Informatics, University of Fribourg, CH 1700 Fribourg,
Switzerland

British Library Cataloguing in Publication Data
Kohlas, J.
   Information Algebras : generic structures for inference. –
   (Discrete mathematics and theoretical computer science)
   1.Algebra 2.Information theory – Mathematics 3. Computer
   science – Mathematics
   I.Title
   512'.001154
   ISBN 1852336897

Library of Congress Cataloging-in-Publication Data
A catalog record for this book is available from the Library of Congress

Discrete Mathematics and Theoretical Computer Science Series ISSN 1439-9911
ISBN 1-85233-689-7 Springer-Verlag London Berlin Heidelberg
a member of BertelsmannSpringer Science+Business Media GmbH
http://www.springer.co.uk

Typesetting: Camera ready by author
Printed and bound at the Athenæum Press Ltd., Gateshead, Tyne & Wear
34/3830-543210 Printed on acid-free paper SPIN 10890758

Preface

What is the book about?

Information is a central concept of science, especially of Computer Science. The well developed fields of statistical and algorithmic information theory focus on measuring information content of messages, statistical data or computational objects. There are however many more facets to information. We mention only three:

1. Information should always be considered relative to a given specific question. And if necessary, information must be focused onto this question. The part of information relevant to the question must be extracted.

2. Information may arise from different sources. There is therefore the need for aggregation or combination of different pieces of information. This is to ensure that the whole information available is taken into account.

3. Information can be uncertain, because its source is unreliable or because it may be distorted. Therefore it must be considered as defeasible and it must be weighed according to its reliability.

The first two aspects induce a natural algebraic structure of information. The two basic operations of this algebra are combination or aggregation of information and focusing of information. Especially the operation of focusing relates information to a structure of interdependent questions. This book is devoted to an introduction into this algebraic structure and its variants. It will demonstrate that many well known, but very different formalisms such as for example relational algebra and probabilistic networks, as well as many others, fit into this abstract framework. The algebra allows for a generic study of the structure of information. Furthermore it permits to construct generic architectures for inference.

The third item above, points to a very central property of real-life information, namely uncertainty. Clearly, probability is the appropriate tool to describe and study uncertainty. Information however is not only numerical. Therefore, the usual concept of random variables is not sufficient to study uncertain information. It turns out that the algebra of information is the natural framework to model uncertain information. This will be another important subject of this book.

Surely the algebra of information has links to statistical and algorithmic information theory. This very important synthesis is yet to be worked out. It promises to be a fruitful area of research.

This book has its root in the past European Basic Research activity no. 6156 DRUMS 2 (Defeasible Reasoning and Uncertainty Management). In this project, scientists from different fields like philosophy, formal logic, probability theory, statistics, evidence theory and fuzzy systems met to discuss approaches to uncertainty. It was in this stimulating environment that the desire for a unifying treatment of apparently quite different formalisms arose. The foundation for it was fortunately given in a basic axiomatic system proposed in (Shenoy & Shafer 1990). This system in turn was derived from a fundamental paper on probabilities on graphical structures (Lauritzen & Spiegelhalter, 1988). For more information about this background we refer to Chapter 1.

Organization

The dependencies of the sections are shown in the chart on the next page. The material covered is organized into seven chapters:

- Chap. 1 gives a more detailed overview of the subject of this book and its background.

- Chap. 2 introduces the basic algebraic structure and presents several important examples.

- Chap. 3 develops the algebraic theory of information.

- Chap. 4 presents generic architectures for inference.

- Chap. 5 discusses the notion of conditional independence in the abstract frame of information algebra.

- Chap. 6 is devoted to a discussion of idempotent information algebras and information systems.

- Chap. 7 treats a general model of uncertain information.

As the dependency graph shows, the book need not be read in a sequential manner. Chap. 2 is basic and needed for all other parts of the book. Readers interested in computational questions may continue with Chapter 4. However, for the more advanced architectures, Chapter 3 is needed. Chapter 3 and 4 are a prerequisite for conditional independence, Chapter 5. Readers interested more in the algebraic theory can concentrate on Chapters 3 and 6. Finally the discussion of uncertain information in Chapter 7 is based on Chapter 6.

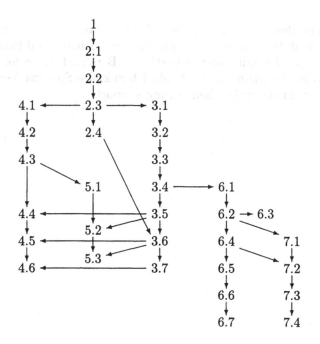

Acknowledgments

The participation to the DRUMS 2 project mentioned above, was made possible by financial support from Swiss Federal Office for Science and Education. I thank the participants in this projects for numerous discussions. They are too numerous for all to be listed. But I would like to mention more particularly Philippe Smets (the project leader), Philippe Besnard, Steffen Lauritzen, Serafin Moral and Nic Wilson. The research started at DRUMS 2 was subsequently supported by several grants from the Swiss National Foundation for Research. I am grateful to the University of Fribourg, Switzerland, which gave me the opportunity and the freedom to write this book. A large part of the book was written during my sabbatical leave at the University of Kansas, Lawrence, USA. This visit was supported by a grant from the Ronald G. Harper Chair of Artificial Intelligence at the University of Kansas Business School. I thank Prof. Prakash Shenoy for offering me this opportunity. He is also one of the founders of the theory developed in this book and I thank him for the many discussions and exchanges of ideas.

I would also like to thank my collaborators who participated in the research leading to this book. At the first place my thanks are due to Robert Staerk, who made essential contributions to Chapter 6, which were not published. Bernhard Anrig, Dritan Berzati, Rolf Haenni, Norbert Lehmann and Paul-André Monney all made important contributions to probabilistic argumentation systems, which were our test field for the generic inference architecture and which laid also the foundation for Chapter 7 on uncertain information. I thank Bernhard Anrig, Dritan Berzati, Norbert Lehmann and Patrick Herte-lendy also for proof reading and in particular Bernhard Anrig for extremely valuable help in typesetting of this book. I thank also Springer-Verlag and in particular Stephen Bailey for their editing support.

Jürg Kohlas

Contents

1 Introduction

There are many different formalisms for representing knowledge or information. However a few elementary features are found with every formalism. Knowledge or information comes in chunks or piecewise. These pieces of information are then somehow combined or aggregated to represent the whole of the information. Inference then usually means to extract from the whole of the knowledge the part relevant to a given problem or question. This can be called focusing of information. This observation leads naturally to considering an algebraic structure, composed of a set of "pieces of information" which can be manipulated by two operations: combination to aggregate pieces of information, and focusing to extract the part of information related to a certain domain. The goal of this book is to formally describe this algebraic structure and to study it in some detail.

Probably the most popular instance of such a structure comes from probability theory. A multidimensional probability density over some set of variables can often be decomposed into a product of densities, defined on smaller sets of variables. This factorization reflects the structure of conditional independence between variables. In many cases these independence structures can be represented graphically. Lauritzen and Spiegelhalter in a pioneering work (Lauritzen & Spiegelhalter, 1988) showed how such a factorization can be used to compute marginals of the multidimensional probability density by so-called "local computation". The main advantage of this approach is that the multidimensional density must never be computed in its full dimension, but that rather the computation can be carried out "locally" on the much smaller dimensions of the factors. Only this possibility may make computation feasible. Shenoy and Shafer (Shenoy, 1989; Shenoy & Shafer 1990) introduced for the first time an abstract, axiomatic system capturing the essence for the type of local propagation introduced by (Lauritzen & Spiegelhalter, 1988). They pointed out, that many other formalisms satisfy also the axioms needed for local propagation. The next well-known example of such a formalism is probably provided by belief functions in the sense of Dempster-Shafer theory of evidence. In numerical analysis, sparse matrix techniques can be subsumed under these axioms. Looking through the literature of inference in artificial intelligence and elsewhere, it seems that the structure defined by Shenoy and Shafer is very often implicitly exploited, but mostly without explicit reference to it. Also the associated computational possibilities seem to be recognized over and over again. Therefore, it seems to be the time to formulate this structure explicitly and to study it in detail as a general and generic structure for inference.

A slightly changed version of the axiomatic formulation of Shenoy and Shafer is the starting point for this book. The mathematical structure defined by these axioms is called here *valuation algebra*. It will be studied both from an algebraic point of view as well as from a computational one. Special attention is paid to the idempotent variant of valuation algebras. These structures are called *information algebras*. Idempotency introduces a lot of additional structure. Indeed it places information algebra in the neighborhood of other theories of "information" like relational algebra, domain theory and the theory of information systems.

The elements of valuation algebras are called "valuations". This term is used in mathematics to denote some generalizations of measures, especially probability measures. Here, valuations are introduced in a similar spirit, although in a strict technical sense they do not correspond exactly to the concept used elsewhere in the theory of valuations.

The overview of the book is as follows: In Chapter 2 the axioms of (labeled) valuation algebras are introduced and a few elementary consequences are presented. Essentially, valuation algebras are commutative semigroups with an additional operation of marginalization, representing focusing. These two operations are linked by a kind of distributive law, which is fundamental for the theory. Several examples or instances of valuation algebras are described. These include systems inspired by probability theory like discrete probability potentials as used for example in Bayesian networks, or Gaussian potentials, motivated by normal regression theory, by Kalman filters and the like. In this view combination is represented by multiplication of densities, including conditional densities, and focusing corresponds to marginalization, i.e. summation or integration of densities. But the examples include also non-probabilistic systems. Relational algebra is an important example, and a very basic one as will be seen later. Here combination is the join and focusing corresponds to projection. Possibility potentials and Spohn potentials represent systems related to fuzzy systems. t-norms provide for a variety of combination operators in this field. This shows that valuation algebras indeed cover a wide range of interesting and useful systems for inference. For later reference another, weaker axiomatic system, allowing for partial focusing only, is also introduced.

In the following Chapter 3 the algebraic theory of valuation algebras is developed to some extent. Some concepts from universal algebra are introduced. In particular, there is an important congruence, which allows to group together valuations representing the "same" information. The corresponding quotient algebra gives us then an alternative way to represent the valuation algebra, in a "domain-free" form. Inversely, from the domain-free version we can reconstruct the original "labeled" algebra. This provides us with two equivalent ways to look at a valuation algebra, which proves very valuable. Essentially, the labeled point of view is more appropriate for computational and also for some semantical issues. The domain-free variant is generally more convenient for theoretical considerations. An important issue is the question of division, which, in general, is not defined in semigroups. But from semigroup theory we know that there are commutative semigroups which are a union of groups. We adapt in Chapter 3 this semigroup theory to *regular* valuation algebras. More generally, there are semigroups which are embedded (as a semigroup) in a semigroup which is a union of groups. This generalizes to *separative* valuation algebras. The issue of division is important for computation, but also for the concept of independence, as will be seen in Chapters 4 and 5. It is shown for example that, depending on the t-norm, possibility potentials may or may not allow for some form of division. And this makes a lot of difference both from a computational as well as from a semantical point of view.

There are several architectures for local computation known from the literature, especially for Bayesian networks. These will be presented in Chapter 4. Some of these architectures are valid for any valuation algebra. Others use some form of division. It is shown that these latter architectures can be used especially for regular valuation algebras. In the case of separative algebras, the additional problem arises that in the embedding union of groups, marginalization is only partially defined. It turns out that this partial marginalization is sufficient to apply the architectures with division also in the case of separative valuation algebras. In these algebras scaling is usually needed for semantical reasons. The architectures with division allow for an efficient organization of scaling.

The local character of computation in probability theory is closely related to conditional independence of variables. This concept can be generalized to valuation algebras in general and to regular and separative algebras in particular. This is discussed in Chapter 5. In probability theory conditional independence is also closely related to conditional probability, that is to division. This indicates that the concept of conditional independence depends very much on the structure of the valuation algebra. In a valuation algebra without division not very much can be said about conditional independence. Regular algebras on the other hand maintain many of the properties of conditional independence known from probability theory. In particular a concept of *conditional* can be defined which resembles a conditional density in probability theory. Separative valuation algebras are somewhere in between. Conditionals may also be defined. But they do not necessarily have all the properties of a conditional density. This explains for example why conditional belief functions are not of much interest in Dempster-Shafer theory of evidence.

The last two chapters are devoted to valuation algebras which are *idempotent*. This means that a valuation combined with a focused version of itself does not change the first valuation. This is an essential ingredient of "information". A piece of information combined with part of it gives nothing new. That is why these idempotent valuation algebras are called *information algebras* (Chapter 6). Idempotency allows to introduce a partial order between pieces of information, representing the relation of more (or less) informative information. This order is very essential for the theory. Information algebras become thus semilattices. With the aid of this partial order we can express the idea of "finite" elements, which serve to approximate general, "non-finite" elements. This leads to *compact* information algebras, which are in fact algebraic lattices (but with an additional operation of focusing). And this brings information algebras into the realm of domain theory. In fact, we show that an *information system*, a concept introduced by Scott into domain theory and adapted here to the needs of our theory, induces an information algebra. Inversely, any information algebra determines an information system. Thus information systems are an alternative way to represent information algebras. And they provide for a very important approach to information algebras, especially in practice. Propositional logic, systems of linear equations or linear inequalities are examples of information systems. In fact, information systems link information algebras to logic. Via the information algebras they induce, they can be treated by architectures of local computation as introduced in Chapter 4. Indeed, since idempotency makes division trivial, the architectures can even be simplified for information algebras. Information algebras can, on the other hand, also be related to relational algebra in general. For this purpose an abstract notion of tuple and relation is introduced. Information algebras can then be embedded into an abstract relational algebra over abstract relations. We call this a *file system*. So file systems provide for a second alternative representation of information algebras. In short: a piece of information may be looked at as a file (set) of tuples, that is as a relation. Or it may be looked at as a set of sentences expressed in some logic. Information can thus alternatively be described in a relational or in a logical way.

Information may be uncertain. So it is natural to ask how uncertainty can be represented in information algebras. This can be done by random variables with values in information algebras (Chapter 7). Random variables represent sources of evidence or information. Accordingly an operation of combination and another one of focusing can be defined. Not surprisingly, this leads to an information algebra of random variables. If we look at the distribution of these random variables, we find belief functions (here called support functions) in the sense of Dempster-Shafer theory of evidence. Therefore, we claim that information algebras are the natural mathematical framework for Dempster-Shafer theory of evidence. The usual set-theoretic framework of this theory is only a particular case of information algebras. But for example belief functions on linear manifolds (or systems of linear equations) are better treated in the framework of information algebra than in a purely set-theoretic setting. If information systems are used to express uncertainty, then this leads to assumption-based reasoning and *probabilistic argumentation systems*. Therefore, this is another approach to Dempster-Shafer theory of evidence, and a very practical one indeed. If an appropriate notion of "independence" between sources of evidence is introduced, then combination becomes the well-known rule of Dempster (expressed in information algebras of course). The corresponding algebra of "independent" belief functions is a valuation algebra.

This book depends on many publications and also on numerous personal discussions during different European research projects and other contacts. I want to give credit to the most important documents which helped to shape this book: The axioms and the first part of Chapter 3 (domain-free algebras) are largely based on an unpublished paper by Shafer (Shafer, 1991). The second part of Chapter 3 related to division has been motivated by the paper (Lauritzen & Jensen, 1997). There the author found the references to semi-group theory which are essential for the development of regular and separative valuation algebras. The chapter on local computation, Chapter 4, is based on the various original papers, especially (Lauritzen & Spiegelhalter, 1988; Jensen, Lauritzen & Olesen, 1990), where the different architectures were presented (for the case of probability networks) and on many personal discussions with Prakash Shenoy. Part of this chapter is also based on a chapter "Computation in Valuation Algebras", written by Prakash Shenoy and the author (Kohlas & Shenoy, 2000), in (Gabbay & Smets, 2000). The chapter on conditional independence, Chapter 5, is motivated by (Shenoy, 1997 a). It has been adapted to the axiomatic system used in this book and makes use of the results about regular and separative valuation algebras, as well as of the concept of valuation algebras with partial marginalization. Although these parts draw heavily on former work, the author hopes that there are sufficient new elements in this book to make these chapters interesting even for the reader which knows already the papers mentioned above.

Chapters 6 and 7 draw largely on unpublished material. Special credit is due to Robert Staerk, who contributed to the development of information algebra, and who, among other things, invented the file systems (Kohlas & Staerk, 1996). We remark that the cylindric algebras treated in (Henkin, Monk & Tarski, 1971) are special classes of information algebras, related to first order logic. Furthermore, classification domains as introduced and discussed in (Barwise & Seligman, 1997) seem to bear interesting connections to information algebras. Structures similar to information algebras are used also to study modules and modularity (Bergstra, et. al., 1990; Renardel de Lavalette, 1992). (Mengin & Wilson, 1999) discuss the use of the structure of information algebras for logical deduction.

For the uncertainty in information systems, Chapter 7, the basic literature on Dempster-Shafer theory of evidence was of course important, especially (Dempster, 1967; Shafer, 1973; Shafer, 1976; Shafer, 1979). We mention that there is an alternative, non-probabilistic approach to evidence theory (Smets, 1998). Partially this chapter is based on some former papers of the author (Kohlas, 1993; Kohlas, 1995; Besnard & Kohlas, 1995; Kohlas, 1997). These papers however were not based on information algebras.

This book is, as far as the author knows, the first systematic treatment of valuation algebras from an algebraic point of view. This does of course not mean that the subject is treated in an exhaustive way. Not nearly so. Many questions remain open. Here are only a few of them: What is the full structure theory of valuation and information algebras (what types of these algebras exist and how are they characterized)? What is the exact relation between information algebras and logic, which logic lead to information algebras? Which valuation algebras, representing uncertainty formalisms, can be induced from an algebra of random variables with values in an information algebra? How is Shannon's theory of information and algorithmic information theory related to information algebra, and especially to Dempster-Shafer theory? It is the author's hope that this book may arouse interest in the subject and serve to unify and promote efforts in developing inference schemes in different fields, using different formalisms.

2 Valuation Algebras

2.1 The Framework

Valuation algebras represent the basic unifying structure behind the inference mechanisms for many different calculi of uncertainty or more generally for many formalisms of treating knowledge or information. In this chapter we introduce this concept and give some examples of valuation algebras. The following chapter is devoted to a more thorough discussion of the algebraic theory of valuation algebras.

We consider reasoning and inference to be concerned with variables with unknown values. Thus, the first ingredients for valuation algebras are *variables, frames and configurations*. Variables will be designated by capitals like $X, Y, \ldots$. The symbol Ω_X is used for the set of possible values of a variable X, the frame of X . Often, but not always, these sets are assumed to be finite. We are concerned with finite sets of variables and we use lower-case letters such as $x, y, \ldots, s, r, t, \ldots$ to denote sets of variables.

Given a nonempty set s of variables, let Ω_s denote the Cartesian product of the frames Ω_X of the variables $X \in s$,

$$\Omega_s = \prod_{X \in s} \Omega_X. \tag{2.1}$$

Ω_s is called the frame of the set of variables s . The elements of Ω_s are called configurations of s. We use lower-case, bold-faced letters such as $\mathbf{x}, \mathbf{y}, \ldots$ to denote configurations .

It is convenient to extend this terminology to the case where s is empty. We adopt the convention that the frame for the empty set consists of a single configuration and we use the symbol $\diamond$ to name that configuration ; $\Omega_\emptyset = \{\diamond\}$.

If $\mathbf{x}$ is a configuration of s and $t \subset s$, then $\mathbf{x}^{\downarrow t}$ denotes the projection of $\mathbf{x}$ to the subset t of variables . That is $\mathbf{x}^{\downarrow t}$ contains only the components corresponding to variables in the set t. If $t = \emptyset$, then $\mathbf{x}^{\downarrow \emptyset} = \diamond$. Sometimes, in order to stress the decomposition of $\mathbf{x}$ into subsets of components belonging to the subsets t and $s - t$ of s we write $\mathbf{x} = (\mathbf{x}^{\downarrow t}, \mathbf{x}^{\downarrow s-t})$.

If A is a subset of Ω_s and again $t \subset s$, then $A^{\downarrow t}$ denotes the projection of A to the subset of variables t ,

$$A^{\downarrow t} \;=\; \{ \mathbf{x}^{\downarrow t} : \mathbf{x} \in A \}. \tag{2.2}$$

On the other hand, if $t \supset s$, then $A^{\uparrow t}$ denotes the cylindric extension of A from Ω_s to Ω_t . That is,

$$A^{\uparrow t} \;=\; \{ \mathbf{x} \in \Omega_t : \mathbf{x}^{\downarrow s} \in A \}. \tag{2.3}$$

The primitive elements of a valuation algebra are *valuations* . Intuitively, a valuation represents some knowledge about the possible values of a set s of variables. So, any valuation refers to some determined set of variables. Given a (possibly empty) set s of variables, there is a set Φ_s of valuations. The elements of Φ_s are called valuations for s. Let r be the set of all variables considered. Then Φ denotes the set of all valuations, i.e.

$$\Phi \;=\; \bigcup_{s \subseteq r} \Phi_s \tag{2.4}$$

We use lower-case Greek letters such as $\phi, \psi, \ldots$ to denote valuations. Let finally D denote the lattice of subsets of r (The powerset of r). We call D also the lattice of domains of the valuations.

If ϕ is a valuation for s, then we write $d(\phi) = s$ and call s the domain for ϕ. d is a mapping of Φ onto the power set D; it is called *labeling* and $d(\phi)$ is also called the *label* of ϕ .

There are two operations assumed to be defined for valuations. The *combination* is a binary operation $\Phi \times \Phi \to \Phi$ denoted by $(\phi, \psi) \mapsto \phi \otimes \psi$. It represents aggregation of knowledge. The combined knowledge bears on the union of the domains of the factors, thus we assume $d(\phi \otimes \psi) = d(\phi) \cup d(\psi)$. This operation is furthermore assumed to be *associative* and *commutative*,

$$\begin{aligned}
(\phi_1 \otimes \phi_2) \otimes \phi_3 &= \phi_1 \otimes (\phi_2 \otimes \phi_3), \\
\phi_1 \otimes \phi_2 &= \phi_2 \otimes \phi_1.
\end{aligned} \tag{2.5}$$

Associativity means that it does not matter in what order the operation of combination is performed. This is an important property one surely wants to be assured for every sensible system to combine knowledge or information. It permits to write simply $\phi_1 \otimes \phi_2 \otimes \cdots \otimes \phi_{n-1} \otimes \phi_n$ without parentheses, when n valuations have to be combined. A structure with an associative operation is called a *semigroup* . Commutativity means that it does not matter whether the first valuation is combined with the second or the second with the first one. Again this seems to be a sensible assumption for combining knowledge or information. We assume thus Φ to be a commutative semigroup under the operation of combination .

Any set of valuations Φ_s with domain s is itself a semigroup, since $d(\phi \otimes \psi) = s$ if $d(\phi) = d(\psi) = s$. It is a sub-semigroup of Φ. In many (but not all) cases there is a neutral element e_s for every domain $s \subseteq r$ which has the property that $\phi \otimes e_s = e_s \otimes \phi = \phi$ for all valuations $\phi \in \Phi_s$. The elements e_s represent intuitively something like empty, vacuous information or ignorance with respect to the variables s. Note however that the elements e_s are not neutral elements in Φ, that is $\phi \otimes e_s \neq \phi$ in general. However we assume that the combination of these neutral elements yield again neutral elements, $e_s \otimes e_t = e_{s \cup t}$, in accordance with the labeling rule for combination.

The second operation is *marginalization* . It is defined for all $\phi \in \Phi$ and $x \in D$ such that $x \subseteq d(\phi)$. It is denoted by $(\phi, x) \mapsto \phi^{\downarrow x}$. This operation of marginalization represents focusing of the knowledge captured by ϕ for $d(\phi)$ to the smaller domain x. It is a reduction or extraction of information. Accordingly we assume that $d(\phi^{\downarrow x}) = x$.

This is the basic structure we are going to study throughout this book. In the next section we shall impose further restrictions on these operations, which are important for simplifying computation. Its usefulness will be proved by the wide variety of examples which correspond to this algebraic structure.

Let's however consider first a small example which illustrates abstract modeling with valuations and which also introduces the basic computational problem of inference with valuations.

Here is the example, which is drawn from (Lauritzen & Spiegelhalter, 1988):

> Shortness of breath (dyspnoea) may be due to tuberculosis, lung cancer or bronchitis or none of them, or more than one of them. A recent visit to Asia increases the chances of tuberculosis, while smoking is known to be a risk factor for both lung cancer and bronchitis. The results of a single chest X-ray do not discriminate between lung cancer and tuberculosis, as neither does the presence or absence of dyspnoea.

In this example there are eight binary variables - A (visit to Asia), S (Smoking), T (Tuberculosis), L (Lung cancer), B (Bronchitis), E (Either tuberculosis or lung cancer), X (positive X-ray) and D (Dyspnoea). Each variable has two possible values, *true* or *false*, which we represent by 1 and 0. The frame of each variables is thus $\{0, 1\}$.

Prior to any observations, there are eight valuations - α for $\{A\}$, σ for $\{S\}$, τ for $\{A, T\}$, λ for $\{S, L\}$, β for $\{S, B\}$, ϵ for $\{T, L, E\}$, ξ for $\{E, X\}$ and δ for $\{E, B, D\}$. These valuations represent the basic medical knowledge. Additional observations related to a specific patient will also be modeled by valuations. Suppose for example, a patient is observed, who visited Asia recently and is suffering from dyspnoea. These two observations can be modeled by two valuations o_A for $\{A\}$ and o_D for $\{D\}$ respectively.

A graphical display of such a set of valuations is shown in Fig. 2.1. There, variables are represented by circular nodes, and valuations are represented by square nodes. Each valuation is connected by an edge to each of the variables of its domain. Such a bipartite graph is called a *valuation network* .

Note that the example has been formulated without selecting specific valuations. We did not say what exactly the eight valuations of the problem are,

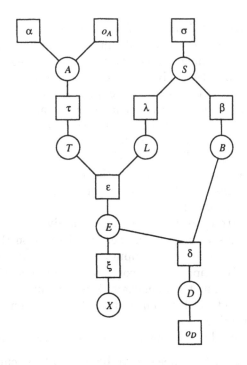

Figure 2.1: The valuation network for the medical example.

we leave it open which specific knowledge or uncertainty formalism will finally be used. It is a generic problem formulation.

Given this knowledge, one wants usually to focus on one or several variables and to know what can be concluded about the possible values of these variables. In the example the variables of interest will be the possible sicknesses, represented by T (tuberculosis), L (lung cancer) and B (bronchitis). Thus one wants for example to compute the marginal of the combination of all valuations to the variable T, that is to focus the whole information on the question whether there is tuberculosis or not,

$$(\alpha \otimes \sigma \otimes \tau \otimes \lambda \otimes \beta \otimes \epsilon \otimes \xi \otimes \delta \otimes o_A \otimes o_D)^{\downarrow\{T\}}. \tag{2.6}$$

We shall see in the examples later, that the complexity of computation for the operations of combination and marginalization as well as the memory space tends to increase exponentially with the size of the domains of the valuations. This limits the possibility of an explicit computation of (2.6) since in combining one valuation after the other the domain of the result increases steadily due to the labeling rule. Its size becomes in general very fast infeasible. So the direct computation of (2.6) will, in most cases, be impossible.

Fortunately there are conditions which are satisfied by many systems, and which will be formulated in the next section, which allow to organize the computation of (2.6) in such a way that all operations take place on domains not essentially larger than the domains of the valuations appearing in (2.6). This is called *local computation* and will be one of the main subjects of this book (see Chapter 4).

2.2 Axioms

We define now formally a valuation algebra by a system of axioms. Thus, let Φ be a set of valuations with domains in D, the lattice of subsets of the set r of variables. Suppose there are three operations defined :

1. *Labeling*: $\Phi \rightarrow D; \phi \mapsto d(\phi)$,

2. *Combination*: $\Phi \times \Phi \rightarrow \Phi; (\phi, \psi) \mapsto \phi \otimes \psi$,

3. *Marginalization*: $\Phi \times D \rightarrow \Phi; (\phi, x) \mapsto \phi^{\downarrow x}$, for $x \subseteq d(\phi)$.

We now impose the following axioms on Φ and D :

1. *Semigroup*: Φ is associative and commutative under combination. For all $s \in D$ there is an element e_s with $d(e_s) = s$ such that for all $\phi \in \Phi$ with $d(\phi) = s, e_s \otimes \phi = \phi \otimes e_s = \phi$.

2. *Labeling*: For $\phi, \psi \in \Phi$,

$$d(\phi \otimes \psi) \;=\; d(\phi) \cup d(\psi). \tag{2.7}$$

3. *Marginalization*: For $\phi \in \Phi, x \in D, x \subseteq d(\phi)$,

$$d(\phi^{\downarrow x}) \;=\; x. \tag{2.8}$$

4. *Transitivity*: For $\phi \in \Phi$ and $x \subseteq y \subseteq d(\phi)$,

$$(\phi^{\downarrow y})^{\downarrow x} \;=\; \phi^{\downarrow x}. \tag{2.9}$$

5. *Combination*: For $\phi, \psi \in \Phi$ with $d(\phi) = x, d(\psi) = y$,

$$(\phi \otimes \psi)^{\downarrow x} \;=\; \phi \otimes \psi^{\downarrow x \cap y}. \tag{2.10}$$

6. *Neutrality*: For $x, y \in D$,

$$e_x \otimes e_y \;=\; e_{x \cup y}. \tag{2.11}$$

Axiom (1) says that Φ is a commutative semigroup under combination and that a neutral element is adjoined for every sub-semigroup Φ_s of valuations for s. Note that the neutral element is unique. If there would be a second one, say e_s', then we have $e_s' = e_s \otimes e_s' = e_s$. The labeling axiom says that under combination the domains of the factors are joined. The marginalization axioms says that marginalization to a domain x yields a valuation for this domain. The transitivity axiom tells us that marginalization to some domain x can be done in two (or more) steps by intermediate marginalization to intermediate domains. The combination axioms assures that in order to marginalize to a domain of one of the factors of a combination it is not necessary to compute first the combination, but that we can as well first marginalize the other factor to the domain of the first one and then combine the two valuations. It means that we need in fact not leave the domains of the two factors to compute the marginal of the combination. It is essentially this axiom which allows local computation. The neutrality axiom finally specifies combination of neutral elements to give neutral elements.

A system (Φ, D) of valuations with the operations of labeling, combination and marginalization satisfying these axioms is called a *valuation algebra* or more precisely, a *labeled* valuation algebra.

We now state a few elementary consequences of these axioms.

Lemma 2.1   *1. If $\phi, \psi \in \Phi$ such that $d(\phi) = x, d(\psi) = y$, then*

$$(\phi \otimes \psi)^{\downarrow x \cap y} \;=\; \phi^{\downarrow x \cap y} \otimes \psi^{\downarrow x \cap y}. \tag{2.12}$$

2. If $\phi, \psi \in \Phi$ such that $d(\phi) = x, d(\psi) = y, z \subseteq x$, then

$$(\phi \otimes \psi)^{\downarrow z} \;=\; (\phi \otimes \psi^{\downarrow x \cap y})^{\downarrow z}. \tag{2.13}$$

3. If $\phi, \psi \in \Phi$ such that $d(\phi) = x, d(\psi) = y, x \subseteq z \subseteq x \cup y$, then

$$(\phi \otimes \psi)^{\downarrow z} \;=\; \phi \otimes \psi^{\downarrow y \cap z}. \tag{2.14}$$

Proof. (1) By the transitivity and combination axioms

$$\begin{aligned}
(\phi \otimes \psi)^{\downarrow x \cap y} &= ((\phi \otimes \psi)^{\downarrow x})^{\downarrow x \cap y} \\
&= (\phi \otimes \psi^{\downarrow x \cap y})^{\downarrow x \cap y} \\
&= \phi^{\downarrow x \cap y} \otimes \psi^{\downarrow x \cap y}.
\end{aligned} \tag{2.15}$$

(2) By the transitivity and combination axioms

$$\begin{aligned}
(\phi \otimes \psi)^{\downarrow z} &= ((\phi \otimes \psi)^{\downarrow x})^{\downarrow z} \\
&= (\phi \otimes \psi^{\downarrow x \cap y})^{\downarrow z}.
\end{aligned} \tag{2.16}$$

(3) Using the neutral element for z we obtain by the marginalization axiom

$$(\phi \otimes \psi)^{\downarrow z} = (\phi \otimes \psi)^{\downarrow z} \otimes e_z. \tag{2.17}$$

Since $z \subseteq x \cup y$ implies $z \cap (x \cup y) = z$, the labeling and the combination axioms together with associativity of combination gives us then,

$$\begin{aligned}
(\phi \otimes \psi)^{\downarrow z} &= (\phi \otimes \psi \otimes e_z)^{\downarrow z} \\
&= (\phi \otimes e_z) \otimes \psi^{\downarrow y \cap z} \\
&= (\phi \otimes \psi^{\downarrow y \cap z}) \otimes e_z.
\end{aligned} \tag{2.18}$$

Now, the first factor in this last formula has domain $x \cup (y \cap z) = z$ since $x \subseteq z \subseteq x \cup y$. Thus we obtain finally

$$(\phi \otimes \psi)^{\downarrow z} = \phi \otimes \psi^{\downarrow y \cap z}. \tag{2.19}$$

$\square$

The computational implications of these statements are clear: each time it is allowed to restrict computation to smaller domains by using the right hand side of the identities.

A word of warning: it may be conjectured that if $x = d(\phi)$, then $\phi^{\downarrow x} = \phi$. This is however not true in general. Consider for example a valuation algebra Φ and double each element, by distinguishing two versions of each element, one marked, one unmarked. Both behave the same way, except that the marginal of each element is marked, and when any valuation in a combination is marked, then so is the combination. This system is still a valuation algebra, but $\phi^{\downarrow x} \neq \phi$, if ϕ is not marked (this example is due to (Shafer, 1991)). This identity holds nevertheless in many, in fact in most, practical examples of valuation algebras. It is implied by an additional axiom which will be introduced later (see Chapter 3). For the time being however, we will stick with the above minimal set of axioms, which are sufficient for our computational purposes.

Instead of marginalization we may introduce another primitive operation in a set of valuations Φ, for a valuation ϕ and a variable $X \in d(\phi)$,

$$\phi^{-X} = \phi^{\downarrow d(\phi) - \{X\}}. \tag{2.20}$$

This is called *variable elimination* . If we replace marginalization with this new operation, then we obtain a system (Φ, D) with the following operations:

1. *Labeling*: $\Phi \to D; \phi \mapsto d(\phi)$,

2. *Combination*: $\Phi \times \Phi \to \Phi; (\phi, \psi) \mapsto \phi \otimes \psi$,

3. *Variable Elimination*: $\Phi \times r \to \Phi; (\phi, X) \mapsto \phi^{-X}$, for $X \in d(\phi)$.

We shall prove below that, if (Φ, D) is a valuation algebra, then the following system of axioms hold :

1. *Semigroup*: Φ is associative and commutative under combination. For all $s \in D$ there is an element e_s with $d(e_s) = s$ such that for all $\phi \in \Phi$ with $d(\phi) = s, e_s \otimes \phi = \phi \otimes e_s = \phi$.

2. *Labeling*: For $\phi, \psi \in \Phi$,

$$d(\phi \otimes \psi) = d(\phi) \cup d(\psi). \tag{2.21}$$

3. *Variable Elimination*: For $\phi \in \Phi, X \in d(\phi)$,

$$d(\phi^{-X}) = d(\phi) - \{X\}. \tag{2.22}$$

4. *Commutativity of Elimination.* For $\phi \in \Phi$, and $X, Y \in d(\phi)$,

$$(\phi^{-X})^{-Y} = (\phi^{-Y})^{-X}. \qquad (2.23)$$

5. *Combination.* For $\phi, \psi \in \Phi$ with $d(\phi) = x, d(\psi) = y$ and $Y \notin x, Y \in y$,

$$(\phi \otimes \psi)^{-Y} = \phi \otimes \psi^{-Y}. \qquad (2.24)$$

6. *Neutrality.* For $x, y \in D$,

$$e_x \otimes e_y = e_{x \cup y}. \qquad (2.25)$$

We denote a system of valuations Φ over subsets of variables of r, with the operation of labeling, combination and variable elimination, satisfying these axioms, by (Φ, r).

Commutativity of elimination allows to define unambiguously the elimination of several variables $X_1, X_2, \ldots, X_n \in d(\phi)$ as

$$\phi^{-\{X_1, X_2, \ldots, X_n\}} = (\cdots ((\phi^{-X_1})^{-X_2}) \cdots)^{-X_n} \qquad (2.26)$$

independently of the actual elimination sequence used.

Here are a few elementary results on variable elimination:

Lemma 2.2   *1. If $x \subseteq d(\phi)$ for some $\phi \in \Phi$, then*

$$d(\phi^{-x}) = d(\phi) - x. \qquad (2.27)$$

2. If x and y are two disjoint subsets of $d(\phi)$ for some $\phi \in \Phi$, then

$$(\phi^{-x})^{-y} = \phi^{-(x \cup y)}. \qquad (2.28)$$

3. If $\phi, \psi \in \Phi$ and y a subset of $d(\psi)$, disjoint to $d(\phi)$, then

$$(\phi \otimes \psi)^{-y} = \phi \otimes \psi^{-y}. \qquad (2.29)$$

Proof. (1) In order to prove the first item, we order the variables of the set x arbitrarily into a sequence, say, $X_1, X_2, \ldots, X_n$ and apply the definition (2.26) as well as axiom (3) of the system above

$$\begin{aligned}
d(\phi^{-x}) &= d((\cdots ((\phi^{-X_1})^{-X_2}) \cdots)^{-X_n}) \\
&= (\cdots ((d(\phi) - \{X_1\}) - \{X_2\}) \cdots) - \{X_n\} \\
&= d(\phi) - x. \qquad (2.30)
\end{aligned}$$

(2) Follows directly from the definition of the elimination of subsets of variables of the domain (2.26).

(3) Follows from (2.26) and a repeated application of axiom (5) of the system above.   □

Marginalization can now conversely be expressed as variable elimination. For $x \subseteq d(\phi)$, we have

$$\phi^{\downarrow x} = \phi^{-(d(\phi)-x)}. \tag{2.31}$$

The following theorem states that the two ways of defining algebras (Φ, D) and (Φ, r) are in fact equivalent.

Theorem 2.3 *If (Φ, D) is a valuation algebra and variable elimination is defined by (2.20), then the system of axioms above is satisfied.*

If (Φ, r) satisfies the axioms above and marginalization is defined by (2.31), then (Φ, D) is a valuation algebra.

Proof. Let (Φ, D) be a valuation algebra. We have only to prove axioms (3) to (5) above.

Axiom (3) follows from the definition of variable elimination and the marginalization axiom of valuation algebras

$$d(\phi^{-X}) = d(\phi^{\downarrow d(\phi)-\{X\}}) = d(\phi) - \{X\}.$$

Axiom (4) follows from the definition of variable elimination, axiom (3) just proved and the transitivity axiom of valuation algebras,

$$\begin{aligned}
(\phi^{-X})^{-Y} &= (\phi^{\downarrow d(\phi)-\{X\}})^{\downarrow (d(\phi)-\{X\})-\{Y\}} \\
&= \phi^{\downarrow d(\phi)-\{X,Y\}} \\
&= (\phi^{\downarrow d(\phi)-\{Y\}})^{\downarrow (d(\phi)-\{Y\})-\{X\}} \\
&= (\phi^{-Y})^{-X}.
\end{aligned}$$

To derive axiom (5) note that $x \subseteq (x \cup y) - \{Y\} \subseteq x \cup y$. Take then $z = (x \cup y) - \{Y\}$ and apply Lemma 2.1 (3)

$$\begin{aligned}
(\phi \otimes \psi)^{-Y} &= (\phi \otimes \psi)^{\downarrow (x \cup y)-\{Y\}} \\
&= \phi \otimes \psi^{\downarrow y \cap ((x \cup y)-\{Y\})}.
\end{aligned}$$

Since $Y \in d(\psi) = y, Y \notin x$ we obtain $y \cap ((x \cup y) - \{Y\}) = y - \{Y\}$ and hence

$$(\phi \otimes \psi)^{-Y} = \phi \otimes \psi^{\downarrow y-\{Y\}} = \phi \otimes \psi^{-Y}.$$

In the same way, if (Φ, r) satisfies the axioms above, we have only to prove axioms (3) to (5) of a valuation algebra.

In order to prove axiom (3) apply Lemma 2.2 (1) to the definition of marginalization,

$$d(\phi^{\downarrow x}) = d(\phi^{-(d(\phi)-x)}) = d(\phi) - (d(\phi) - x) = x.$$

To prove the transitivity axiom (4) of valuation algebras, use the definition of marginalization and apply Lemma 2.2 (1) and (2),

$$(\phi^{\downarrow x})^{\downarrow y} \;=\; (\phi^{-(d(\phi)-x)})^{-(x-y)} \;=\; \phi^{d(\phi)-y} \;=\; \phi^{\downarrow y}.$$

To prove the combination axiom (5) of valuation algebras, note that $(x \cup y) - x = y - x$, use the definition of marginalization and apply Lemma 2.2 (3)

$$\begin{aligned}(\phi \otimes \psi)^{\downarrow x} &\;=\; (\phi \otimes \psi)^{-(x \cup y)-x} \;=\; (\phi \otimes \psi)^{-(y-x)} \\ &\;=\; \phi \otimes \psi^{-(y-x)} \;=\; \phi \otimes \psi^{\downarrow x \cap y}.\end{aligned}$$

$\square$

We remark here that there is a slightly weaker system of axioms which is still sufficient for the computational theory developed in the following chapter. In fact, the existence of the neutral elements e_s can be dispensed with. Then the neutrality axiom (6) is dropped. However, without neutral elements it is no more possible to prove Lemma 2.1 (3). Instead, the combination axiom (5) must be strengthened to:

(5) *Combination.* For $\phi, \psi \in \Phi$ with $d(\phi) = x, d(\psi) = y$, and $z \in D$ such that $x \subseteq z \subseteq x \cup y$,

$$(\phi \otimes \psi)^{\downarrow z} \;=\; \phi \otimes \psi^{\downarrow z \cap y}. \tag{2.32}$$

Variable elimination can in this system of axioms be introduced just as before and it can still be shown to be equivalent to marginalization. We will however use throughout the book in general the former system of axioms. Nevertheless, examples of valuation algebras without neutral elements exist and will be encountered below.

After this abstract discussion of valuation algebras, it is time to look for examples of valuation algebras. In the next section several examples will be presented. The systems presented in the following section are abstracted from different well-known formalisms to treat information and uncertainty. Only their structure as valuation algebras will be highlighted. It will not be possible to present the full fledged semantic background and formal theory of all these examples.

2.3   Examples of Valuation Algebras

2.3.1   Indicator Functions

Functions $i : \Omega_s \to \{0, 1\}$ are called *indicator functions* for s. Indicator functions for $s, s \subseteq r$ will be the valuations in the first example. If Ω_s is *finite*, as we shall assume here, then an indicator function for s can simply be represented by a $|s|$-dimensional table with $|\Omega_s|$ zero-one entries. So, the valuations for $s = \{X_1, \ldots, X_n\}$ are n-dimensional tables of size $|\Omega_{X_1}| \times \cdots \times |\Omega_{X_n}|$ and entries 0 or 1. Each such table represents an indicator function $i(\mathbf{x}), \mathbf{x} \in \Omega_s$ for s.

Combination of indicator functions is defined by multiplication. More precisely, if i_1, i_2 are indicator functions for s and t respectively, then, for $\mathbf{x} \in \Omega_{s \cup t}$,

$$i_1 \otimes i_2(\mathbf{x}) \;=\; i_1(\mathbf{x}^{\downarrow s}) i_2(\mathbf{x}^{\downarrow t}). \tag{2.33}$$

The neutral valuation for s is simply given by $e_s(\mathbf{x}) = 1$ for all $\mathbf{x} \in \Omega_s$.

Marginalization of an indicator function i for s to some domain $t \subseteq s$ is defined as follows . For $\mathbf{x} \in \Omega_t$,

$$i^{\downarrow t}(\mathbf{x}) \;=\; \max_{\mathbf{y} \in \Omega_s : \mathbf{y}^{\downarrow t} = \mathbf{x}} i(\mathbf{y}). \tag{2.34}$$

This can also be expressed alternatively, more in the style of variable elimination: if $\mathbf{y}$ is a configuration of Ω_s, decompose it into components belonging to t and not belonging to t, $(\mathbf{y}_t, \mathbf{y}_{s-t})$. Then, for $\mathbf{x} \in \Omega_t$,

$$i^{\downarrow t}(\mathbf{x}) \;=\; \max_{\mathbf{y}_{s-t} \in \Omega_{s-t}} i(\mathbf{x}, \mathbf{y}_{s-t}). \tag{2.35}$$

Indicator functions for s correspond one-to-one to subsets of Ω_s. If C is such a subset, then the corresponding indicator function, denoted by i_C is defined as follows:

$$i_C(\mathbf{x}) \;=\; \begin{cases} 1 \text{ for } \mathbf{x} \in C, \\ 0 \text{ for } \mathbf{x} \notin C. \end{cases}$$

Inversely, an indicator function i for s defines a set $\{\mathbf{x} \in \Omega_s : i(\mathbf{x}) = 1\}$.

The operations of combination and marginalization correspond to set intersection, in a sense, which becomes clear below. In fact, if i_{C_1}, i_{C_2} are indicator functions for s, t respectively, and if C_1, C_2, refer to the corresponding sets, we have

$$i_{C_1} \otimes i_{C_2} \;=\; i_{C_1^{\uparrow s \cup t} \cap C_2^{\uparrow s \cup t}}. \tag{2.36}$$

Furthermore, if i_C is an indicator function for s representing a subset C, and if $t \subseteq s$, then

$$i_C^{\downarrow t} \;=\; i_{C^{\downarrow t}}. \tag{2.37}$$

Each set, represented by its indicator function, can be considered as a restriction of the unknown value of the variables considered. In this sense the operations of combination and marginalization correspond clearly to the combination of knowledge and to the focusing of knowledge to specified variables.

Do these indicator functions satisfy the axioms for valuation algebras? Commutativity and associativity of combination follow easily from the definition of combination, since multiplication of numbers is commutative and associative. Suppose that i_1, i_2, i_3 are indicator functions for s, t, u. Then, for $\mathbf{x} \in \Omega_{s \cup t \cup u}$,

$$(i_1 \otimes i_2) \otimes i_3(\mathbf{x}) \;=\; i_1 \otimes i_2(\mathbf{x}^{\downarrow s \cup t}) i_3(\mathbf{x}^{\downarrow u}) \;=\; (i_1(\mathbf{x}^{\downarrow s}) i_2(\mathbf{x}^{\downarrow t})) i_3(\mathbf{x}^{\downarrow u})$$

$$=\; i_1(\mathbf{x}^{\downarrow s})(i_2(\mathbf{x}^{\downarrow t}) i_3(\mathbf{x}^{\downarrow u})) \;=\; i_1(\mathbf{x}^{\downarrow s}) i_2 \otimes i_3(\mathbf{x}^{\downarrow t \cup u})$$

$$=\; i_1 \otimes (i_2 \otimes i_3)(\mathbf{x}).$$

The existence of neutral elements for domains s has also been shown above. The labeling axiom as well as the neutrality axiom follow directly from the definition of combination.

The marginalization axiom is part of the definition. The transitivity axiom follows from a corresponding transitivity of maximizing. In fact, if f is a function of two variables x, y, then $\max_{x,y} f(x,y) = \max_x(\max_y f(x,y))$ as is well known. Hence, if i is an indicator function for s and $x \subseteq y \subseteq s$, for $\mathbf{x} \in \Omega_x$ we have

$$
\begin{aligned}
(i^{\downarrow y})^{\downarrow x}(\mathbf{x}) &= \max_{\mathbf{y} \in \Omega_{y-x}} i^{\downarrow y}(\mathbf{x}, \mathbf{y}) &= \max_{\mathbf{y} \in \Omega_{y-x}} \left(\max_{\mathbf{z} \in \Omega_{s-y}} i(\mathbf{x}, \mathbf{y}, \mathbf{z}) \right) \\
&= \max_{(\mathbf{y},\mathbf{z}) \in \Omega_{s-x}} i(\mathbf{x}, \mathbf{y}, \mathbf{z})) &= i^{\downarrow x}(\mathbf{x}).
\end{aligned}
$$

It remains finally to prove the combination axiom, which is usually the most difficult to verify. Here we see that, if we decompose a configuration of $\Omega_{s \cup t}$ according to the subsets $s \cap t, s - t, t - s$, that

$$
\begin{aligned}
(i_1 \otimes i_2)^{\downarrow s}(\mathbf{x}_{s \cap t}, \mathbf{x}_{s-t}) &= \max_{\mathbf{x}_{t-s}} i_1 \otimes i_2(\mathbf{x}_{s \cap t}, \mathbf{x}_{s-t}, \mathbf{x}_{t-s}) \\
&= \max_{\mathbf{x}_{t-s}} i_1(\mathbf{x}_{s \cap t}, \mathbf{x}_{s-t}) i_2(\mathbf{x}_{s \cap t}, \mathbf{x}_{t-s}) \\
&= i_1(\mathbf{x}_{s \cap t}, \mathbf{x}_{s-t}) \max_{\mathbf{x}_{t-s}} i_2(\mathbf{x}_{s \cap t}, \mathbf{x}_{t-s}) \\
&= i_1(\mathbf{x}_{s \cap t}, \mathbf{x}_{s-t}) i_2^{\downarrow s \cap t}(\mathbf{x}_{s \cap t}) \\
&= i_1 \otimes i_2^{\downarrow s \cap t}(\mathbf{x}_{s \cap t}, \mathbf{x}_{s-t}).
\end{aligned} \tag{2.38}
$$

Hence the indicator functions form indeed a valuation algebra. This algebra covers the calculus of constraints in finite frames. Clearly, the algebra can be extended to constraints on more general frames. We mention only, that for example linear manifolds, convex sets and convex polyhedra in Euclidean spaces $\mathcal{R}^n$ form valuation algebras too (see Sections 6.5.3 and 6.5.4) .

2.3.2   Relations

This example is closely related to the previous one. In fact it uses simply another representation of the subsets of Ω_s. Instead of representing a subset by an indicator function, we work with the set itself. For later reference and also for the interest in the example itself, we present the example in the language and notation of *relational databases* .

Let $\mathcal{A}$ be a set of symbols, called *attributes*. For each $\alpha \in \mathcal{A}$ let U_α be a non-empty set, the set of all possible values of the attribute α. For example, if $\mathcal{A} = \{\text{name}, \text{age}, \text{income}\}$, then U_{name} could be the set of strings, whereas U_{age} and U_{income} are both sets of non-negative integers. Clearly, the set of attributes corresponds to the set of variables in our original terminology and the U_α are the frames of the variables.

Let $x \subseteq \mathcal{A}$. A x-*tuple* is a function f with domain x and $f(\alpha) \in U_\alpha$ for each $\alpha \in x$. The set of all x-tuples is called E_x. For any x-tuple f and a subset

$y \subseteq x$ the restriction $f[y]$ is defined to be the y-tuple g so that $g(\alpha) = f(\alpha)$ for all $\alpha \in y$. Tuples correspond to configurations and the restriction of a tuple to its projection.

A *relation R over x* is a set of x-tuples, i.e. a subset of E_x . The set of attributes x is called the *domain* of R. As usual, we denote it by $d(R)$. Relations are thus simply subsets of frames of a group of variables. For $y \subseteq d(R)$ the *projection* of R onto y is defined as follows:

$$\pi_y(R) = \{f[y] : f \in R\}. \tag{2.39}$$

The *join* of a relation R over x and a relation S over y is defined as follows:

$$R \bowtie S = \{f : f \in E_{x \cup y}, f[x] \in R, f[y] \in S\}. \tag{2.40}$$

Clearly, the idea is that projection is marginalization and join is combination. It is easy to see that relational databases satisfy the axioms for valuation algebras. Here the axioms are expressed in the notation of relational databases:

1. $(R_1 \bowtie R_2) \bowtie R_3 = R_1 \bowtie (R_2 \bowtie R_3)$ (Associativity),

2. $R_1 \bowtie R_2 = R_2 \bowtie R_1$ (Commutativity),

3. $d(E_x) = x$, and if $d(R) = x$, then $R \bowtie E_x = R$ (Neutral element),

4. $d(R \bowtie S) = d(R) \cup d(S)$ (Labeling),

5. If $y \subseteq d(R)$, then $d(\pi_y(R)) = y$ (Marginalization),

6. If $x \subseteq y \subseteq d(R)$, then $\pi_x(\pi_y(R)) = \pi_x(R)$ (Transitivity),

7. If $d(R) = x$ and $d(S) = y$, then $\pi_x(R \bowtie S) = R \bowtie \pi_{x \cap y}(S)$ (Combination; the right hand side is also called *semi-join* in database theory, see (Maier, 1983)),

8. $E_x \bowtie E_y = E_{x \cup y}$ (Neutrality).

Thus relational databases have the structure of a valuation algebra. In fact it is not essentially different from the valuation algebra of the previous example. It uses simply another representation of subsets. But the question of an appropriate representation may be important for computations.

Surely, relational databases have a lot more structure than that of a simple valuation algebra. For example we have that $\pi_x(E_y) = E_x$, if $x \subseteq y$. This property is called *stability*. (see Chapter 3). Not all valuation algebras possess this property (see the example of probability potentials below). Furthermore the following holds also

$$\text{If } x \subseteq d(R), \text{ then } R \bowtie \pi_x(R) = R. \tag{2.41}$$

This property is called *idempotency* . It says that an information, combined with a part of it, gives no new information. Valuation algebras satisfying this additional axiom are numerous and very important. We shall study them in Chapter 6 in some depth. Besides joins, unions of relations and complements can be defined too. In relational database theory this leads to a relational algebra, which is richer than valuation algebras.

2.3.3   Probability Potentials

This example is motivated by discrete probability theory . The frames are still assumed to be finite sets. A valuation for s is here then a non-negative function $p : \Omega_s \to \mathbf{R}^+$. Any such valuation can be represented, just as indicator functions, by a n-dimensional table of size $|\Omega_s| = |\Omega_{X_1}| \times \cdots |\Omega_{X_n}|$, if $s = \{X_1, \ldots X_n\}$. If

$$\sum_{\mathbf{x} \in \Omega_s} p(\mathbf{x}) = 1,$$

then the potential is said to be *normalized* or *scaled*. It corresponds to a discrete *probability distribution*.

Combination is defined as multiplication . If p_1 is a potential for s and p_2 is a potential for t, then for $\mathbf{x} \in \Omega_{s \cup t}$,

$$p_1 \otimes p_2(\mathbf{x}) = p_1(\mathbf{x}^{\downarrow s}) p_2(\mathbf{x}^{\downarrow t}).$$

The neutral element for s is $e(\mathbf{x}) = 1$ for all $\mathbf{x} \in \Omega_s$.

Marginalization consists of summing out all variables to be eliminated . That is, if p is a potential for s, and $t \subseteq s$, decompose configurations in Ω_s according to the subsets t and $s - t$ of s. Then if $\mathbf{x} \in \Omega_t$

$$p^{\downarrow t}(\mathbf{x}) = \sum_{\mathbf{y} \in \Omega_{s-t}} p(\mathbf{x}, \mathbf{y}). \tag{2.42}$$

It is evident that combination is associative and commutative, since these laws hold for multiplication of numbers. The neutral elements for s are also identified. The axioms of labeling and marginalization follow directly from the respective definitions of the operations.

The transitivity axiom says simply that we may sum out variables in two steps. That is, if $t \subseteq s \subseteq d(p) = x$

$$\begin{aligned} (p^{\downarrow s})^{\downarrow t}(\mathbf{x}) &= \sum_{\mathbf{y} \in \Omega_{s-t}} p^{\downarrow s}(\mathbf{x}, \mathbf{y}) = \sum_{\mathbf{y} \in \Omega_{s-t}} \sum_{\mathbf{z} \in \Omega_{x-s}} p(\mathbf{x}, \mathbf{y}, \mathbf{z}) \\ &= \sum_{(\mathbf{y}, \mathbf{z}) \in \Omega_{x-t}} p(\mathbf{x}, \mathbf{y}, \mathbf{z}) = p^{\downarrow t}(\mathbf{x}). \end{aligned}$$

The combination axiom follows equally very easily. Suppose p_1 is a potential for s, p_2 is a potential for t and $\mathbf{x} \in \Omega_s$. Then we have

$$(p_1 \otimes p_2)^{\downarrow s}(\mathbf{x}) = \sum_{\mathbf{y} \in \Omega_{(s \cup t) - s}} p_1 \otimes p_2(\mathbf{x}, \mathbf{y}) = \sum_{\mathbf{y} \in \Omega_{t-s}} p_1(\mathbf{x}) p_2(\mathbf{x}^{\downarrow s \cap t}, \mathbf{y})$$

$$= p_1(\mathbf{x}) \sum_{\mathbf{y} \in \Omega_{t-s}} p_2(\mathbf{x}^{\downarrow s \cap t}, \mathbf{y}) = p_1(\mathbf{x}) p_2^{\downarrow s \cap t}(\mathbf{x}^{\downarrow s \cap t})$$

$$= p_1 \otimes p_2^{\downarrow s \cap t}(\mathbf{x}). \tag{2.43}$$

The neutrality axiom follows finally directly from the definition of neutral elements and the definition of combination.

So, the potentials form a valuation algebra. Note that the marginal of the neutral element for s to a subset t of s is not the neutral element for t! Hence, in contrast to relational algebra, the algebra of potentials is not stable.

For an in-depth discussion how this algebra relates to probability theory and the problems to be solved there, we refer to (Shafer, 1996; Cowell et. al., 1999).

We remark furthermore that the combination of normalized potentials yields in general no normalized potentials. However there is an alternative definition of combination for normalized potentials . Let p_1 be a normalized potential for s and p_2 be a normalized potential for t and $\mathbf{x} \in \Omega_{s \cup t}$. Define

$$p_1 \otimes p_2(\mathbf{x}) = \frac{1}{K} p_1(\mathbf{x}^{\downarrow s}) p_2(\mathbf{x}^{\downarrow t}) \tag{2.44}$$

where

$$K = \sum_{\mathbf{x} \in \Omega_{s \cup t}} p_1(\mathbf{x}^{\downarrow s}) p_2(\mathbf{x}^{\downarrow t}). \tag{2.45}$$

If $K = 0$ define $p_1 \otimes p_2(\mathbf{x}) = 0$. It can be verified that, if $K \neq 0$, then the combined potential is again normalized. If we add the null-potential $p(\mathbf{x}) = 0$ to the set of normalized potentials, then this set is both closed under the new combination and ordinary marginalization. The neutral element on s is now $e_s(\mathbf{x}) = 1/|\Omega_s|$. It can be shown as above that the axioms of valuations algebras are satisfied for normalized potentials . They give us thus a further example of a valuation algebra, which in fact is stable. We remark also, that functions $f : \Omega_s \to \mathcal{R}$, without being restricted to be nonnegative values, form still a valuation algebra with combination and marginalization defined as with probability potentials .

2.3.4 Possibility Potentials

Possibility theory has been proposed by Zadeh (Zadeh, 1978; Zadeh, 1979) as an alternative approach to uncertainty, emphasizing different aspects of uncertainty than probability theory. A possibility distribution on a frame Ω_s is determined by a function $p : \Omega_s \to [0, 1]$. The value $p(\mathbf{x})$ is the degree of possibility of the configuration $\mathbf{x}$. Sometimes it is assumed that

$$\max_{\mathbf{x} \in \Omega_s} p(\mathbf{x}) = 1. \tag{2.46}$$

Then p is called *normalized* or *scaled* . These functions p are the valuations and form the set Φ. We call these elements *possibility potentials* . Like a probability potential they are represented by $|s|$-dimensional tables of size $|\Omega_s|$. If p is a possibility potential for the set of variables s, then its label $d(p)$ equals s.

Marginalization is defined as *maximizing* . If p is a possibility potential on s, and $t \subseteq s$, then decompose configurations of Ω_s into a part corresponding to the variables in t and a part corresponding to those in $s - t$. Then, if $\mathbf{x} \in \Omega_t$ and $\mathbf{y} \in \Omega_{s-t}$, we define

$$p^{\downarrow t}(\mathbf{x}) = \max_{\mathbf{y} \in \Omega_{s-t}} p(\mathbf{x}, \mathbf{y}). \tag{2.47}$$

If p is scaled, then so is $p^{\downarrow t}$.

Combination is defined using so-called *t-norms* . A *t*-norm is a function $T : [0,1] \times [0,1] \rightarrow [0,1]$, which satisfies the following conditions:

1. *Boundary:* For all $a \in [0,1]$,

$$T(1,a) = a, \qquad T(0,a) = 0. \tag{2.48}$$

2. *Monotonicity:* For $a_1, a_2, b_1, b_2 \in [0,1]$ such that $a_1 \le a_2$ and $b_1 \le b_2$,

$$T(a_1, b_1) \le T(a_2, b_2). \tag{2.49}$$

3. *Commutativity, Associativity:* For $a, b, c \in [0,1]$,

$$T(a,b) = T(b,a), \qquad T(T(a,b),c) = T(a, T(b,c)). \tag{2.50}$$

If p_1 is a possibility potential on s and p_2 a potential on t, then their combination $p_1 \otimes p_2$ is a potential on $s \cup t$ defined, for $\mathbf{x} \in \Omega_{s \cup t}$ by

$$p_1 \otimes p_2(\mathbf{x}) = T(p_1(\mathbf{x}^{\downarrow s}), p_2(\mathbf{x}^{\downarrow t})). \tag{2.51}$$

We claim that the set Φ of possibility potentials, together with the lattice of subsets D of a set r of variables, furnished with the operations of labeling, marginalization, defined by (2.47), and combination defined by (2.51), forms a valuation algebra for any *t*-norm T . Commutativity and Associativity of combination follow from the corresponding properties of *t*-norms. Also, from the boundary property of *t*-norms it follows that $e(\mathbf{x}) = 1$ for all $\mathbf{x} \in \Omega_s$ is the neutral element of combination on the domain s. The labeling and marginalization axioms follow directly from the definition of combination and marginalization. The transitivity axiom for marginalization follows from the corresponding property of maximizing. The neutrality axiom, follows from the boundary property of *t*-norms and the definition of combination.

As usual, the most difficult axiom to verify is the combination axiom: let p_1 and p_2 be possibility potentials on domains x and y respectively, and assume $\mathbf{x} \in \Omega_x$ and $\mathbf{z} \in \Omega_{y-x}$. Then we have

$$(p_1 \otimes p_2)^{\downarrow x}(\mathbf{x}) = \max_{\mathbf{z} \in \Omega_{y-x}} T(p_1(\mathbf{x}), p_2(\mathbf{x}^{\downarrow x \cap y}, \mathbf{z})).$$

We see that by the monotonicity of t-norms

$$\max_{\mathbf{z} \in \Omega_{y-x}} T(p_1(\mathbf{x}), p_2(\mathbf{x}^{\downarrow x \cap y}, \mathbf{z})) \geq T(p_1(\mathbf{x}), \max_{\mathbf{z} \in \Omega_{y-x}} p_2(\mathbf{x}^{\downarrow x \cap y}, \mathbf{z}))$$

$$\geq T(p_1(\mathbf{x}), p_2(\mathbf{x}^{\downarrow x \cap y}, \mathbf{z})).$$

On the other hand, from the last inequality above we deduce also that

$$T(p_1(\mathbf{x}), \max_{\mathbf{z} \in \Omega_{y-x}} p_2(\mathbf{x}^{\downarrow x \cap y}, \mathbf{z})) \geq \max_{\mathbf{z} \in \Omega_{y-x}} T(p_1(\mathbf{x}), p_2(\mathbf{x}^{\downarrow x \cap y}, \mathbf{z})).$$

So, we conclude that

$$\max_{\mathbf{z} \in \Omega_{y-x}} T(p_1(\mathbf{x}), p_2(\mathbf{x}^{\downarrow x \cap y}, \mathbf{z})) = T(p_1(\mathbf{x}), \max_{\mathbf{z} \in \Omega_{y-x}} p_2(\mathbf{x}^{\downarrow x \cap y}, \mathbf{z})).$$

But the right hand side of this identity equals $p_1 \otimes p_2^{\downarrow x \cap y}(\mathbf{x})$. Thus, this identity proves the combination axiom. Possibility potentials with max-marginalization and t-norm combination form thus indeed a valuation algebra for any t-norm.

There are many t-norms, hence many variants of possibility valuation algebras. The most frequently used t-norms are:

1. *Product t-norm:* $T(a,b) = a \cdot b$.

2. *Gödel's t-norm:* $T(a,b) = \min\{a,b\}$.

3. *Lukasiewicz's t-norm:* $T(a,b) = \max\{0, a+b-1\}$.

Remark, that the valuation algebra of possibility potentials with the Gödel t-norm is *idempotent*, but not with the other two t-norms. This shows already that the choice of a t-norm may make a lot of difference. We shall see later other examples of such structural differences between t-norms. We note also that the combination of scaled potentials is under none of these t-norms guaranteed to remain scaled. Therefore, we may be tempted to define a new combination between scaled potentials, whose result is still normalized, for example by

$$p_1 \otimes p_2(\mathbf{x}) = \frac{1}{K} T(p_1(\mathbf{x}^{\downarrow s}), p_2(\mathbf{x}^{\downarrow t})),$$

with

$$K = \max_{\mathbf{x} \in \Omega_{s \cup t}} T(p_1(\mathbf{x}^{\downarrow s}), p_2(\mathbf{x}^{\downarrow t}))$$

Then, indeed $p_1 \otimes p_2$ is scaled. Unfortunately however this combination operation is not associative. So in this way we do not get a valuation algebra.

2.3.5 Spohn Potentials

Spohn (Spohn, 1988) proposed a dynamic theory of graded belief states based on ordinal numbers . For a configuration $\mathbf{x} \in \Omega_s$ its degree of disbelief is expressed by a nonnegative integer. Thus, a Spohn disbelief potential is a function $p : \Omega_s \to \mathbf{N}^+$. Spohn assumes that

$$\min_{\mathbf{x} \in \Omega_s} p(\mathbf{x}) \; = \; 0.$$

For our purposes, it is not necessary to make this assumption, but a Spohn potential, which satisfies this condition is called *normalized* or *scaled* . If p is a Spohn potential, defined on Ω_s, then its label is $d(p) = s$.

Combination is defined as addition . If p_1 is a potential for s and p_2 one for t, then, for $\mathbf{x} \in \Omega_{s \cup t}$,

$$p_1 \otimes p_2(\mathbf{x}) \; = \; p_1(\mathbf{x}^{\downarrow s}) + p_2(\mathbf{x}^{\downarrow t}). \tag{2.52}$$

Marginalization is minimization . If p is a potential on s and $t \subseteq s$, then decompose configurations in Ω_s into a part corresponding to variables $s \cap t$ and the rest, corresponding to $s - t$. The marginal of p relative to t is then, for $\mathbf{x} \in \Omega_t$, defined as

$$p^{\downarrow t}(\mathbf{x}) \; = \; \min_{\mathbf{y} \in \Omega_{s-t}} p(\mathbf{x}, \mathbf{y}). \tag{2.53}$$

It is easy to verify that Spohn potentials, furnished with these operations of labeling, combination and marginalization, form a valuation algebra.

Scaled potentials remain so under marginalization, but not necessarily under combination. But we may here change the definition of combination as follows,

$$p_1 \otimes p_2(\mathbf{x}) \; = \; p_1(\mathbf{x}^{\downarrow s}) + p_2(\mathbf{x}^{\downarrow t}) - \min_{\mathbf{x} \in \Omega_{s \cup t}} (p_1(\mathbf{x}^{\downarrow s}) + p_2(\mathbf{x}^{\downarrow t})). \tag{2.54}$$

This transforms normalized potentials in a normalized combination. Scaled potentials with this modified combination form still a valuation algebra.

2.3.6   Set Potentials

The examples we introduce in this subsection are related to belief functions. For the semantical background and a formal theory of belief functions we refer to (Shafer, 1976; Kohlas & Monney, 1995) . Here we content ourselves to look at the related valuations in a more or less abstract way.

As before all frames are assumed finite. A valuation for s is a set function $m : \mathcal{P}(\Omega_s) \to R$, where $\mathcal{P}(A)$ denotes the power set of a set A. The $m(A)$ are thus real numbers for all subsets A of Ω_s. We call such a function a *quasi-set potential* for s . We also consider especially functions m into the set of non-negative real numbers, such that $m(A) \geq 0$ for all subsets A of Ω_s. Such non-negative functions will be called *set potentials* for s . Finally we consider set functions m such that

$$m(\emptyset) \; = \; 0, \qquad m(A) \; \geq \; 0, \qquad \sum_{A \subseteq \Omega_s} m(A) \; = \; 1. \tag{2.55}$$

Such a function will be called a *normalized* or *scaled set potential* for s. In the theory of belief functions this is called a *basic probability assignment*.

If m is a (quasi-) set potential for s, then we put $d(m) = s$. This is the labeling function.

The combination of (quasi-) set potentials is defined as follows: if m_1 is a (quasi-) set potential for s and m_2 is a (quasi-) set potential for t, then for subsets $A \subseteq \Omega_{s \cup t}$

$$m_1 \otimes m_2(A) \;=\; \sum_{A_1^{\uparrow s \cup t} \cap A_2^{\uparrow s \cup t} = A} m_1(A_1) m_2(A_2). \tag{2.56}$$

The function v_s defined by $v_s(A) = 0$ for all proper subsets of Ω_s and $v_s(\Omega_s) = 1$ is clearly the neutral element for s.

Marginalization of a (quasi-) set potential m for s to a domain $t \subseteq s$ is defined as follows: for subsets A of Ω_t

$$m^{\downarrow t}(A) \;=\; \sum_{B : B^{\downarrow t} = A} m(B). \tag{2.57}$$

Before we verify that the (quasi-) set potentials form indeed a valuation algebra, we consider some transformations of the m- functions. Thus define

$$b_m(A) \;=\; \sum_{B \cdot B \subseteq A} m(A), \qquad q_m(A) \;=\; \sum_{B : B \supseteq A} m(A). \tag{2.58}$$

Both b_m as well as q_m are set functions as m is. We are going to show that these are one-to-one transformations. We can therefore carry the operations from m- functions both to b- and q-functions by

$$\begin{aligned} b_{m_1} \otimes b_{m_2} &:= b_{m_1 \otimes m_2}, & b_m^{\downarrow t} &:= b_{m^{\downarrow t}}, \\ q_{m_1} \otimes q_{m_2} &:= q_{m_1 \otimes m_2}, & q_m^{\downarrow t} &:= q_{m^{\downarrow t}}. \end{aligned} \tag{2.59}$$

Since the mapping is one-to-one, if the system of m-function is a valuation algebra, so are the systems of the b- and q-functions and vice-versa. Also, we may verify any axiom of valuation algebras in any of these systems and then be assured that it is valid in the other systems. That is useful, since some axioms are indeed verified more easily in one system than in the other one. We have here another example of different, but *isomorph* systems, as we had with indicator functions and relational databases.

In order to prove that these mappings are indeed one-to-one we need two well-known combinatorial lemmas.

Lemma 2.4 *If $B \subseteq A$, then*

$$\sum_{C \cdot B \subseteq C \subseteq A} (-1)^{|C|} \;=\; \begin{cases} (-1)^{|A|} & \text{if } A = B, \\ 0 & \text{otherwise.} \end{cases} \tag{2.60}$$

Proof. Suppose first that $B = \emptyset$. Then, if $|A| = n > 0$

$$\sum_{C.C \subseteq A} (-1)^{|C|} = 1 - |\{C: C \subseteq A, |C| = 1\}|$$

$$+ |\{C: C \subseteq A, |C| = 2\}| - \cdots + (-1)^n$$

$$= \binom{n}{0} - \binom{n}{1} + \binom{n}{2} - \cdots + (-1)^n \binom{n}{n}$$

$$= (1-1)^n = 0.$$

Here we use the binomial theorem. If on the other hand $A = \emptyset$, then

$$\sum_{C:C \subseteq A} (-1)^{|C|} = 1.$$

Now, in general, for $B \subseteq A$ we have

$$\sum_{C.B \subseteq C \subseteq A} (-1)^{|C|} = \sum_{D.D \subseteq A-B} (-1)^{|B \cup D|}$$

$$= (-1)^{|B|} \sum_{D.D \subseteq A-B} (-1)^{|D|}.$$

The last sum, according to the first part of the proof, is either 0 if $A - B \neq \emptyset$ or equal to 1, if $A = B$. This proves the lemma. $\square$

This lemma is used to prove the next one.

Lemma 2.5 *Suppose f and g are functions on a finite set Ω. Then*

$$f(A) = \sum_{B \subseteq A} g(B) \tag{2.61}$$

for all $A \subseteq \Omega$ if, and only if,

$$g(A) = \sum_{B \subseteq A} (-1)^{|A-B|} f(B). \tag{2.62}$$

for all $A \subseteq \Omega$.
    Also,

$$f(A) = \sum_{B \supseteq A} g(B) \tag{2.63}$$

for all $A \subseteq \Omega$ if and only if

$$g(A) = \sum_{B \supseteq A} (-1)^{|B-A|} f(B). \tag{2.64}$$

for all $A \subseteq \Omega$.

Proof. Suppose that (2.61) holds for all $A \subseteq \Omega$. Then

$$
\begin{aligned}
\sum_{B \subseteq A} (-1)^{|A-B|} f(B) &= (-1)^{|A|} \sum_{B \subseteq A} (-1)^{|B|} f(B) \\
&= (-1)^{|A|} \sum_{B \subseteq A} (-1)^{|B|} \sum_{C \subseteq B} g(C) \\
&= (-1)^{|A|} \sum_{C \subseteq A} g(C) \sum_{C \subseteq B \subseteq A} (-1)^{|B|} \\
&= (-1)^{|A|} g(A) (-1)^{|A|} = g(A).
\end{aligned}
$$

Here we use Lemma 2.4.

If on the other hand (2.62) holds for all $A \subseteq \Omega$, then, again by using Lemma 2.4,

$$
\begin{aligned}
\sum_{B \subseteq A} g(B) &= \sum_{B \subseteq A} \sum_{C \subseteq B} (-1)^{|B-C|} f(C) \\
&= \sum_{C \subseteq A} (-1)^{|C|} f(C) \sum_{C \subseteq B \subseteq A} (-1)^{|B|} \\
&= (-1)^{|A|} f(A) (-1)^{|A|} = f(A).
\end{aligned}
$$

The second part of the lemma is proved exactly the same way. $\square$

The transformations in Lemma 2.5 are called *Möbius transformations* . We see now that the m- and the b-functions are Möbius transformations, as well as the m- and the q-functions. In particular it follows that

$$
\begin{aligned}
m(A) &= \sum_{B \subseteq A} (-1)^{|A-B|} b_m(B), \\
m(A) &= \sum_{B \supseteq A} (-1)^{|B-A|} q_m(B).
\end{aligned}
\tag{2.65}
$$

Lemma 2.5 shows that these transformations are one-to-one.

Marginalization is now particularly elegant in the b-system as the following theorem shows.

Theorem 2.6 *If b_m is for s and $t \subseteq s$, then for all $A \subseteq \Omega_t$*

$$
b_m^{\downarrow t}(A) = b_m(A^{\uparrow s}).
\tag{2.66}
$$

Proof. By the definition of $b_m^{\downarrow t}$ in terms of m and the marginalization for m we obtain

$$
\begin{aligned}
b_m^{\downarrow t}(A) = b_{m^{\downarrow t}}(A) &= \sum_{B : B \subseteq A} m^{\downarrow t}(B) = \sum_{B : B \subseteq A} \sum_{C : C^{\downarrow t} = B} m(C) \\
&= \sum_{C : C^{\downarrow t} \subseteq A} m(C) = \sum_{C : C \subseteq A^{\uparrow s}} m(C) = b_m(A^{\uparrow s}).
\end{aligned}
$$

$\square$

Similarly, combination is elegantly expressed in the q-system.

Theorem 2.7 *If q_{m_1} is for s and q_{m_2} is for t, then, for $A \subseteq \Omega_{s \cup t}$*

$$q_{m_1} \otimes q_{m_2}(A) = q_{m_1}(A^{\downarrow s})q_{m_2}(A^{\downarrow t}). \tag{2.67}$$

Proof. We use the definition of $q_{m_1} \otimes q_{m_2}$ in terms of the m-function and the definition of combination for m-functions.

$$
\begin{aligned}
q_{m_1} \otimes q_{m_2}(A) &= q_{m_1 \otimes m_2}(A) = \sum_{B \supseteq A} m_1 \otimes m_2(B) \\
&= \sum_{B \supseteq A} \sum_{B_1^{\uparrow s \cup t} \cap B_2^{\uparrow s \cup t} = B} m_1(B_1)m_2(B_2) \\
&= \sum_{B_1^{\uparrow s \cup t} \cap B_2^{\uparrow s \cup t} \supseteq A} m_1(B_1)m_2(B_2) \\
&= \sum_{B_1 \supseteq A^{\downarrow s}} m_1(B_1) \sum_{B_2 \supseteq A^{\downarrow t}} m_2(B_2) \\
&= q_{m_1}(A^{\downarrow s})q_{m_2}(A^{\downarrow t}). \tag{2.68}
\end{aligned}
$$

$\square$

We are now free to switch between the m-, b- or q- system at our convenience.

In order to verify the axioms of a valuation algebra note first from the combination of q-functions as defined in Theorem 2.7 that associativity and commutativity of combination hold, since these laws are valid for the multiplication of numbers and also since projection of sets is transitive. So, if q_1 is for s, q_2 is for t, q_3 is for u, and $A \subseteq \Omega_{s \cup t \cup u}$, then

$$
\begin{aligned}
(q_1 \otimes q_2) \otimes q_3(A) &= (q_1 \otimes q_2)(A^{\downarrow s \cup t})q_3(A^{\downarrow u}) \\
&= q_1((A^{\downarrow s \cup t})^{\downarrow s})q_2((A^{\downarrow s \cup t})^{\downarrow t})q_3(A^{\downarrow u}) \\
&= q_1(A^{\downarrow s})q_2(A^{\downarrow t})q_3(A^{\downarrow u}). \tag{2.69}
\end{aligned}
$$

In the same way, we obtain exactly the same result for $q_1 \otimes (q_2 \otimes q_3)(A)$. This proves associativity. Commutativity is evident from the definition of combination.

If v_s is the neutral m-set potential for s, then clearly $q_{v_s}(A) = 1$ for all subsets A of Ω_s. It is evident that this is the neutral q-function for s. Thus the semigroup axiom is satisfied.

The labeling and the marginalization axioms are evident from the definitions of combination and marginalization for the m-functions.

Transitivity follows easily from Theorem 2.6. Since $(A^{\uparrow s})^{\uparrow t} = A^{\uparrow t}$ if $s \subseteq t$, we obtain, if $s \subseteq t \subseteq d(b) = x$ and $A \in \Omega_s$

$$(b^{\downarrow t})^{\downarrow s}(A) = b^{\downarrow t}(A^{\uparrow t}) = b((A^{\uparrow t})^{\uparrow x}) = b(A^{\uparrow x}) = b^{\downarrow s}(A). \tag{2.70}$$

The combination axiom is best verified in the m-system. We use here the combination axiom for subsets, as it has been verified in the example for indicator functions (see Subsection 2.3.1). We have then, if m_1 is a m- function for s and m_2 is a m- function for t and $A \subseteq \Omega_s$,

$$
\begin{aligned}
(m_1 \otimes m_2)^{\downarrow s}(A) &= \sum_{B \cdot B^{\downarrow s} = A} m_1 \otimes m_2(B) \\
&= \sum_{B : B^{\downarrow s} = A} \sum_{C_1^{\uparrow s \cup t} \cap C_2^{\uparrow s \cup t} = B} m_1(C_1) m_2(C_2) \\
&= \sum_{(C_1^{\uparrow s \cup t} \cap C_2^{\uparrow s \cup t})^{\downarrow s} = A} m_1(C_1) m_2(C_2) \\
&= \sum_{C_1 \cap (C_2^{\downarrow s \cap t})^{\uparrow s} = A} m_1(C_1) m_2(C_2) \\
&= \sum_{C_1 \cap C_{12}^{\uparrow s} = A} m_1(C_1) \sum_{C_2^{\downarrow s \cap t} = C_{12}} m_2(C_2) \\
&= \sum_{C_1 \cap C_{12}^{\uparrow s} = A} m_1(C_1) m_2^{\downarrow s \cup t}(C_{12}) \\
&= m_1 \otimes m_2^{\downarrow s \cup t}(A).
\end{aligned}
\tag{2.71}
$$

The neutrality axiom finally follows easily both in the m- and in the q-system.

Let then M, B and Q denote the systems of m-, b- and q-functions with their corresponding definitions of labeling, combination and marginalization . We have just shown that $(M, D), (B, D)$ and (Q, D) are all valuation algebras . In fact they are all *isomorph*, that is essentially the same. Both for theoretical as well as for computational purposes we may switch between the different representations at convenience.

Let M^+ denote the system of non-negative m-functions and B^+, Q^+ the images of M^+ under the corresponding Möbius transformations. Clearly, for $b \in B^+$ and for $q \in Q^+$ we have $b(A) \geq 0$ and $q(A) \geq 0$. However, not any non-negative b-function or non-negative q- function transforms back into an element of M^+. That is B^+ and Q^+ are not simply the set of non-negative set functions as M^+ is. Elements $b \in B^+$ are called *belief functions* or *support functions* and elements $q \in Q^+$ are called *commonality functions*.

It follows from the definition of combination and marginalization of m-functions, that M^+ is closed under both operations, i.e. $m_1, m_2, m \in M^+$ and $s \subseteq d(m)$ imply that $m_1 \otimes m_2 \in M^+$ and $m^{\downarrow s} \in M^+$. That is, (M^+, D) is a *sub-valuation algebra* of (M, D) and so are $(B^+, D), (Q^+, D)$ sub-valuation algebras of (B, D) and (Q, D) respectively.

Let M^n be the set of normalized m-functions . Although marginals of normalized m-functions are still normalized, this is not true in general for the combination. Thus under the actual definition of combination (M^n, D) is not a sub-valuation algebra of (M, D), it is not a valuation algebra at all.

But there is an alternative definition of combination for normalized m-functions. If m_1 is for s, m_2 is for t, both are normalized and $A \subseteq \Omega_{s\cup t}, A \neq \emptyset$, then define

$$m_1 \otimes m_2(A) = \frac{1}{K} \sum_{A_1^{\uparrow s\cup t} \cap A_2^{\uparrow s\cup t} = A} m_1(A_1)m_2(A_2). \qquad (2.72)$$

where

$$K = \sum_{A_1^{\uparrow s\cup t} \cap A_2^{\uparrow s\cup t} \neq \emptyset} m_1(A_1)m_2(A_2). \qquad (2.73)$$

Furthermore we define $m_1 \otimes m_2(\emptyset) = 0$. If $K = 0$, then, $m_1 \otimes m_2(A) = 0$ for all $A \neq \emptyset$ and $m_1 \otimes m_2(\emptyset) = 1$. This way to combine scaled set potentials is called Dempster's rule (Shafer, 1976).

If we add the function $m(A) = 0$ for $A \neq \emptyset$, $m(\emptyset) = 1$, to M^n, then M^n is closed under this new variant of combination. Of course we can define the corresponding sets of B^n and Q^n with this new combination induced by the transformations. It can be shown that all the axioms are satisfied under this variant of combination. So (M^n, D), as well as $(B^n, D), (Q^n, D)$ are new examples of valuation algebras. We refer to Section 3.7 for a discussion of normalized set potentials.

2.3.7   Gaussian Potentials

Gaussian or multivariate normal distributions are very popular and important probability models. We are going to establish here two valuation algebras related to these distributions. In the first one multivariate normal density functions are the valuations to be considered. In the second, more interesting and more general one, Gaussian set potentials are considered.

We consider here variables with the set of real numbers $\mathbf{R}$ as domains. The frame of a set s of variables is then $\Omega_s = \mathbf{R}^n$, if $n = |s|$. A normal distribution on Ω_s is then defined by the density function

$$f(\mathbf{x}) = (2\pi)^{-n/2}(\det \Sigma)^{-1/2}e^{-(1/2)(\mathbf{x}-\mu)'\Sigma^{-1}(\mathbf{x}-\mu)}. \qquad (2.74)$$

Here $\mathbf{x}$ represents configurations in Ω_s, μ is a vector in Ω_s and Σ is a symmetric, positive definite matrix with elements in $\Omega_s \times \Omega_s$. μ is the *mean value vector* and Σ the *variance-covariance matrix* of the distribution. The matrix

$$\mathbf{K} = \Sigma^{-1} \qquad (2.75)$$

is called the *concentration matrix*. It is also symmetric and positive definite.

The normal densities for sets s of variables are the valuations we are going to consider first. Their domains are the label of the valuations. It is well known, that multiplication of normal densities results (up to renormalization) again in normal densities, and the marginals of normal densities are still normal. To facilitate computations, we represent a normal density like (2.74) by the mean vector and the concentration matrix $(\mu, \mathbf{K})$. These two parameters specify the density fully. We call such a pair of a vector μ in Ω_s and a symmetric, positive definite matrix in $\Omega_s \times \Omega_s$ a *Gaussian potential* on the domain s. More precisely, these Gaussian potentials form the set Φ of the valuations considered here.

If μ is a vector from Ω_s, $\mathbf{K}$ a matrix from $\Omega_s \times \Omega_s$ and $t \subseteq s$, then we denote by $\mu^{\downarrow t}$ and $\mathbf{K}^{\downarrow t}$ the vector, respective matrix obtained by selecting only the elements in Ω_t or $\Omega_t \times \Omega_t$. $\mu^{\downarrow \emptyset}$ and $\mathbf{K}^{\downarrow \emptyset}$ equal both $\diamond$. If, on the contrary, $t \supseteq s$, then $\mu^{\uparrow t}$ and $\mathbf{K}^{\uparrow t}$ represent the vector or matrix obtained by adding to μ and to $\mathbf{K}$ 0-entries for all variables in $t - s$.

Now, if f_1 and f_2 are two normal densities, corresponding to the potentials $(\mu_1, \mathbf{K}_1)$ and $(\mu_2, \mathbf{K}_2)$ on s and t respectively, then, neglecting the normalization constants, we have for $\mathbf{x} \in \Omega_{s \cup t}$

$$f_1(\mathbf{x}^{\downarrow s}) \cdot f_2(\mathbf{x}^{\downarrow t}) \;=\; e^{-(1/2)(\mathbf{x}^{\downarrow s}-\mu_1)'\mathbf{K}_1(\mathbf{x}^{\downarrow s}-\mu_1)-(1/2)(\mathbf{x}^{\downarrow t}-\mu_2)'\mathbf{K}_2(\mathbf{x}^{\downarrow t}-\mu_2)}.$$

It is easy to verify that

$$(\mathbf{x}^{\downarrow s} - \mu_1)'\mathbf{K}_1(\mathbf{x}^{\downarrow s} - \mu_1) + (1/2)(\mathbf{x}^{\downarrow t} - \mu_2)'\mathbf{K}_2(\mathbf{x}^{\downarrow t} - \mu_2)$$
$$= \quad (\mathbf{x} - \mu)'\mathbf{K}(\mathbf{x} - \mu)$$

with

$$\mathbf{K} \;=\; \mathbf{K}_1^{\uparrow s \cup t} + \mathbf{K}_2^{\uparrow s \cup t},$$
$$\mu \;=\; \mathbf{K}^{-1}\left(\mathbf{K}_1^{\uparrow s \cup t} \cdot \mu_1^{\uparrow s \cup t} + \mathbf{K}_2^{\uparrow s \cup t} \cdot \mu_2^{\uparrow s \cup t}\right) \qquad (2.76)$$

Therefore, we define an operation of combination between Gaussian potentials as follows:

$$(\mu_1, \mathbf{K}_1) \otimes (\mu_2, \mathbf{K}_2) \;=\; (\mu, \mathbf{K}), \qquad\qquad (2.77)$$

where μ and $\mathbf{K}$ are defined as in (2.76).

Further, it is well known that the marginal relative to t of a normal distribution for s with mean μ and the covariance matrix Σ is a normal distribution with mean $\mu^{\downarrow t}$ and covariance matrix $\Sigma^{\downarrow t}$. Therefore, we define marginalization of Gaussian potentials by

$$(\mu, \mathbf{K})^{\downarrow t} \;=\; (\mu^{\downarrow t}, ((\mathbf{K}^{-1})^{\downarrow t})^{-1}. \qquad\qquad (2.78)$$

Clearly, multiplication (and renormalization) of normal densities is associative and commutative. So Gaussian potentials form a semigroup under combination. But this time there are no neutral elements. Marginalization, that is integration, is transitive and the combination axiom holds also for normal densities, hence Gaussian potentials (we refer to (Monney, 2000), see also below). So Gaussian potentials form a valuation algebra, without neutral elements.

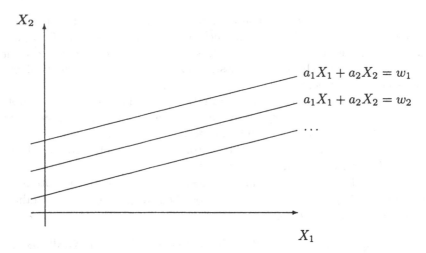

Figure 2.2: Geometric representation of a Gaussian hint: parallel linear manifolds as focal sets of Gaussian variables.

We are now going to generalize this system. Consider a linear system

$$\mathbf{AX} \;=\; \mathbf{W}, \tag{2.79}$$

where $\mathbf{A}$ is an $m \times n$ matrix of rank m, $\mathbf{X}$ a vector of variables of dimension n and $\mathbf{W}$ a vector of variables of dimension m. For any fixed value for the vector $\mathbf{W}$, system (2.79) determines a *linear manifold* of dimension m in the n-dimensional space $\mathcal{R}^n$. For different values of the vector $\mathbf{W}$, the manifolds are *parallel* (see Fig. 2.2 for a schematic representation of the situation in two dimensions). We assume now, that $\mathbf{W}$ are random vectors with a normal distribution of dimension m with mean value vector μ and a positive definite covariance matrix Σ. Then system (2.79) defines a multivalued mapping from the sample space of the vectors $\mathbf{W}$ to the frame of the variables of the vector $\mathbf{X}$. Each sample value $\mathbf{w}$ of the random vector $\mathbf{W}$ maps to the linear manifold defined by the system of linear equations $\mathbf{AX} = \mathbf{w}$. The manifold

$$\Gamma(\mathbf{w}) \;=\; \{\mathbf{x} \in \Omega_s : \mathbf{Ax} = \mathbf{w}\}$$

is called the focal manifold associated to $\mathbf{w}$. This is an instance of a more general situation described by (Dempster, 1967). (Kohlas & Monney, 1995) call such a system a *hint*, or, more precisely, since the underlying probability structure is defined by a normal distribution, a *Gaussian hint*. Hints or multivalued mappings lead to *belief functions*, in our case to *Gaussian belief functions* . Gaussian hints have been treated by (Liu L., 1996; Monney, 2000). We sketch here the approach of (Monney, 2000).

A Gaussian hint on Ω_s, with $|s| = n$ is specified by a triple $(\mathbf{A}, \mu, \Sigma)$, where $\mathbf{A}$ is a $m \times n$ matrix of rank m, μ a m-dimensional vector and Σ a positive definite $m \times m$ matrix. The matrix μ and Σ determine the normal distribution of the vector $\mathbf{W}$ (here Σ is the covariance matrix of the normal distribution) and $\mathbf{A}$ fixes the family of linear manifolds. Note that a Gaussian potential of the former kind is now given by $(\mathbf{I}_n, \mu, \Sigma)$, where $\mathbf{I}_n$ is the $n \times n$ identity matrix. The focal manifolds reduce to points, and the linear system associated with the hint is in this case simply $\mathbf{X} = \mathbf{W}$. Such Gaussian hints are called *precise*.

We remark, that there are many equivalent representations of the same Gaussian hint. If $\mathbf{T}$ is a $m \times m$ matrix of rank m, then the hint $(\mathbf{T} \cdot \mathbf{A}, \mathbf{T} \cdot \mu, \mathbf{T}\Sigma\mathbf{T}')$ represents the same system as $(\mathbf{A}, \mu, \Sigma)$.

We allow also for the degenerate normal distribution, where $\mathbf{W} = 0$ with probability one. In this case we put $\mu = 0$ and $\Sigma = 0$ (here 0 represents first a vector and then a matrix of the appropriate dimensions with zero entries). This degenerate normal distribution is used to define the *vacuous Gaussian hint* $(0, 0, 0)$ where the only focal manifold is the frame Ω_s itself (again 0 is to be taken as the matrix or vector of the appropriate dimension).

Marginalization of a Gaussian hint on s to t means to project their focal manifolds in Ω_s to the subspace Ω_t. The projections $\Gamma(\mathbf{w})^{\downarrow t}$ of the linear manifolds $\Gamma(\mathbf{w})$ are linear manifolds in Ω_t. The focal manifolds of different sample vectors $\mathbf{w}$ may project to the same linear manifold. This means that the normal density of $\mathbf{W}$ has to be marginalized accordingly. In terms of the triple $(\mathbf{A}, \mu, \Sigma)$ these operations amount to the following:

Decompose $\mathbf{A}$ into the columns $\mathbf{A}_1$ corresponding to the variables in t and the columns $\mathbf{A}_2$ associated with the variables in $s - t$,

$$\mathbf{A} = [\mathbf{A}_1\mathbf{A}_2].$$

If the rank of $\mathbf{A}_2$ equals the rank of $\mathbf{A}$, then the projection of the focal manifolds $\Gamma(\mathbf{w})$ equal all Ω_t. Thus, in this case the marginal of $(\mathbf{A}, \mu, \Sigma)$ is the vacuous Gaussian hint on t. If the rank of $\mathbf{A}_2$ is strictly smaller than the rank of $\mathbf{A} = m$, say $p < m$, then there exists a $(m - p) \times m$ matrix $\mathbf{P}$ of rank $m - p$ such that

$$\mathbf{P} \cdot \mathbf{A}_2 = 0.$$

Then the system manifolds $\mathbf{A} \cdot \mathbf{X} = \mathbf{W}$ are transformed by the projection matrix $\mathbf{P}$ into

$$\mathbf{P} \cdot \mathbf{A} \cdot \mathbf{X} = \mathbf{P} \cdot \mathbf{W}.$$

The variables $\mathbf{W}' = \mathbf{P} \cdot \mathbf{W}$ are still normally distributed, namely with mean values $\mathbf{P} \cdot \mu$ and covariance matrix $\mathbf{P} \cdot \Sigma \cdot \mathbf{P}'$. Therefore, marginalization of a Gaussian hint $(\mathbf{A}, \mu, \Sigma)$ on s to $t \subseteq s$ is defined as follows:

$$(\mathbf{A}, \mu, \Sigma)^{\downarrow t} = \begin{cases} (0, 0, 0), & \text{if } rank(\mathbf{A}_2) = rank(\mathbf{A}), \\ (\mathbf{P} \cdot \mathbf{A}, \mathbf{P} \cdot \mu, \mathbf{P} \cdot \Sigma \cdot \mathbf{P}'), & \text{otherwise.} \end{cases}$$

We note that for precise hints, the second case applies and we obtain

$$(\mathbf{I}_n, \mu, \Sigma)^{\downarrow t} = (\mathbf{I}_m, \mu^{\downarrow t}, \Sigma^{\downarrow t}). \tag{2.80}$$

This is exactly the marginalization defined for Gaussian densities above.

Combination of two Gaussian hints $(\mathbf{A}_1, \mu_1, \Sigma_1)$ and $(\mathbf{A}_2, \mu_2, \Sigma_2)$ means to intersect focal manifolds $\Gamma_1(w_1)$ and $\Gamma_2(w_2)$ from the two hints. This is schematically illustrated in Fig. 2.3. Here we assume the normal distributed vectors $\mathbf{W}_1$ and $\mathbf{W}_2$ of the two hints to be *stochastically independent*. Remark however, that there may be sample vectors $\mathbf{w}_1$ and $\mathbf{w}_2$ such that $\Gamma_1(\mathbf{w}_1)$ and $\Gamma_2(\mathbf{w}_2)$ do not intersect (see Fig. 2.3 b). The common normal distribution of $\mathbf{W}_1$ and $\mathbf{W}_2$ must be conditioned on the condition that the focal manifolds have a non-empty intersection. This corresponds to Dempster's rule (Dempster, 1967; Kohlas & Monney, 1995).

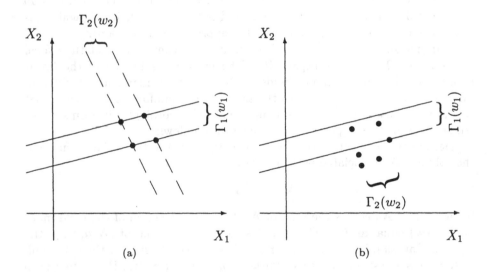

Figure 2.3: Combining Gaussian hints by intersecting their focal linear manifolds (a). Sometimes this intersection may be empty (b).

Intersecting manifolds means to take the union of the system of linear equations

$$\mathbf{A}_1 \cdot \mathbf{X}_1 = \mathbf{W}_1$$
$$\mathbf{A}_2 \cdot \mathbf{X}_2 = \mathbf{W}_2.$$

Here $\mathbf{X}_1$ denotes the vector of variables in the domain s of the first Gaussian hint and $\mathbf{X}_2$ the vector of variables in the domain t of the second Gaussian hint. Let $\mathbf{X}$ be the vector of variables in $s \cup t$ and $\mathbf{A}$ the matrix and $\mathbf{W}$ the right hand side of the combined system above, such that

$$\mathbf{A} \cdot \mathbf{X} = \mathbf{W}. \qquad (2.81)$$

$\mathbf{A}$ is an $(m_1 + m_2) \times n$ matrix, if $n = |s \cup t|$. $\mathbf{W}$ is an $m_1 + m_2$ dimensional random vector, which is normally distributed with mean and covariance

$$\mu = \begin{bmatrix} \mu_1 \\ \mu_2 \end{bmatrix}, \qquad \Sigma = \begin{bmatrix} \Sigma_1 & 0 \\ 0 & \Sigma_2 \end{bmatrix}.$$

We have now to distinguish four cases:

(1) $rank(\mathbf{A}) = m_1 + m_2$: In this case $(\mathbf{A}, \mu, \Sigma)$ is a Gaussian hint, and

$$(\mathbf{A}_1, \mu_1, \Sigma_1) \otimes (\mathbf{A}_2, \mu_2, \Sigma_2) = (\mathbf{A}, \mu, \Sigma). \qquad (2.82)$$

(2) $rank(\mathbf{A}) < m_1 + m_2 < n$: In this case $(\mathbf{A}, \mu, \Sigma)$ represents not a Gaussian hint, since $\mathbf{A}$ has not the full rank. Let

$$\mathbf{A} = \begin{bmatrix} \mathbf{B}_1 \\ \mathbf{B}_2 \end{bmatrix}$$

such that $rank(\mathbf{B}_1) = rank(\mathbf{A}) = k$. Then there is a matrix Λ such that

$$\mathbf{B}_2 = \Lambda \cdot \mathbf{B}_1.$$

This linear dependence between the equations of the combined system (2.81) has as consequence, that not all sample values $\mathbf{w}$ of $\mathbf{W}$ are admissible, that is, allow for a solution of the system (2.81). In fact, if we decompose $\mathbf{W}$ into $\mathbf{W}_1$ and $\mathbf{W}_2$ according to the decomposition of $\mathbf{A}$, then we see that

$$\Lambda \cdot \mathbf{W}_1 - \mathbf{W}_2 = 0$$

must hold. So the normal distribution of $\mathbf{W}$ has to be conditioned on this relation. Decompose then the mean μ and covariance Σ of $\mathbf{W}$ according to the decomposition of $\mathbf{A}$,

$$\mu = \begin{bmatrix} \mu_1 \\ \mu_2 \end{bmatrix}, \qquad \Sigma = \begin{bmatrix} \Sigma_{11} & \Sigma_{12} \\ \Sigma_{21} & \Sigma_{22} \end{bmatrix}.$$

This leads then to the following new mean and covariance of the conditional distribution:

$$\begin{aligned} \mu &= \mu_1 - (\Sigma_{11} \cdot \Lambda' - \Sigma_{12}) \cdot (\Lambda \cdot \Sigma_{11} \cdot \Lambda' - \Lambda \cdot \Sigma_{12} - \Sigma_{21} \cdot \Lambda' + \Sigma_{22})^{-1} \\ &\quad \cdot (\Lambda \cdot \mu_1 - \mu_2), \\ \Sigma &= \Sigma_{11} - (\Sigma_{11} \cdot \Lambda' - \Sigma_{12}) \cdot (\Lambda \cdot \Sigma_{11} \cdot \Lambda' - \Lambda \cdot \Sigma_{12} - \Sigma_{21} \cdot \Lambda' + \Sigma_{22})^{-1} \\ &\quad \cdot (\Lambda \cdot \Sigma_{11} - \Sigma_{21}). \end{aligned} \qquad (2.83)$$

Therefore, in this case we obtain

$$(\mathbf{A}_1, \mu_1, \Sigma_1) \otimes (\mathbf{A}_2, \mu_2, \Sigma_2) = (\mathbf{B}_1, \mu, \Sigma), \qquad (2.84)$$

where $\mathbf{B}_1$ is as defined above and μ and Σ are given in (2.83).

(3) $rank(\mathbf{A}) = n < m_1 + m_2$: $(\mathbf{A}, \mu, \Sigma)$ represents again not a Gaussian hint. In this case we decompose the matrix $\mathbf{A}$ into

$$\mathbf{A} = \begin{bmatrix} \mathbf{B}_1 \\ \mathbf{B}_2 \end{bmatrix},$$

where this time $\mathbf{B}_1$ is an $n \times n$ matrix of rank n. If we decompose the combined system (2.81) and the vector $\mathbf{W}$ accordingly, then we obtain

$$\mathbf{B}_1 \cdot \mathbf{X} = \mathbf{W}_1,$$
$$\mathbf{B}_2 \cdot \mathbf{X} = \mathbf{W}_2.$$

From this we conclude that $\mathbf{X} = \mathbf{B}_1^{-1} \cdot \mathbf{W}_1$. If we introduce this into the second equation above, we obtain

$$\mathbf{B}_2 \cdot \mathbf{B}_1^{-1} \cdot \mathbf{W}_1 = \mathbf{W}_2.$$

This is again a relation onto which we have to condition the normal distribution of $\mathbf{W}$. Also, clearly, in this case the focal manifolds of the two hints intersect into one-dimensional manifolds (if they intersect). So the combined hint becomes a precise Gaussian hint. We refer to (Monney, 2000), which shows that in this case the combined hint is as follows:

$$(\mathbf{A}_1, \mu_1, \Sigma_1) \otimes (\mathbf{A}_2, \mu_2, \Sigma_2) \qquad (2.85)$$
$$= (\mathbf{I}_n, (\mathbf{A}' \cdot \Sigma^{-1} \cdot \mathbf{A})^{-1} \cdot \mathbf{A}' \cdot \Sigma^{-1} \cdot \mu, (\mathbf{A}' \cdot \Sigma^{-1} \cdot \mathbf{A})^{-1}).$$

We remark that for example the combination of two precise hints leads to this case. After some simple algebraic transformations, one can verify that this rule of combination is identical to the one defined for Gaussian potentials above.

(4) $rank(\mathbf{A}) < n \le m_1 + m_2$: Once again, $(\mathbf{A}, \mu, \Sigma)$ is not a Gaussian hint. We decompose $\mathbf{A}$ into

$$\mathbf{A} = [\mathbf{B}_1 \mathbf{B}_2],$$

where this time $rank(\mathbf{B}_1) = rank(\mathbf{A}) = k$. Then there is a matrix Λ such that $\mathbf{B}_2 = \mathbf{B}_1 \cdot \Lambda$. Decompose $\mathbf{X}$ into $\mathbf{X}_1$ and $\mathbf{X}_2$ according to the decomposition of $\mathbf{A}$. Define $\mathbf{Y} = \mathbf{X}_1 + \Lambda \cdot \mathbf{X}_2$. Then the combined system (2.81) can be written as

$$\mathbf{B}_1 \cdot \mathbf{Y} = \mathbf{W}.$$

Here we are essentially back to the former case with the matrix $\mathbf{B}_1$ at the place of $\mathbf{A}$ (except, that the system is in a lower dimensional space than $\Omega_{s \cup t}$). (Monney, 2000) shows that in this case the combined Gaussian hint is

$$(\mathbf{A}_1, \mu_1, \Sigma_1) \otimes (\mathbf{A}_2, \mu_2, \Sigma_2) \qquad (2.86)$$
$$= ([\mathbf{I}_k, \Lambda], (\mathbf{B}_1' \cdot \Sigma^{-1} \cdot \mathbf{B}_1)^{-1} \cdot \mathbf{B}_1' \cdot \Sigma^{-1} \mu, (\mathbf{B}_1' \cdot \Sigma^{-1} \cdot \mathbf{B}_1)^{-1}).$$

In this way, operations of labeling, marginalization and combination are defined among Gaussian hints. (Monney, 2000) proves that this system forms a valuation algebra. Contrary to Gaussian potentials considered above, this algebra has neutral elements, the vacuous Gaussian hints. It is also clear that Gaussian potentials, if we consider them as precise Gaussian hints, are contained in the algebra of Gaussian hints, form a subalgebra.

2.4 Partial Marginalization

There are interesting systems, where marginalization is not defined completely. Consider variables with real numbers as frames, such that the frames Ω_s equal $n = |s|$-dimensional spaces $\mathbf{R}^n$. Consider nonnegative functions $f : \Omega_s \to \mathbf{R}^+$. Marginalization should again correspond to integration. Already, integration is only defined for, say, Lebesgue-measurable functions. But even within this class of functions, integration may only be possible over certain variables, but not all of them (we interpret integration over a variable as possible, only if the integral is almost everywhere finite). Density functions of absolutely continuous probability distributions on Ω_s belong to this class. They are integrable over all variables. But for example, if f is such a positive density on Ω_s, $f(\mathbf{z}) > 0$ almost everywhere, where $\mathbf{z} = (\mathbf{x}, \mathbf{y})$ and $\mathbf{y} \in \Omega_t$, then the quotient $f(\mathbf{x}, \mathbf{y})/f^{\downarrow t}(\mathbf{y})$ belongs to the class, but is not integrable over the variables in t. More precisely,

$$\left(\frac{f(\mathbf{x}, \mathbf{y})}{f^{\downarrow t}(\mathbf{y})} \right)^{\downarrow t} = \frac{f^{\downarrow t}(\mathbf{y})}{f^{\downarrow t}(\mathbf{y})} = 1 \tag{2.87}$$

for almost all $\mathbf{y}$. The quotient $f(\mathbf{x}, \mathbf{y})/f^{\downarrow t}(\mathbf{y})$ represents a family of conditional distributions associated with the density f. So already this important function is only partially integrable (over the variables in $s-t$). Therefore, we are going to generalize valuation algebras to account for systems with only partially defined marginalization.

Let Φ be a set of valuations over domains $s \subseteq r$ and D the lattice of subsets of r. Suppose that there are three operations defined:

1. *Labeling*: $\Phi \to D; \phi \mapsto d(\phi)$,

2. *Combination*: $\Phi \times \Phi \to \Phi; (\phi, \psi) \mapsto \phi \otimes \psi$,

3. *Variable Elimination*: $\Phi \times \mathcal{P}(D) \to \Phi; (\phi, t) \mapsto \phi^{-t}$ for subsets t in some family of subsets $\mathcal{V}(\phi) \in D$.

We take here the point of view of variable elimination, rather than the one of marginalization. But this is, as we know, only a matter of perspective. We can freely move between the point of view of variable elimination and the one of marginalization.

We impose now the following axioms on Φ and D:

1. *Semigroup*: Φ is associative and commutative under combination. For all $s \in D$ there is an element e_s with $d(e_s) = s$ such that for all $\phi \in \Phi$ with $d(\phi) = s$, $e_s \otimes \phi = \phi \otimes e_s = \phi$.

2. *Labeling*: For $\phi, \psi \in \Phi$,

$$d(\phi \otimes \psi) \;=\; d(\phi) \cup d(\psi). \tag{2.88}$$

3. *Variable Elimination*: For $\phi \in \Phi$, $t \in \mathcal{V}(\phi)$,

$$d(\phi^{-t}) \;=\; d(\phi) - t. \tag{2.89}$$

4. *Transitivity*: For $\phi \in \Phi$, $t \in \mathcal{V}(\phi)$, $u \subseteq t$ implies $u \in \mathcal{V}(\phi)$, $t - u \in \mathcal{V}(\phi^{-u})$, and

$$\phi^{-t} \;=\; (\phi^{-u})^{-(t-u)} \tag{2.90}$$

5. *Combination*: For $\phi, \psi \in \Phi$, and $d(\phi) = x$, $d(\psi) = y$, $u \subseteq y - x \in \mathcal{V}(\psi)$ implies $u \in \mathcal{V}(\phi \otimes \psi)$, and

$$(\phi \otimes \psi)^{-u} \;=\; \phi \otimes \psi^{-u}. \tag{2.91}$$

6. *Neutrality*: For $x, y \in D$,

$$e_x \otimes e_y \;=\; e_{x \cup y}. \tag{2.92}$$

As usual, we define marginalization of $\phi \in \Phi$ by

$$\phi^{\downarrow t} \;=\; \phi^{-(d(\phi)-t)},$$

if $d(\phi) - t \in \mathcal{V}(\phi)$. Let $\mathcal{M}(\phi) = \{t : d(\phi) - t \in \mathcal{V}(\phi)\}$. The transitivity axiom translates then as follows: if $t \in \mathcal{M}(\phi)$, then $t \subseteq s \subseteq d(\phi)$ implies $s \in \mathcal{M}(\phi)$ and

$$(\phi^{\downarrow s})^{\downarrow t} \;=\; \phi^{\downarrow t}. \tag{2.93}$$

It is like the usual transitivity axiom, when the marginalization to t is possible. The (extended) combination axiom, in terms of marginalization, becomes: let $d(\phi) = x$, $d(\psi) = y$, $x \subseteq z \subseteq x \cup y$. Then $y \cap z \in \mathcal{M}(\psi)$ implies $z \in \mathcal{M}(\phi \otimes \psi)$, and

$$(\phi \otimes \psi)^{\downarrow z} \;=\; \phi \otimes \psi^{\downarrow y \cap z}. \tag{2.94}$$

Again this is the usual combination axiom, or rather, its extended form as stated in Lemma 2.1 (3), provided that the marginal of ψ relative to $y \cap z$ exists.

It is evident, that any valuation algebra with full marginalization or variable elimination, satisfies these axioms. We call a system satisfying these new axioms therefore a *valuation algebra with partial marginalization*.

As an example we consider real valued variables X and Lebesgue-measurable, nonnegative functions $f : \Omega_s \to \mathbf{R}$. Such a function obtains the label $d(f) = s$. Two functions f and g are considered as equal, if they differ only on a set of measure zero (are almost everywhere equal). Lebesgue-measurable functions are closed under multiplication (Halmos, 1950). They form therefore a commutative semigroup under this combination operation. And $e(\mathbf{x}) = 1$ almost everywhere are the neutral elements. So the semigroup axiom holds. Also labeling and neutrality axioms are satisfied.

If $t \subseteq s$, then decompose configuration in Ω_s into a part corresponding to variables in t and a part corresponding to $s - t$. We consider then f with $d(f) = s$ as integrable over $s - t$, if

$$\int f(\mathbf{x}, \mathbf{y}) dy \ < \ \infty \tag{2.95}$$

for almost all $\mathbf{x} \in \Omega_t$. This integral is defined as the marginal $f^{\downarrow t}(\mathbf{x})$, if it exists. This function is Lebesgue-measurable. Then $d(f^{\downarrow t}) = t$ and the marginalization (or the variable-elimination) axiom holds. Assume now that $f^{\downarrow t}$ exists and $t \subseteq s \subseteq d(f)$. Let $\mathbf{x}, \mathbf{y}$ and $\mathbf{z}$ be configurations for t, $s - t$ and $d(f) - s$ respectively. Then Fubini's theorem says that

$$f^{\downarrow s}(\mathbf{x}, \mathbf{y}) \ = \ \int f(\mathbf{x}, \mathbf{y}, \mathbf{z}) dz$$

exists (almost everywhere) and that

$$f^{\downarrow t}(\mathbf{x}) \ = \ \int f(\mathbf{x}, \mathbf{y}, \mathbf{z}) dz dy \ = \ \int (\int f(\mathbf{x}, \mathbf{y}, \mathbf{z}) dz) dy$$
$$= \ \int f^{\downarrow s}(\mathbf{x}, \mathbf{y}) dy \ = \ (f^{\downarrow s})^{\downarrow t}.$$

So the transitivity axiom holds too.

Let finally f and g be Lebesgue-measurable functions on x and y respectively, and suppose that $x \subseteq z \subseteq x \cup y$. Let $\mathbf{x}, \mathbf{y}, \mathbf{z}$ and $\mathbf{u}$ be configurations for $x - y$, $x \cap y$, $z - x$ and $y - z$ respectively. Assume that

$$g^{\downarrow y \cap z}(\mathbf{y}, \mathbf{z}) \ = \ \int g(\mathbf{y}, \mathbf{z}, \mathbf{u}) du$$

exists. Then

$$f \otimes g^{\downarrow y \cap z}(\mathbf{x}, \mathbf{y}, \mathbf{z}) \ = \ f(\mathbf{x}, \mathbf{y}) \cdot \int g(\mathbf{y}, \mathbf{z}, \mathbf{u}) du$$
$$= \ \int f(\mathbf{x}, \mathbf{y}) \cdot g(\mathbf{y}, \mathbf{z}, \mathbf{u}) du \ = \ (f \otimes g)^{\downarrow z}(\mathbf{x}, \mathbf{y}, \mathbf{z}).$$

Thus, finally, the combination axiom is verified too. The Lebesgue-measurable functions on frames Ω_s for $s \subseteq r$ form a valuation algebra with partial marginalization.

Note that Gaussian potentials form a subalgebra of this valuation algebra with full marginalization, but without neutral elements.

This concludes our introductory discussion of valuation algebras and of examples of them. In the next chapter we examine the algebraic theory of these structures.

3 Algebraic Theory

3.1 Congruences

In this chapter we study valuation algebras from an algebraic point of view. That is, we consider them as algebraic structures in their own right. We shall see that this gives interesting insight, not only into the computational aspects, but also into the way valuations represent uncertain, partial or other knowledge. The approach is motivated both by universal algebra (many algebraic techniques and results are not so much linked to specific structures such as groups, rings, etc., but are generic) as well as by the specific problem setting and semantic intuition behind valuation algebras.

An important construction in any part of mathematics is the *quotient construction* associated with an *equivalence relation* on a set. An equivalence relation θ on a set A is a relation which is *binary, reflexive, symmetric* and *transitive* . This means that θ is a subset of $A \times A$, and if we write $a \equiv b$ (mod θ) for $(a, b) \in \theta$, it satisfies for any $a, b \in A$ the following properties:

1. $a \equiv a$ (mod θ) (reflexivity),

2. $a \equiv b$ (mod θ) implies $b \equiv a$ (mod θ) (symmetry),

3. $a \equiv b$ (mod θ), $b \equiv c$ (mod θ) imply $a \equiv c$ (mod θ).

Given an equivalence relation θ, we can define for each element a of A the class of equivalent elements $[a]_\theta = \{b \in A : b \equiv a \pmod{\theta}\}$. It is well known (and easy to verify) that these equivalence classes form a partition of A (the classes are disjoint and cover A). The family of equivalence classes is called the *quotient* or *factor set* of A and is written A/θ.

In algebra, equivalence relations which are compatible with the operations of the algebra are particularly important. Such equivalence relations are called *congruences*. In the case of a *valuation algebra* (Φ, D) consider an equivalence relation θ in Φ such that for all $\phi, \psi \in \Phi$

1. $\phi \equiv \psi$ (mod θ) implies $d(\phi) = d(\psi)$,

2. $\phi \equiv \psi \pmod{\theta}$ implies $\phi^{\downarrow x} \equiv \psi^{\downarrow x} \pmod{\theta}$ if $x \subseteq d(\phi) = d(\psi)$,

3. $\phi_1 \equiv \psi_1 \pmod{\theta}, \phi_2 \equiv \psi_2 \pmod{\theta}$ imply $\phi_1 \otimes \phi_2 \equiv \psi_1 \otimes \psi_2 \pmod{\theta}$.

Such an equivalence relation is a congruence in the valuation algebra (Φ, D).

An intuitive motivation for such a congruence is that different valuations represent sometimes the same knowledge or situation. In the case of probability potentials for example, we are ultimately interested in normalized potentials, which represent discrete probability distributions. Since potentials which differ only by a multiplicative constant have the same normalized version, they represent essentially the same information. If we consider potentials which differ only by a multiplicative constant as equivalent, then we have a congruence.

If we have a congruence θ on valuations, then we may define labeling, combination and marginalization in Φ/θ as follows:

1. *Labeling*: $d([\phi]_\theta) = d(\phi)$.

2. *Combination*: $[\phi]_\theta \otimes [\psi]_\theta = [\phi \otimes \psi]_\theta$.

3. *Marginalization*: $[\phi]_\theta^{\downarrow x} = [\phi^{\downarrow x}]_\theta$ for $x \subseteq d([\phi]_\theta)$.

These operations are well defined if θ is a congruence, since the result does not depend on which representative of the equivalence class is chosen for the definition. So we have:

$$\phi \equiv \psi \pmod{\theta} \Rightarrow d([\phi]_\theta) = d([\psi]_\theta),$$
$$\phi_1 \equiv \psi_1 \pmod{\theta}, \phi_2 \equiv \psi_2 \pmod{\theta} \Rightarrow [\phi_1]_\theta \otimes [\phi_2]_\theta = [\psi_1]_\theta \otimes [\psi_2]_\theta,$$
$$\phi \equiv \psi \pmod{\theta} \Rightarrow [\phi]_\theta^{\downarrow x} = [\psi]_\theta^{\downarrow x}.$$

It can easily be verified that $(\Phi/\theta, D)$ is still a valuation algebra.

In a valuation algebra, we need not only to restrict valuations to smaller domains (marginalization), but also to extend them sometimes to larger ones. This extension has to be done without adding information. Therefore we define for $\phi \in \Phi$ and $y \supseteq d(\phi)$

$$\phi^{\uparrow y} = \phi \otimes e_y. \tag{3.1}$$

$\phi^{\uparrow y}$ is called the vacuous extension of ϕ to the domain y. It is of course only defined in valuation algebras with neutral elements. Since ϕ is combined with the neutral, vacuous valuation e_y only the domain is changed to y, because, thanks to the labeling axiom, we have $d(\phi^{\uparrow y}) = y$, but intuitively no information is added. This would be neatly underlined if, for $d(\phi) = x$, we would have $(\phi^{\uparrow y})^{\downarrow x} = \phi$. However this is not true in general under the axioms imposed on valuations algebras (look for example at the case of probability potentials). It is a consequence of the fact that different valuations (even for the same domain) may represent the same knowledge. We shall see below how to remove this ambiguity.

The following lemma summarizes a few useful results on vacuous extension and related issues.

Lemma 3.1

1. If $x \subseteq y$ then $e_x^{\uparrow y} = e_y$.

2. If $d(\phi) = x$, then for all $y \in D$ we have $\phi \otimes e_y = \phi^{\uparrow x \cup y}$.

3. If $d(\phi) = x$, then $\phi^{\uparrow x} = \phi$.

4. If $d(\phi) = x$, and $y \subseteq x$, then $\phi \otimes e_y = \phi$.

5. If $d(\phi) = x$, and $x \subseteq y \subseteq z$, then $(\phi^{\uparrow y})^{\uparrow z} = \phi^{\uparrow z}$.

6. If $d(\phi) = x, d(\psi) = y$, and $x, y \subseteq z$, then $(\phi \otimes \psi)^{\uparrow z} = \phi^{\uparrow z} \otimes \psi^{\uparrow z}$.

7. If $d(\phi) = x, d(\psi) = y$, then $\phi \otimes \psi = \phi^{\uparrow x \cup y} \otimes \psi^{\uparrow x \cup y}$.

8. If $d(\phi) = x$, then $(\phi^{\uparrow x \cup y})^{\downarrow y} = (\phi^{\downarrow x \cap y})^{\uparrow y}$.

Proof. (1) By definition of vacuous extension and the neutrality axiom and since $x \subseteq y$ implies $x \cup y = y$ we have $e_x^{\uparrow y} = e_x \otimes e_y = e_{x \cup y} = e_y$.

(2) Again by the neutrality axiom, $\phi \otimes e_y = \phi \otimes e_x \otimes e_y = \phi \otimes e_{x \cup y} = \phi^{\uparrow x \cup y}$.

(3) Since e_x is the neutral valuation for x, we have $\phi^{\uparrow x} = \phi \otimes e_x = \phi$.

(4) Since $x \cup y = x$, the neutrality axiom implies that $\phi \otimes e_y = \phi \otimes e_x \otimes e_y = \phi \otimes e_{x \cup y} = \phi \otimes e_x = \phi$.

(5) Once more by the neutrality axiom, since $y \cup z = z$, we have $(\phi^{\uparrow y})^{\uparrow z} = (\phi \otimes e_y) \otimes e_z = \phi \otimes e_{y \cup z} = \phi \otimes e_z = \phi^{\uparrow z}$.

(6) By definition of vacuous extension, and since $e_z = e_z \otimes e_z$, we obtain $\phi^{\uparrow z} \otimes \psi^{\uparrow z} = \phi \otimes e_z \otimes \psi \otimes e_z = (\phi \otimes \psi) \otimes e_z = (\phi \otimes \psi)^{\uparrow z}$.

(7) By (3) and the labeling axiom we have $\phi \otimes \psi = (\phi \otimes \psi)^{\uparrow x \cup y}$. The rest follows from (6), since $x, y \subseteq x \cup y$.

(8) Using (2) above and the combination axiom, we obtain $(\phi^{\uparrow x \cup y})^{\downarrow y} = (\phi \otimes e_y)^{\downarrow y} = \phi^{\downarrow x \cap y} \otimes e_y = (\phi^{\downarrow x \cap y})^{\uparrow y}$. □

Even if, for $d(\phi) = x$, the identity $(\phi^{\uparrow y})^{\downarrow x} = \phi$ does not hold in general, it seems reasonable to assume that somehow $(\phi^{\uparrow y})^{\downarrow x}$ and ϕ represent the same knowledge. Therefore we extend the idea of a congruence θ by adding the following requirement: for $d(\phi) = x$ and $x \subseteq y$

$$(\phi^{\uparrow y})^{\downarrow x} \equiv \phi \pmod{\theta}. \tag{3.2}$$

Extending a valuation to a larger domain and marginalizing back should give, if not the initial valuation, then at least a valuation equivalent to the initial one. Since $(\phi^{\uparrow y})^{\downarrow x} = (\phi \otimes e_y)^{\downarrow x} = \phi \otimes e_y^{\downarrow x}$, (3.2) is equivalent to

$$e_y^{\downarrow x} \equiv e_x \pmod{\theta} \tag{3.3}$$

if $x \subseteq y$. A congruence on a valuation algebra (Φ, D) which satisfies (3.3) is called a *stable congruence*. Note for example, that proportionality of probability potentials (see above) defines a stable congruence on the valuation algebra of probability potentials.

If θ is a stable congruence on (Φ, D), then the valuation algebra $(\Phi/\theta, D)$ has the following property: If $d([\phi]_\theta) = x \subseteq y$, then

$$([\phi]_\theta^{\uparrow y})^{\downarrow x} = [\phi]_\theta. \tag{3.4}$$

This follows from (3.2).

From now on we consider *stable* valuation algebras in this section. That is, we add the following additional axiom to the six old ones (see Section 2.2):

7. *Stability*: For $x \subseteq y$,

$$e_y^{\downarrow x} = e_x. \tag{3.5}$$

A valuation algebra which satisfies this additional axiom is called *stable*. This implies for ϕ such that $d(\phi) = x \subseteq y$

$$(\phi^{\uparrow y})^{\downarrow x} = (\phi \otimes e_y)^{\downarrow x} = \phi \otimes e_y^{\downarrow x} = \phi \otimes e_x = \phi. \tag{3.6}$$

Furthermore, we obtain from stability the following result: If $d(\phi) = x$, then, by the combination axiom,

$$\phi^{\downarrow x} = (\phi \otimes e_x)^{\downarrow x} = \phi \otimes e_x^{\downarrow x} = \phi \otimes e_x = \phi. \tag{3.7}$$

The stability axiom is satisfied in many examples of valuation algebras, especially for constraint systems, such as relational algebras, for possibility and Spohn potentials, and also for set potentials and Gaussian hints, but not for probability potentials. In the latter case however the stable congruence defined by proportionality reduces this algebra to a stable quotient algebra. Also normalized potentials with normalized combination were seen to be stable. So stable valuation algebras are important and also very interesting. But, of course, stability makes no sense for valuation algebras without neutral elements, like for example Gaussian potentials.

On stable valuation algebras we may also consider congruences where the labeling compatibility is weakened. This is motivated by the observation that even valuations on different domains may represent the same knowledge or situation. Consider for example subsets C_s of Ω_s which are cylindric, that is $C_s = C_t \times \Omega_{s-t}$ for some $t \subseteq s$ and subset C_t of Ω_t. The subset C_s for s and C_t for t, when looked at as constraints on the variables in s, represent finally the same information for the possible values of the variables $X \in r$, since, although C_s is on a larger domain, it gives no information about the possible values of variables $X \in s - t$. So it adds nothing to C_t.

Motivated by this example we introduce on a stable valuation algebra (Φ, D) the following relation σ. If ϕ, ψ are two valuations with $d(\phi) = x, d(\psi) = y$, then we define

$$\phi \equiv \psi \pmod{\sigma} \quad \text{if, and only if,} \quad \phi^{\uparrow x \cup y} = \psi^{\uparrow x \cup y}. \tag{3.8}$$

We note first, that this is an equivalence relation. Clearly reflexivity and symmetry are satisfied. In order to verify transitivity, assume that $d(\phi) =$

$x, d(\psi) = y, d(\gamma) = z$ and $\phi \equiv \psi \pmod{\sigma}, \psi \equiv \gamma \pmod{\sigma}$. This means that $\phi^{\uparrow x \cup y} = \psi^{\uparrow x \cup y}$ and $\psi^{\uparrow y \cup z} = \gamma^{\uparrow y \cup z}$. By the transitivity of vacuous extension (see Lemma 3.1, (5)), we obtain that $\phi^{\uparrow x \cup y \cup z} = \psi^{\uparrow x \cup y \cup z} = \gamma^{\uparrow x \cup y \cup z}$. Stability implies then

$$\phi^{\uparrow x \cup z} = (\phi^{\uparrow x \cup y \cup z})^{\downarrow x \cup z} = (\gamma^{\uparrow x \cup y \cup z})^{\downarrow x \cup z} = \gamma^{\uparrow x \cup z}.$$

Thus, we have that $\phi \equiv \gamma \pmod{\sigma}$.

We remark that $\phi \equiv \psi \pmod{\sigma}$ if, and only if,

$$\phi^{\uparrow r} = \psi^{\uparrow r}.$$

In fact, assume $d(\phi) = x, d(\psi) = y$. Then we have $\phi^{\uparrow x \cup y} = ((\phi^{\uparrow x \cup y})^{\uparrow r})^{\downarrow x \cup y}$, $\psi^{\uparrow x \cup y} = ((\psi^{\uparrow x \cup y})^{\uparrow r})^{\downarrow x \cup y}$ by stability. So we obtain $\phi^{\uparrow x \cup y} = (\phi^{\uparrow r})^{\downarrow x \cup y} = (\psi^{\uparrow r})^{\downarrow x \cup y} = \psi^{\uparrow x \cup y}$.

If $\phi \equiv \psi \pmod{\sigma}$, then stability implies also, if $d(\phi) = x, d(\psi) = y$, that

$$(\psi^{\uparrow x \cup y})^{\downarrow x} = \phi, \quad (\phi^{\uparrow x \cup y})^{\downarrow y} = \psi. \tag{3.9}$$

We can thus start with the valuation ψ, extend it to $x \cup y$, marginalize to x to obtain ϕ and vice versa. So, equivalent valuations modulo σ really represent the same information. To exploit this result further we introduce a new operation $(\phi, y) \mapsto \phi^{\to y}$ by defining, if $d(\phi) = x$,

$$\phi^{\to y} = (\phi^{\uparrow x \cup y})^{\downarrow y}. \tag{3.10}$$

By Lemma 3.1 (8) we have also

$$\phi^{\to y} = (\phi^{\downarrow x \cap y})^{\uparrow y}. \tag{3.11}$$

With this notation we may say that $\phi \equiv \psi \pmod{\sigma}$ implies that $\psi^{\to x} = \phi$ and $\phi^{\to y} = \psi$, if $d(\phi) = x, d(\psi) = y$. We call the operation $\to$ *transportation* and say for $\phi^{\to y}$ that ϕ is transported to domain y.

Let's collect a few elementary results about this new operation in the following lemma.

Lemma 3.2

1. *If $d(\phi) = x$, then $\phi^{\to x} = \phi$.*

2. *If $d(\phi) = x$, $x, y, \subseteq z$, then $\phi^{\to y} = (\phi^{\uparrow z})^{\downarrow y}$.*

3. *If $d(\phi) = x$, $x \subseteq y$, then $\phi^{\to y} = \phi^{\uparrow y}$.*

4. *If $d(\phi) = x$, $x \supseteq y$, then $\phi^{\to y} = \phi^{\downarrow y}$.*

5. *If $y \subseteq z$, then $\phi^{\to y} = (\phi^{\to z})^{\to y}$.*

6. *For each ϕ, and all x, y, $(\phi^{\to x})^{\to y} = (\phi^{\to x \cap y})^{\to y}$.*

7. *If $d(\phi) = x, d(\psi) = y$, then $(\phi \otimes \psi)^{\to x} = \phi \otimes \psi^{\to x}$.*

Proof. (1) Follows from Lemma 3.1 (3) and the stability axiom, since $\phi^{\rightarrow x} = (\phi^{\uparrow x})^{\downarrow x} = \phi^{\downarrow x} = \phi$.

(2) We use first transitivity of vacuous extension (Lemma 3.1 (5)) and transitivity of marginalization, and then stability to obtain

$$(\phi^{\uparrow z})^{\downarrow y} \;=\; (((\phi^{\uparrow x \cup y})^{\uparrow z})^{\downarrow x \cup y})^{\downarrow y} \;=\; (\phi^{\uparrow x \cup y})^{\downarrow y} \;=\; \phi^{\rightarrow y}.$$

(3) Here we use stability and obtain, since $x \cup y = y$,

$$\phi^{\rightarrow y} \;=\; (\phi^{\uparrow y})^{\downarrow y} \;=\; \phi^{\uparrow y}.$$

(4) We have by definition and Lemma 3.1 (3)

$$\phi^{\rightarrow y} \;=\; (\phi^{\uparrow x \cup y})^{\downarrow y} \;=\; (\phi^{\uparrow x})^{\downarrow y} \;=\; \phi^{\downarrow y}.$$

(5) Here we apply first point (4) just proved, then transitivity of marginalization, followed by transitivity of vacuous extension (Lemma 3.1 (5)), and finally stability. Assume $d(\phi) = x$. This gives us the following derivation, since $x \cup y \subseteq x \cup z$,

$$(\phi^{\rightarrow z})^{\rightarrow y} \;=\; ((\phi^{\uparrow x \cup z})^{\downarrow z})^{\downarrow y} \;=\; (\phi^{\uparrow x \cup z})^{\downarrow y}$$
$$=\; (((\phi^{\uparrow x \cup y})^{\uparrow x \cup z})^{\downarrow x \cup y})^{\downarrow y} \;=\; (\phi^{\uparrow x \cup y})^{\downarrow y} \;=\; \phi^{\rightarrow y}.$$

(6) To prove this result we apply first points (5) and (3) proved above, then point (4) and finally Lemma 3.1 (8). Thus,

$$(\phi^{\rightarrow x \cap y})^{\rightarrow y} \;=\; ((\phi^{\rightarrow x})^{\rightarrow x \cap y})^{\uparrow y} \;=\; ((\phi^{\rightarrow x})^{\downarrow x \cap y})^{\uparrow y} \;=\; (\phi^{\rightarrow x})^{\rightarrow y}.$$

(7) Here, point (4) above, and the combination axiom imply that

$$(\phi \otimes \psi)^{\rightarrow x} \;=\; (\phi \otimes \psi)^{\downarrow x} \;=\; \phi \otimes \psi^{\downarrow x \cap y} \;=\; \phi \otimes e_x \otimes \psi^{\downarrow x \cap y}$$
$$=\; \phi \otimes (\psi^{\downarrow x \cap y})^{\uparrow x} \;=\; \phi \otimes \psi^{\rightarrow x}.$$

$\square$

This lemma shows, among other things, that the transport operation is both a generalization of marginalization and vacuous extension.

As we have seen, for equivalent valuations modulo σ the one transported to the domain of the other one, equals the other one. The next lemma shows that this is not only necessary for the equivalence σ but also sufficient, which gives us a second definition of this equivalence.

Lemma 3.3 *If ϕ, ψ are two valuations with $d(\phi) = x, d(\psi) = y$, then $\phi \equiv \psi$ (mod σ) if and only if*

$$\phi^{\rightarrow y} \;=\; \psi, \qquad \psi^{\rightarrow x} \;=\; \phi. \qquad\qquad (3.12)$$

Proof. The necessity of the condition has already been shown above. Assume thus that (3.12) holds. Then, Lemma 3.1 (8), and stability imply

$$\phi^{\downarrow x \cap y} = (\psi^{\to x})^{\downarrow x \cap y} = ((\psi^{\downarrow x \cap y})^{\uparrow x})^{\downarrow x \cap y} = \psi^{\downarrow x \cap y}. \tag{3.13}$$

From this we obtain furthermore, again using Lemma 3.1 (8),

$$\psi = \phi^{\to y} = (\phi^{\downarrow x \cap y})^{\uparrow y} = (\psi^{\downarrow x \cap y})^{\uparrow y}.$$

In the same way we obtain also

$$\phi = (\phi^{\downarrow x \cap y})^{\uparrow x}.$$

This implies, using (3.13), and transitivity of vacuous extension, Lemma 3.1 (5),

$$\phi^{\uparrow x \cup y} = ((\phi^{\downarrow x \cap y})^{\uparrow x})^{\uparrow x \cup y} = ((\psi^{\downarrow x \cap y})^{\uparrow x})^{\uparrow x \cup y} = (\psi^{\downarrow x \cap y})^{\uparrow x \cup y}$$
$$= ((\psi^{\downarrow x \cap y})^{\uparrow y})^{\uparrow x \cup y} = \psi^{\uparrow x \cup y}.$$

But this says that $\phi \equiv \psi \pmod{\sigma}$. □

(3.13) is itself a remarkable result. It says that if $\phi \equiv \psi \pmod{\sigma}$ and $d(\phi) = x, d(\psi) = y$, then $\phi^{\downarrow x \cap y} = \psi^{\downarrow x \cap y}$, that is the two equivalent valuations are identical on their largest common domain.

The equivalence relation σ is now shown to be a congruence relative to the operations of *combination* and *transport*.

Theorem 3.4

1. *If $\phi_1 \equiv \psi_1 \pmod{\sigma}$ and $\phi_2 \equiv \psi_2 \pmod{\sigma}$, then*

$$\phi_1 \otimes \phi_2 \equiv \psi_1 \otimes \psi_2 \pmod{\sigma}. \tag{3.14}$$

2. *If $\phi \equiv \psi \pmod{\sigma}$, then for all $z \in D$,*

$$\phi^{\to z} \equiv \psi^{\to z} \pmod{\sigma}. \tag{3.15}$$

Proof. (1) Let x_1, x_2, y_1, y_2 be the domains of $\phi_1, \phi_2, \psi_1, \psi_2$ respectively. The equivalences assumed in the theorem imply that (see Lemma 3.1 (2))

$$\phi_1 \otimes e_{y_1} = \psi_1 \otimes e_{x_1} \qquad \phi_2 \otimes e_{y_2} = \psi_2 \otimes e_{x_2}.$$

Thus, we have

$$
\begin{aligned}
(\phi_1 \otimes \phi_2)^{\uparrow (x_1 \cup x_2) \cup (y_1 \cup y_2)} &= (\phi_1 \otimes \phi_2) \otimes e_{y_1 \cup y_2} \\
&= (\phi_1 \otimes e_{y_1}) \otimes (\phi_2 \otimes e_{y_2}) \\
&= (\psi_1 \otimes e_{x_1}) \otimes (\psi_2 \otimes e_{x_2}) \\
&= (\psi_1 \otimes \psi_2) \otimes e_{x_1 \cup x_2} \\
&= (\psi_1 \otimes \psi_2)^{\uparrow (x_1 \cup x_2) \cup (y_1 \cup y_2)}.
\end{aligned}
$$

But this shows that $\phi_1 \otimes \phi_2 \equiv \psi_1 \otimes \psi_2 \pmod{\sigma}$.

(2) We show first that $\phi \equiv \psi \pmod{\sigma}$ implies $\phi^{\downarrow z} = \psi^{\downarrow z}$ if $z \subseteq d(\phi) = x, z \subseteq d(\psi) = y$. In fact, we have by the transitivity axiom, and (3.13)

$$\phi^{\downarrow z} = (\phi^{\downarrow x \cap y})^{\downarrow z} = (\psi^{\downarrow x \cap y})^{\downarrow z} = \psi^{\downarrow z}.$$

Now, since for any z we have $\phi^{\rightarrow z} = (\phi \otimes e_z)^{\downarrow z}, \psi^{\rightarrow z} = (\psi \otimes e_z)^{\downarrow z}$, and since $z \subseteq d(\phi) \cup z, d(\psi) \cup z$ and since finally by point (1) of the theorem just proved we have $\phi \otimes e_z \equiv \psi \otimes e_z \pmod{\sigma}$ we see that $\phi^{\rightarrow z} \equiv \psi^{\rightarrow z}$. In fact, we have proved more, namely $\phi^{\rightarrow z} = \psi^{\rightarrow z}$. □

If (Φ, D) is a stable valuation algebra, then we may consider the quotient algebra $(\Phi/\sigma, D)$ relative to the congruence σ. The elements of this algebra, the equivalence classes $[\phi]_\sigma$, group all valuations together, which represent the same knowledge or information, even on different domains. Therefore, we call $[\phi]_\sigma$ *domain-free valuations*. The quotient algebra has two operations, combination, which we denote as before by the symbol $\otimes$ and a second operation, which we call *focusing* and which we denote by $\Rightarrow$. The definitions of these two operations are as follows:

$$[\phi]_\sigma \otimes [\psi]_\sigma = [\phi \otimes \psi]_\sigma,$$
$$[\phi]_\sigma^{\Rightarrow x} = [\phi^{\rightarrow x}]_\sigma. \tag{3.16}$$

The fact that σ is a congruence relative to the operations of combination and transport in the algebra (Φ, D) (Theorem 3.4) guarantees that these operations are defined unambiguously. Combination is also commutative and associative and $[e_s]_\sigma$ is the neutral element.

The two operations in $(\Phi/\sigma, D)$ have interesting and basic properties, which are announced in the next theorem.

Theorem 3.5   *1. For all $x, y \in D$*

$$([\phi]_\sigma^{\Rightarrow x})^{\Rightarrow y} = [\phi]_\sigma^{\Rightarrow x \cap y}. \tag{3.17}$$

2. For all $x \in D$,

$$([\phi]_\sigma^{\Rightarrow x} \otimes [\psi]_\sigma)^{\Rightarrow x} = [\phi]_\sigma^{\Rightarrow x} \otimes [\psi]_\sigma^{\Rightarrow x}. \tag{3.18}$$

3. For all $x \in D$

$$[e_s]_\sigma^{\Rightarrow x} = [e_s]_\sigma \tag{3.19}$$

4. For all $[\phi]_\sigma \in \Phi/\sigma$

$$[\phi]_\sigma^{\Rightarrow r} = [\phi]_\sigma. \tag{3.20}$$

Proof. (1) By Lemma 3.2 (6) $(\phi^{\to x})^{\to y} = (\phi^{\to x \cap y})^{\to y}$. But we have $(\phi^{\to x \cap y})^{\to y} = (\phi^{\to x \cap y})^{\uparrow y}$, hence $(\phi^{\to x \cap y})^{\to y} \equiv \phi^{\to x \cap y}$ (mod σ). Therefore we obtain $([\phi]_\sigma^{\Rightarrow x})^{\Rightarrow y} = [(\phi^{\to x})^{\to y}]_\sigma = [\phi^{\to x \cap y}]_\sigma = [\phi]_\sigma^{\Rightarrow x \cap y}$.

(2) Since $d(\phi^{\to x}) = x$, we obtain, using Lemma 3.2 (7),

$$
\begin{aligned}
([\phi]_\sigma^{\Rightarrow x} \otimes [\psi]_\sigma)^{\Rightarrow x} &= [(\phi^{\to x} \otimes \psi)^{\to x}]_\sigma = [\phi^{\to x} \otimes \psi^{\to x}]_\sigma \\
&= [\phi]_\sigma^{\Rightarrow x} \otimes [\psi]_\sigma^{\Rightarrow x}.
\end{aligned}
$$

(3) Follows from the stability axiom, since $[e_s]_\sigma^{\Rightarrow x} = [e_s^{\to x}]_\sigma = [(e_s^{\uparrow x \cup s})^{\downarrow x}]_\sigma = [e_{x \cup s}^{\downarrow x}]_\sigma = [e_x]_\sigma = [e_s]_\sigma$.

(4) Follows since for all $\phi \in \Phi$ the equivalence $\phi \equiv \phi^{\to r}$ (mod σ) holds. □

These domain-free valuations are interesting in themselves and will be studied in detail in the next section.

3.2 Domain-Free Valuation Algebras

Motivated by the construction in the previous section, we introduce here a variant of valuation algebras, which we call domain-free. So let Ψ be a set of elements, called domain-free valuations and D the lattice of subsets of a set r. Suppose that two operations are defined:

1. *Combination*: $\Psi \times \Psi \to \Psi; (\phi, \psi) \mapsto \phi \otimes \psi$,

2. *Focusing*: $\Psi \times D \to \Psi; (\psi, x) \mapsto \psi^{\Rightarrow x}$.

Intuitively, the elements of Ψ are representations of knowledge and information, which however, this time are not related directly to a subset s of variables, but to all variables of r. Combination is, as before, aggregation of information, and focusing is extracting the information in a valuation, which pertains to a subset x of variables, and removing the part of information pertaining to the variables in $r - x$. In fact, we shall see in a moment, that focusing is closely related to variable elimination, as marginalization was.

We impose the following axioms on Ψ and D:

1. *Semigroup*: Ψ is associative and commutative under combination. There is a neutral element $e \in \Psi$ such that $e \otimes \psi = \psi \otimes e = \psi$ for all $\psi \in \Psi$.

2. *Transitivity*: For $\psi \in \Psi$ and $x, y \in D$,

$$
(\psi^{\Rightarrow x})^{\Rightarrow y} = \psi^{\Rightarrow x \cap y}. \tag{3.21}
$$

3. *Combination*: For $\psi, \phi \in \Psi$ and $x \in D$

$$
(\psi^{\Rightarrow x} \otimes \phi)^{\Rightarrow x} = \psi^{\Rightarrow x} \otimes \phi^{\Rightarrow x}. \tag{3.22}
$$

4. *Neutrality*: For $x \in D$

$$e^{\Rightarrow x} = e. \tag{3.23}$$

5. *Support*: For $\psi \in \Psi$,

$$\psi^{\Rightarrow r} = \psi. \tag{3.24}$$

The last axiom essentially states that any valuation must have a support (see below). In the case of lattices of domains D with a top element, this must necessarily be the top domain. The theory could be developed without this axiom. But then the full correspondence between domain-free and labeled valuation algebras (see below) would be lost.

A system (Ψ, D) with the operations of combination and focusing, satisfying these axioms, is called a *domain-free valuation algebra*. This is in contrast to the formerly defined labeled valuation algebra with labeled valuations. If (Φ, D) is a stable, labeled valuation algebra, then Theorem 3.5 shows that $(\Phi/\sigma, D)$ is a domain-free valuation algebra. And this construction is also the main source of examples for domain-free valuation algebras.

If (Φ, D) is a stable, labeled valuation algebra, then the associated domain-free valuation algebra $(\Phi/\sigma, D)$ is best viewed by taking the representative $\phi^{\uparrow r}$ of the class $[\phi]_\sigma$. Remember that $\phi \equiv \psi \pmod{\sigma}$ if and only if $\phi^{\uparrow r} = \psi^{\uparrow r}$. Thus, we may consider Φ_r with the usual combination defined in it and define focusing by

$$\phi^{\Rightarrow x} = (\phi^{\downarrow x})^{\uparrow r}.$$

Clearly this corresponds to the domain-free algebra.

So, consider in a first example the subsets of Ω_r as representatives of the classes of equivalent subsets in the labeled valuation algebra of subsets. Combination is then simply set-intersection,

$$C_1 \otimes C_2 = C_1 \cap C_2.$$

Focusing of C on a subset x of r is to take the cylindric extension of the projection of C to x,

$$C^{\Rightarrow x} = C^{\downarrow x} \times \Omega_{r-x}.$$

This is a domain-free valuation algebra. There is no need to verify the axioms; they follow from Theorem 3.5 since this is the quotient algebra modulo σ of the labeled algebra of subsets on domains s. We may note here that subsets have more structure than what is seized by the valuation algebra.   For example, there is not only intersection, but also union and complements; subsets form a Boolean algebra and not only a semigroup as indicated here. On the other hand, focusing is an additional operation in Boolean algebra. We refer to (Henkin, Monk & Tarski, 1971) for a study of Boolean algebras with an additional operation of variable elimination, representing existential qualifier elimination in predicate logic.

It is of course also possible to consider the corresponding indicator functions on r. If i_1, i_2 are indicator functions on $r, \mathbf{x} \in \Omega_r$, then in this representation, combination is defined as

$$i_1 \otimes i_2(\mathbf{x}) \quad = \quad i_1(\mathbf{x})i_2(\mathbf{x}).$$

Focusing of an indicator function i on a subset x of variables is defined for all $\mathbf{x} \in \Omega_r$ by,

$$i^{\Rightarrow x}(\mathbf{x}) \quad = \quad i^{\downarrow x}(\mathbf{x}^{\downarrow x}).$$

Probability potentials are not stable. So there is no domain-free version of the algebra of probability potentials. However, the quotient algebra modulo proportionality is stable and possesses thus a domain-free version. Its elements may simply be represented by potentials on Ω_r, interpreted modulo proportionality.

Possibility potentials on the other hand are stable. The elements of the corresponding domain-free algebra are possibility potentials $p : \Omega_r \to [0,1]$. Such potentials may also be interpreted as *fuzzy subsets* of Ω_r (Zadeh, 1978). Combination is defined by a t-norm T,

$$p_1 \otimes p_2(\mathbf{x}) \quad = \quad T(p_1(\mathbf{x}), p_2(\mathbf{x})).$$

This defines a variant of intersection of fuzzy subsets. Focusing is obtained by

$$p^{\Rightarrow x}(\mathbf{x}, \mathbf{y}) \quad = \quad \max_{\mathbf{y} \in \Omega_{r-x}} p(\mathbf{x}, \mathbf{y}) \quad \dot{=} \quad p^{\downarrow x}(\mathbf{x})$$

This is the fuzzy version of set projection.

Domain-free (quasi-) set potentials are given by m, b or q-functions on the power-set $\mathcal{P}(\Omega_r)$. Combination can be defined by q- functions,

$$q_1 \otimes q_2(A) \quad = \quad q_1(A)q_2(A)$$

for all subsets $A \subseteq \Omega_r$. Alternatively, we could also use m-functions,

$$m_1 \otimes m_2(A) \quad = \quad \sum_{A_1 \cap A_2 = A} m_1(A_1)m_2(A_2).$$

Focusing can be defined in terms of m-functions

$$m^{\Rightarrow x}(A) \quad = \quad \sum_{B:B^{\Rightarrow x}=A} m(B).$$

Again, the axioms of a domain-free valuation algebra are granted by Theorem 3.5 since these m-, q- and b-functions are simply the representatives of the σ-equivalence classes $[m]_\sigma, [q]_\sigma, [b]_\sigma$ on r.

As these examples show, the labeled version of valuations is to be preferred for computational purposes, because they limit the memory requirement to what is needed, whereas the domain-free versions of a valuation waste memory, because they contain a lot of redundancy.

A domain $x \in D$ is called a *support* of $\psi \in \Psi$, if

$$\psi^{\Rightarrow x} = \psi. \tag{3.25}$$

According to the support axiom, r is a support of every element of Ψ. And the neutrality axiom says that e is supported by all $x \in D$ (including the empty set). The following lemma collects some properties of support.

Lemma 3.6

1. x *is support of* $\psi^{\Rightarrow x}$.

2. *If* x *is support of* ψ, *then* x *is support of* $\psi^{\Rightarrow y}$.

3. *If* x *and* y *are supports of* ψ, *then* $x \cap y$ *is support of* ψ.

4. *If* x *is support of* ψ, *then* $x \cap y$ *is support of* $\psi^{\Rightarrow y}$.

5. *If* x *is support of* ψ, *then* $\psi^{\Rightarrow y} = \psi^{\Rightarrow x \cap y}$.

6. *If* x *is support of* ψ *and* $x \subseteq y$, *then* y *is support of* ψ.

7. *If* x *is support of* ψ *and* ϕ, *then* x *is support of* $\psi \otimes \phi$.

8. *If* x *is support of* ψ *and* y *support of* ϕ, *then* $x \cup y$ *is support of* $\psi \otimes \phi$.

Proof. (1) The transitivity axiom implies that $(\psi^{\Rightarrow x})^{\Rightarrow x} = \psi^{\Rightarrow x \cap x} = \psi^{\Rightarrow x}$.

(2) Again, by the transitivity axiom, $(\psi^{\Rightarrow y})^{\Rightarrow x} = \psi^{\Rightarrow x \cap y} = (\psi^{\Rightarrow x})^{\Rightarrow y} = \psi^{\Rightarrow y}$, since $\psi^{\Rightarrow x} = \psi$.

(3) Still by the transitivity axiom, $\psi^{\Rightarrow x \cap y} = (\psi^{\Rightarrow x})^{\Rightarrow y} = \psi^{\Rightarrow y} = \psi$.

(4) Once more, by the transitivity axiom, $(\psi^{\Rightarrow y})^{\Rightarrow x \cap y} = \psi^{\Rightarrow x \cap y}$ since $y \cap (x \cap y) = x \cap y$.

(5) Here we use the transitivity axiom to obtain $\psi^{\Rightarrow y} = (\psi^{\Rightarrow x})^{\Rightarrow y} = \psi^{\Rightarrow x \cap y}$.

(6) We have $\psi = \psi^{\Rightarrow x} = \psi^{\Rightarrow x \cap y}$ since $x = x \cap y$. But then the transitivity axiom shows that $\psi^{\Rightarrow x \cap y} = (\psi^{\Rightarrow x})^{\Rightarrow y} = \psi^{\Rightarrow y}$. This proves that $\psi = \psi^{\Rightarrow y}$.

(7) By the combination axiom, $(\psi \otimes \phi)^{\Rightarrow x} = (\psi^{\Rightarrow x} \otimes \phi)^{\Rightarrow x} = \psi^{\Rightarrow x} \otimes \phi^{\Rightarrow x} = \psi \otimes \phi$.

(8) Follows from (6) and (7). □

Every valuation ψ has a support, namely at least r. If it has two supports x and y, then it has also the support $x \cap y$ (Lemma 3.6 (3)). Hence it has a unique *least* support z, such that for any other support x, we have $z \subseteq x$. The neutral element, for example has least support $\emptyset$. Denote the least support set of ψ by $\Delta \psi$. The cardinality $|\Delta \psi|$ of the least support set is called the *dimension* of ψ.

Here are two basic properties of least supports.

Lemma 3.7

1. For each $\psi \in \Psi, x \in D$,

$$\Delta \psi^{\Rightarrow x} \subseteq \Delta \psi. \tag{3.26}$$

2. For $\phi, \psi \in \Psi$,

$$\Delta(\phi \otimes \psi) \subseteq \Delta \phi \cup \Delta \psi. \tag{3.27}$$

Proof. (1) $\Delta \psi$ is a support of ψ. By Lemma 3.6 (2) it is then also a support of $\psi^{\Rightarrow x}$, hence $\Delta \psi^{\Rightarrow x} \subseteq \Delta \psi$.

(2) $\Delta \phi$ and $\Delta \psi$ are supports of ϕ and ψ respectively. By Lemma 3.6 (8), $\Delta \phi \cup \Delta \psi$ is a support of $\phi \otimes \psi$, hence $\Delta(\phi \otimes \psi) \subseteq \Delta \phi \cup \Delta \psi$. $\qquad \square$

It is easy to give examples which show that in Lemma 3.7 in general, equality does not hold.

Just as in labeled valuation algebras, we may introduce in domain-free valuation algebras (Ψ, D) the operation of *variable elimination*. For $\psi \in \Psi$ and $X \in r$ define

$$\psi^{-X} = \psi^{\Rightarrow r - \{X\}}. \tag{3.28}$$

We prove first the following basic set of results on variable elimination.

Lemma 3.8

1. Commutativity of Variable Elimination: For $\psi \in \Psi$ and $X, Y \in r$,

$$(\psi^{-X})^{-Y} = (\psi^{-Y})^{-X}. \tag{3.29}$$

2. Combination: For $\phi, \psi \in \Psi$ and $X \in r$,

$$(\psi^{-X} \otimes \phi)^{-X} = \psi^{-X} \otimes \phi^{-X}. \tag{3.30}$$

3. Neutrality: For $X \in r$,

$$e^{-X} = e. \tag{3.31}$$

Proof. (1) By definition of variable elimination and the transitivity axiom for domain-free valuation algebras, we obtain

$$(\psi^{-X})^{-Y} = (\psi^{\Rightarrow r - \{X\}})^{\Rightarrow r - \{Y\}} = \psi^{\Rightarrow (r - \{X\}) \cap (r - \{Y\})}$$
$$= (\psi^{\Rightarrow r - \{Y\}})^{\Rightarrow r - \{X\}} = (\psi^{-Y})^{-X}.$$

(2) Here we use the combination axiom for domain-free algebras, together with the definition of variable elimination:

$$(\psi^{-X} \otimes \phi)^{-X} = (\psi^{\Rightarrow r - \{X\}} \otimes \phi)^{\Rightarrow r - \{X\}}$$

$$= \quad \psi^{\Rightarrow r - \{X\}} \otimes \phi^{\Rightarrow r - \{X\}} \quad = \quad \psi^{-X} \otimes \phi^{-X}.$$

(3) This follows from the neutrality axiom of domain-free valuation algebras, $e^{-X} = e^{\Rightarrow r - \{X\}} = e$.   □

As in the case of labeled valuation algebras, the commutativity of variable elimination allows us to define the operation of the elimination of a subset of variables $\{X_1, X_2, \ldots, X_n\} \subseteq r$ unambiguously by

$$\psi^{-\{X_1, X_2, \ldots, X_n\}} \quad = \quad (\cdots ((\psi^{-X_1})^{-X_2}) \cdots)^{-X_n}. \tag{3.32}$$

The commutativity of variable elimination guarantees that the left hand side in this definition is independent of the actual sequence of variable elimination used on the right hand side.

We can take the three points of Lemma 3.8 as axioms together with the old semigroup axiom of domain-free valuation algebras to obtain an equivalent definition of domain free algebras. In fact, we may recover focusing from variable elimination by defining

$$\psi^{\Rightarrow x} \quad = \quad \psi^{-(r-x)}. \tag{3.33}$$

We leave this as an exercise.

From a domain-free valuation algebra (Ψ, D) we can construct a stable, labeled valuation algebra. For this purpose consider the set of pairs

$$\Psi^* \quad = \quad \{(\psi, x) : \psi \in \Psi, \psi^{\Rightarrow x} = \psi\}. \tag{3.34}$$

These pairs can be considered as valuations, labeled by their supports. Therefore, we define the following operations:

1. *Labeling*: For $(\psi, x) \in \Psi^*$ define

$$d(\psi, x) \quad = \quad x. \tag{3.35}$$

2. *Combination*: For $(\phi, x), (\psi, y) \in \Psi^*$ define

$$(\phi, x) \otimes (\psi, y) \quad = \quad (\phi \otimes \psi, x \cup y) \tag{3.36}$$

3. *Marginalization*: For $(\psi, x) \in \Psi^*$ and $y \subseteq x$ define

$$(\psi, x)^{\downarrow y} \quad = \quad (\psi^{\Rightarrow y}, y). \tag{3.37}$$

Note that combination and marginalization give indeed results in Ψ^* (see Lemma 3.6). It can be verified that this defines a stable, labeled valuation algebra. We leave this as an exercise for the reader. From this stable, labeled valuation algebra we may construct again a domain-free algebra, as in Section 3.1. We shall show in the next section that this is essentially (up to isomorphism) the same algebra as the original algebra (Ψ, D). Similarly, we may start with a stable, labeled algebra, construct the associated domain-free valuation algebra and then reconstruct by the procedure just introduced again a labeled valuation algebra. And this labeled algebra is again essentially (up to isomorphism) the same as the original one.

3.3 Subalgebras, Homomorphisms

3.3.1 Subalgebras

In this section we examine a few important concepts of universal algebra with respect to valuation algebras. The first one is the notion of a *subalgebra*. In terms of valuation algebras, this is a subset of a valuation algebra which is closed under the operations of combination and focusing (in the case of domain-free valuation algebras). More precisely, let (Ψ, D) be a domain-free valuation algebra. (Ψ', D) is said to be a subalgebra of (Ψ, D), if $\Psi' \subseteq \Psi$, and

1. $\phi, \psi \in \Psi'$ implies $\phi \otimes \psi \in \Psi'$.

2. $e \in \Psi'$

3. $\psi \in \Psi'$ implies $\psi^{\Rightarrow x} \in \Psi'$ for all $x \in D$.

It is immediately clear, that a subalgebra of a domain-free valuation algebra is itself a valuation algebra.

A similar notion holds also for labeled valuation algebras. If (Φ, D) is a labeled valuation algebra, then (Φ', D) is a subalgebra of (Φ, D) if $\Phi' \subseteq \Phi$ and

1. $\phi, \psi \in \Phi'$ implies $\phi \otimes \psi \in \Phi'$.

2. $e_s \in \Phi'$ for all $s \subseteq r$.

3. $\psi \in \Phi'$ implies $\psi^{\downarrow x} \in \Phi'$ for all $x \in D$.

As before, a subalgebra of a labeled valuation algebra is itself a labeled valuation algebra. The subalgebra inherits all axioms. In particular it is stable, if its superalgebra is so.

If (Ψ', D) is a subalgebra of a domain-free valuation algebra (Ψ, D), then, since (Ψ', D) is itself a domain-free valuation algebra, we may derive the labeled information algebra (Ψ'^*, D) from it (see Section 3.2). It is clearly a subalgebra of the labeled valuation algebra (Ψ^*, D) derived form (Ψ, D). Inversely, if (Φ', D) is a subalgebra of the stable, labeled valuation algebra (Φ, D), then the congruence σ defined in (3.8) on (Φ, D) induces also a congruence $\sigma \cap \Phi' \times \Phi'$ in the subalgebra (Φ', D). For the sake of simplicity we call this congruence still σ. It is not difficult to see that the quotient algebra $(\Phi'/\sigma, D)$ is a subalgebra of the domain-free algebra $(\Phi/\sigma, D)$.

Let's illustrate the notion of subalgebras of valuation algebras with a few examples. The simplest subalgebra of a domain-free valuation algebra (Ψ, D) is $\{e\}$. The corresponding stable, labeled version is simply $\{e_s : s \subseteq r\}$. This subalgebra is idempotent. Note however, that for a non-stable valuation algebra $\{e_s : s \subseteq r\}$ is not a subalgebra, since it is not closed under marginalization.

Less trivial examples of subalgebras of a domain-free valuation algebra (Ψ, D) are the subsets of valuations with support $x \in D$,

$$\Psi^{\Rightarrow x} = \{\psi \in \Psi : \psi = \psi^{\Rightarrow x}\}.$$

We have $\Psi^{\Rightarrow r} = \Psi$. According to Lemma 3.6 (7), $\Psi^{\Rightarrow x}$ is closed under combination. Point (2) of the same Lemma 3.6 assures us that $\Psi^{\Rightarrow x}$ is also closed under focusing. Finally, according to the neutrality axiom, e belongs to $\Psi^{\Rightarrow x}$. So $(\Psi^{\Rightarrow x}, D)$ is indeed a subalgebra of (Ψ, D).

Indicator functions can be considered as a special kind of possibility potential, taking only the values of zero and one. Marginalization of indicator functions is exactly as marginalization of possibility potentials. Note furthermore, that for potentials with only zero- and one-values, any t-norm gives the same result $(T(0,0) = T(0,1) = T(1,0) = 0, T(1,1) = 1)$. Hence combination of indicator functions corresponds to combination of possibility potentials with any t-norm. Finally, the neutral elements of indicator functions and possibility potentials are identical. Therefore, indicator functions are a subalgebra of possibility potentials for any t-norm. In other words, ordinary subsets (represented by indicator functions) are special cases of fuzzy subsets. In this sense, fuzzy subsets generalize ordinary subsets.

For other examples, consider quasi-set-potentials represented by m-functions. To fix ideas we consider the domain-free version of the valuation algebra (see Section 3.2). Since set-potentials are both closed under combination and focusing or marginalization, and since the neutral element $e(\Omega_r) = 1, e(A) = 0$ for $A \subset \Omega_r$ is a set-potential, the set-potentials form a subalgebra of the valuation algebra of quasi-set potentials.

The normalized set-functions are also a subset of (quasi-) set potentials. However, this set is not closed under combination, but only under normalized combination. So, although the normalized set potentials form a valuation algebra under the normalized combination, they are *not* a subalgebra of (quasi-) set potentials under the non-normalized combination. This is an important point: a subalgebra of a valuation algebra must be a valuation algebra with the *same* operations as the larger algebra.

Consider the set of m-functions, which for subsets $B \in \Omega_r$, including the empty set, are defined by $m_B(A) = 1$, if $A = B, m_B(A) = 0$ otherwise. This is a subset of set-potentials and it is clearly closed under combination and focusing, and contains the neutral element e. So this is a subalgebra both of the valuation algebra of (quasi-) set potentials as well as of the valuation algebra of normalized set-potentials with normalized combination. This subalgebra represents again subsets. Subsets are therefore not only particular cases of fuzzy subsets, but also of belief functions. Or, the other way round, not only fuzzy sets are generalizations of subsets, but also belief functions.

If $(\Psi_i, D), i \in I$ is any family (finite or not) of subalgebras of a valuation algebra (Ψ, D), then $(\cap_{i \in I} \Psi_i, D)$ is also a subalgebra of (Ψ, D). This is so, because, if $\phi, \psi \in \Psi_i$ for all $i \in I$, then $\phi \otimes \psi \in \Psi_i$ and $\psi^{\Rightarrow x} \in \Psi_i$ as well as $e \in \Psi_i$ for all $i \in I$. So $\phi \otimes \psi, \psi^{\Rightarrow x}, e \in \cap_{i \in I} \Psi_i$, which shows that $(\cap_{i \in I} \Psi_i, D)$ is indeed a subalgebra of (Ψ, D).

If $\Upsilon \subseteq \Psi$ is any subset of valuations of a valuation algebra (Ψ, D), then $(\langle \Upsilon \rangle_\Psi, D)$ denotes the smallest subalgebra of (Ψ, D) containing Υ. It is called the subalgebra in (Ψ, D), *generated* by Υ. It can easily be seen that

$$\langle \Upsilon \rangle_\Psi \;=\; \bigcap \{\Psi' : \Upsilon \subseteq \Psi', (\Psi', D) \text{ is a subalgebra of } (\Psi, D)\}.$$

3.3.2 Homomorphisms and Isomorphisms

The second important concept from universal algebra is the notion of *homomorphism* and related notions. Homomorphisms are mappings between valuation algebras which maintain operations. Thus, let (Ψ_1, D) and (Ψ_2, D) be two domain-free valuation algebras. Let e_1 and e_2 be the neutral elements in Ψ_1 and Ψ_2. A mapping $h : \Psi_1 \to \Psi_2$ is called a *homomorphism* if, for $\phi, \psi \in \Psi_1$ and $x \in D$,

1. $h(\phi \otimes_1 \psi) = h(\phi) \otimes_2 h(\psi)$,

2. $h(\psi^{\Rightarrow_1 x}) = h(\psi)^{\Rightarrow_2 x}$,

3. $h(e_1) = e_2$.

We have indexed the operations and neutral elements with the index of the corresponding valuation algebra to make clear that two *different* operations of the same type are involved. In the sequel however, we shall drop this index, since it will always be clear from the context which operation or neutral element is meant.

We denote the image of Ψ_1 under h by $h(\Psi_1)$. Note, that $(h(\Psi_1), D)$ is a subalgebra of (Ψ_2, D), where of course $h(\Psi_1) = \Psi_2$ is not excluded. In this latter case, (Ψ_2, D) is said to be a *homomorphic image* of (Ψ_1, D).

If h is *bijective* (that is one-to-one and $h(\Psi_1) = \Psi_2$), then h is called an *isomorphism*. In this case the inverse mapping h^{-1} exists and is itself an isomorphism. Two valuation algebras, which are isomorph are, from an algebraic point of view, essentially identical. If h is *injective* (that is one-to-one, but $h(\Psi_1) \subset \Psi_2$) then h is called a *monomorphism* or an *embedding*; the valuation algebra (Ψ_1, D) is said to be *embedded* in (Ψ_2, D). In this case the valuation algebras (Ψ_1, D) and $(h(\Psi_1), D)$ are identical from the algebraic point of view. If finally, $\Psi_1 = \Psi_2$, then homomorphisms are called *endomorphisms* and isomorphisms are called *automorphisms*. The identity mapping $h(\psi) = \psi$ is the simplest example of an automorphism.

A similar notion of homomorphism holds for labeled valuation algebras. If (Φ_1, D) and (Φ_2, D) are two labeled algebras, then a mapping $h : \Phi_1 \to \Phi_2$ is a homomorphism, if, for $\phi, \psi \in \Phi_1$ and $x \in D, x \subseteq d(\phi)$,

1. $h(\phi \otimes \psi) = h(\phi) \otimes h(\psi)$,

2. $h(\psi^{\downarrow x}) = h(\psi)^{\downarrow x}$,

3. $h(e_x) = e_x$ for all $x \in D$.

As a consequence of 2., if in a stable valuation algebra $d(\psi) = x$, then $h(\psi) = h(\psi^{\downarrow x}) = h(\psi)^{\downarrow x}$. This implies $d(h(\psi)) = d(\psi)$. In stable valuation algebras, homomorphisms maintain domain.

If h is a homomorphism (or isomorphism) between two domain-free valuation algebras (Ψ_1, D) and (Ψ_2, D), we may take the associated labeled valuation algebras (Ψ_2^*, D) and (Ψ_2^*, D) and define an induced mapping h^* between these two algebras by

$$h^*(\psi, x) = (h(\psi), x)$$

for any labeled valuation $(\psi, x) \in \Psi_1^*$. If x is a support of ψ, then we have $h(\psi)^{\Rightarrow x} = h(\psi^{\Rightarrow x}) = h(\psi)$. So x is a support for $h(\psi)$. Therefore $(h(\psi), x) \in \Psi_2^*$ which justifies the definition above. It can easily be verified that this is a homomorphism between the two labeled valuation algebras.

Inversely, if (Φ_1, D) is a stable, labeled valuation algebra, (Φ_2, D) a second labeled valuation algebra, then we have for $x \subseteq y$

$$e_x = h(e_x) = h(e_y^{\downarrow x}) = h(e_y)^{\downarrow x} = e_y^{\downarrow x}.$$

Thus the valuation algebra (Φ_2, D) is also stable. Furthermore, the homomorphism h respects the congruence relation σ. That is, if $\phi \equiv \psi \pmod{\sigma}$ in (Φ_1, D), then $h(\phi) \equiv h(\psi) \pmod{\sigma}$ in (Φ_2, D). In fact, $\phi^{\uparrow r} = \psi^{\uparrow r}$ implies

$$
\begin{aligned}
h(\phi)^{\uparrow r} &= h(\phi) \otimes e_r = h(\phi) \otimes h(e_r) = h(\phi \otimes e_r) = h(\phi^{\uparrow r}) \\
&= h(\psi^{\uparrow r}) = h(\psi \otimes e_r) = h(\psi) \otimes h(e_r) = h(\psi) \otimes e_r \\
&= h(\psi)^{\uparrow r},
\end{aligned}
$$

hence $h(\phi) \equiv h(\psi) \pmod{\sigma}$. Thus a homomorphism h between (Φ_1, D) and (Φ_2, D) induces a mapping h/σ between $(\Phi_1/\sigma, D)$ and $(\Phi_2/\sigma, D)$, defined by

$$h/\sigma([\phi]_\sigma) = [h(\phi)]_\sigma.$$

Since h maintains the congruence σ, this definition is unambiguous, and does not depend on the representative ϕ of the class $[\phi]_\sigma$ chosen. It can be easily verified, that h/σ is a homomorphism between $(\Phi_1/\sigma, D)$ and $(\Phi_2/\sigma, D)$. So homomorphisms between domain-free and stable, labeled valuation algebras may be freely moved from one type of algebra to the associated alternative type. The same is true of isomorphisms and embeddings.

A first example of an isomorphism is the mapping from indicator-function i on Ω_r to subsets of Ω_r defined by $C(i) = \{\mathbf{x} \in \Omega_r : i(\mathbf{x}) = 1\}$. Already in Chapter 2 we noted this one-to-one relation and identified intuitively the two algebras. Here we have the rigorous background justifying it. Also, the algebra of relations in isomorph to the algebra of indicator functions.

In the realm of (quasi-) set potentials we moved freely between m-, q- and b-representations. This can be justified now, since the Möbius transforms are isomorphisms between m-, q- and b-systems.

Let (Ψ, D) be a domain-free valuation algebra, (Ψ^*, D) its associated, derived stable, labeled valuation algebra and finally $(\Psi^*/\sigma, D)$ the domain-free quotient algebra corresponding to (Ψ^*, D). We define a mapping h from (Ψ, D) onto $(\Psi^*/\sigma, D)$ as follows. For $\psi \in \Psi$ let

$$h(\psi) \;=\; [(\psi, r)]_\sigma$$

First of all, h is one-to-one, since $(\psi, r) \equiv (\phi, r) \pmod \sigma$ implies $\psi = \phi$. It is a mapping onto, since for each element $[(\psi, x)]_\sigma$ in $(\Psi^*/\sigma, D)$, we have $[(\psi, x)]_\sigma = [(\psi, r)]_\sigma$. And this is the image of ψ in Ψ. Finally, the mapping is a homomorphism, since, for $\phi, \psi \in \Psi$ and $x \in D$

$$
\begin{aligned}
h(\phi \otimes \psi) &= [(\phi \otimes \psi, r)]_\sigma &= [(\phi, r) \otimes (\psi, r)]_\sigma \\
&= [(\phi, r)]_\sigma \otimes [(\psi, r)]_\sigma &= h(\phi) \otimes h(\psi),
\end{aligned}
$$

$$
\begin{aligned}
h(\psi^{\Rightarrow x}) &= [(\psi^{\Rightarrow x}, r)]_\sigma &= [(\psi^{\Rightarrow x}, x)]_\sigma \\
&= [(\psi, r)^{\downarrow x}]_\sigma &= [(\psi, r)]_\sigma^{\Rightarrow x} &= h(\psi)^{\Rightarrow x},
\end{aligned}
$$

$$h(e) \;=\; [(e, r)]_\sigma.$$

The mapping h is thus an *isomorphism*. We proved here that, if we start from a domain-free algebra, associate the derived labeled algebra and take the quotient algebra modulo σ of this stable, labeled algebra, we obtain essentially the original valuation algebra back. The same is true, if we start with a stable, labeled valuation algebra, take the quotient modulo σ then the associated, derived labeled algebra, it will be isomorph to the original one.

As an example of an embedding we consider the mapping h, which associates with a subset C of Ω_r the m-function defined by $m(C) = 1, m(A) = 0$, if $A \neq C$. Let's denote this m-function by m_C. This mapping is injective into the valuation algebra of m-functions, that is of set potentials. We have, if C_1 and C_2 are two subsets of Ω_r

$$m_{C_1} \otimes m_{C_2}(A) \;=\; \begin{cases} 1, & \text{if } A = C_1 \cap C_2, \\ 0, & \text{otherwise.} \end{cases}$$

This shows that $m_{C_1 \cap C_2} = m_{C_1} \otimes m_{C_2}$. Similarly, we have

$$m_C^{\Rightarrow x}(A) \;=\; \begin{cases} 1, & \text{if } A = C^{\Rightarrow x}, \\ 0, & \text{otherwise.} \end{cases}$$

This shows that $m_C^{\Rightarrow x} = m_{C \Rightarrow x}$. And furthermore,

$$m_{\Omega_r}(\Omega_r) \;=\; 1, \qquad m_{\Omega_r}(A) \;=\; 0, \quad \text{if } A \subset \Omega_r$$

is the neutral m-function on Ω_r. So this injection is a homomorphism, that is an embedding of *subsets* into *set potentials*. In a similar way, subsets are embedded in the valuation algebras of possibility potentials.

Any *congruence* θ in a valuation algebra (Φ, D) induces a homomorphism between (Φ, D) and $(\Phi/\theta, D)$ defined by

$$pr_\theta(\phi) \;=\; [\phi]_\theta. \tag{3.38}$$

It is called the *projection*. An example is the congruence of proportional potentials whose homomorphism essentially maps a potential to its normalized version, since we may take this potential as representative of the equivalence class. Another example is the congruence σ, where pr_σ maps ϕ essentially to $\phi^{\uparrow r}$, since we may take the latter valuation as representative of the class $[\phi]_\sigma$.

Inversely, a homomorphism h between two valuation algebras (Φ_1, D) and (Φ_2, D) (whether domain-free or labeled) defines a congruence θ_h in (Φ_1, D) as follows

$$\phi \equiv \psi \ (\mathrm{mod} \ \theta_h) \quad \text{if, and only if,} \quad h(\phi) = h(\psi).$$

This is a congruence, since (for example in the case of domain-free valuation algebras) we see that $\phi_1 \equiv \psi_1, \phi_2 \equiv \psi_2 \ (\mathrm{mod} \ \theta_h)$ implies that $h(\phi_1) = h(\psi_1), h(\phi_2) = h(\psi_2)$, hence $h(\phi_1 \otimes \phi_2) = h(\phi_1) \otimes h(\phi_2) = h(\psi_1) \otimes h(\psi_2) = h(\psi_1 \otimes \psi_2)$ and therefore $\phi_1 \otimes \phi_2 \equiv \psi_1 \otimes \psi_2 \ (\mathrm{mod} \ \theta_h)$. Similarly, if $\phi \equiv \psi$ $(\mathrm{mod} \ \theta_h)$, that is $h(\phi) = h(\psi)$ we obtain $h(\phi^{\Rightarrow x}) = h(\phi)^{\Rightarrow x} = h(\psi)^{\Rightarrow x} = h(\psi^{\Rightarrow x})$ and thus $\phi^{\Rightarrow x} \equiv \psi^{\Rightarrow x} \ (\mathrm{mod} \ \theta_h)$. The congruence θ_h of a homomorphism h is called its *kernel*. It induces in the usual way a quotient valuation algebra $(\Phi/\theta_h, D)$.

The following important theorem is an example of a result of universal algebra, but which is stated here for the case of valuation algebras, indeed for domain-free valuation algebras. The reader can formulate the theorem for labeled valuation algebras himself.

Theorem 3.9 Homomorphism Theorem. *Let (Ψ_1, D) and (Ψ_2, D) be two domain-free valuation algebras and $h : \Psi_1 \to \Psi_2$ a homomorphism such that $h(\Psi_1) = \Psi_2$, Then there exists an isomorphism $k : \Psi_1/\theta_h \to \Psi_2$ such that*

$$k \circ pr_h \ = \ h, \tag{3.39}$$

where pr_h is the projection associated with the congruence θ_h induced by the homomorphism h.

Proof. Define k by $k([\psi]_{\theta_h}) = h(\psi)$. If $[\psi]_{\theta_h} = [\phi]_{\theta_h}$, then $h(\psi) = h(\phi)$, therefore $k([\psi]_{\theta_h})$ is uniquely defined. To check, that k is a homomorphism, consider $[\psi]_{\theta_h}, [\phi]_{\theta_h}$ and $x \in D$,

$$
\begin{aligned}
k([\psi]_{\theta_h} \otimes [\phi]_{\theta_h}) \ &= \ k([\psi \otimes \phi]_{\theta_h}) \ &= \ h(\psi \otimes \phi) \\
&= \ h(\psi) \otimes h(\phi) \ &= \ k([\psi]_{\theta_h}) \otimes k([\phi]_{\theta_h}),
\end{aligned}
$$

$$
\begin{aligned}
k([\psi]_{\theta_h}^{\Rightarrow x}) \ &= \ k([\psi^{\Rightarrow x}]_{\theta_h}) \ &= \ h(\psi^{\Rightarrow x}) \\
&= \ h(\psi)^{\Rightarrow x} \ &= \ k([\psi]_{\theta_h})^{\Rightarrow x}.
\end{aligned} \tag{3.40}
$$

So k is a homomorphism. Since h is surjective, k is surjective too. Furthermore, if $[\psi]_{\theta_h} \neq [\phi]_{\theta_h}$, then $h(\psi) \neq h(\phi)$ and so $k([\psi]_{\theta_h}) \neq k([\phi]_{\theta_h})$. So k is an isomorphism.

Finally $k(pr_h(\psi)) = k([\psi]_{\theta_h}) = h(\psi)$ so that $k \circ pr_h = h$. $\qquad \square$

As an application, we consider labeled probability potentials. Let p be a potential for s. We define a mapping of the labeled valuation algebra of potentials onto the valuation algebra of *normalized* potentials, by

$$p^{\downarrow}(\mathbf{x}) = \frac{1}{K}p(\mathbf{x}), \qquad K = \sum_{\mathbf{x} \in \Omega_s} p(\mathbf{x}),$$

if $K > 0$. If $p(\mathbf{x}) = 0$ for all $\mathbf{x} \in \Omega_s$, then we define $p^{\downarrow}(\mathbf{x}) = 0$ for all $\mathbf{x} \in \Omega_s$. Remember that combination in the valuation algebra of normalized potentials involves normalization (see Section 2.3.3, (2.44) and (2.45)).

We show that this is a homomorphism. Let p_1 be a potential for s and p_2 a potential for t, such that $p_1 \otimes p_2 \neq 0$. Then, if $\mathbf{x} \in \Omega_{s \cup t}$,

$$(p_1 \otimes p_2)^{\downarrow}(\mathbf{x}) = \frac{p_1(\mathbf{x}^{\downarrow s})p_2(\mathbf{x}^{\downarrow t})}{\sum_{\mathbf{x} \in \Omega_{s \cup t}} p_1(\mathbf{x}^{\downarrow s})p_2(\mathbf{x}^{\downarrow t})}.$$

On the other hand, we have

$$\begin{aligned}
p_1^{\downarrow} \otimes p_2^{\downarrow}(\mathbf{x}) &= \frac{p_1^{\downarrow}(\mathbf{x}^{\downarrow s})p_2^{\downarrow}(\mathbf{x}^{\downarrow t})}{\sum_{\mathbf{x} \in \Omega_{s \cup t}} p_1^{\downarrow}(\mathbf{x}^{\downarrow s})p_2^{\downarrow}(\mathbf{x}^{\downarrow t})} \\
&= \frac{\frac{1}{K_1}p_1(\mathbf{x}^{\downarrow s})\frac{1}{K_2}p_2(\mathbf{x}^{\downarrow t})}{\sum_{\mathbf{x} \in \Omega_{s \cup t}} \frac{1}{K_1}p_1(\mathbf{x}^{\downarrow s})\frac{1}{K_2}p_2(\mathbf{x}^{\downarrow t})} \\
&= \frac{p_1(\mathbf{x}^{\downarrow s})p_2(\mathbf{x}^{\downarrow t})}{\sum_{\mathbf{x} \in \Omega_{s \cup t}} p_1(\mathbf{x}^{\downarrow s})p_2(\mathbf{x}^{\downarrow t})}.
\end{aligned}$$

Here K_1 and K_2 are the normalization constants associated with p_1 and p_2. And we see that $(p_1 \otimes p_2)^{\downarrow} = p_1^{\downarrow} \otimes p_2^{\downarrow}$ if neither $K_1 = 0$ nor $K_2 = 0$. If one or both of the two constants are zero, then one or both of p_1 or p_2 are zero, hence both $(p_1 \otimes p_2)^{\downarrow}$ and $p_1^{\downarrow} \otimes p_2^{\downarrow}$ are zero. If $p_1 \otimes p_2 = 0$, then $(p_1 \otimes p_2)^{\downarrow} = 0$ and $p_1^{\downarrow} \otimes p_2^{\downarrow} = 0$ such that $(p_1 \otimes p_2)^{\downarrow} = p_1^{\downarrow} \otimes p_2^{\downarrow}$ in this case too. So $(p_1 \otimes p_2)^{\downarrow} = p_1^{\downarrow} \otimes p_2^{\downarrow}$ holds always.

Furthermore, let p be a potential for s, not equal to zero, and $t \subseteq s$, then, for $\mathbf{x} \in \Omega_t$

$$(p^{\downarrow t})^{\downarrow}(\mathbf{x}) = \frac{\sum_{\mathbf{y} \in \Omega_{s-t}} p(\mathbf{x}, \mathbf{y})}{\sum_{\mathbf{x} \in \Omega_t}\sum_{\mathbf{y} \in \Omega_{s-t}} p(\mathbf{x}, \mathbf{y})}.$$

On the other we obtain

$$(p^{\downarrow})^{\downarrow t}(\mathbf{x}) = \sum_{\mathbf{y} \in \Omega_{s-t}} \frac{p(\mathbf{x}, \mathbf{y})}{\sum_{\mathbf{x} \in \Omega_t, \mathbf{y} \in \Omega_{s-t}} p(\mathbf{x}, \mathbf{y})}. \tag{3.41}$$

So, $(p^{\downarrow t})^{\downarrow} = (p^{\downarrow})^{\downarrow t}$ if p is not identically equal to zero. But in this latter case both $(p^{\downarrow t})^{\downarrow}$ and $(p^{\downarrow})^{\downarrow t}$ are zero. So, $(p^{\downarrow t})^{\downarrow} = (p^{\downarrow})^{\downarrow t}$ holds in any case.

If finally $e_s(\mathbf{x}) = 1$ for all $\mathbf{x} \in \Omega_s$ is the neutral element on s, then $e_s^{\downarrow}(\mathbf{x}) = 1/|\Omega_s|$. This is the neutral element for normalized combination.

The mapping defined by normalization is therefore a homomorphism. Its induced congruence in the valuation algebra of potentials is the equivalence relation defined by proportionality which we referred to several times above. The quotient algebra induced by proportionality of potentials is according to the homomorphism theorem above isomorph to the valuation algebra of normalized potentials. Such scaling or normalization homomorphisms will be discussed in a more general setting in Section 3.7.

3.3.3   Weak Subalgebras and Homomorphisms

We have seen in Section 2.2 that a weaker form of labeled valuation algebra exists without neutral elements, but with a modified, strengthened combination axiom. There is a corresponding weak notion of subalgebra and of homomorphism. Let (Φ, D) be a labeled valuation algebra satisfying this weaker system of axioms. (Φ', D) is said to be a *weak subalgebra* of (Φ, D), if $\Phi' \subseteq \Phi$ and

1. $\phi, \psi \in \Psi'$ implies $\phi \otimes \psi \in \Psi'$,

2. $\psi \in \Psi'$ implies $\psi^{\downarrow x} \in \Psi'$ for all $x \in D$.

In a similar spirit, if (Φ_1, D) and (Φ_2, D) are two labeled valuation algebras, then a mapping $h : \Phi_1 \to \Phi_2$ is called a *weak homomorphism*, if

1. $h(\phi \otimes \psi) = h(\phi) \otimes h(\psi)$,

2. $h(\psi^{\downarrow x}) = h(\psi)^{\downarrow x}$.

As an example we consider set potentials and in particular commonality functions q with the following property:

$$q(A) \;=\; 0, \qquad \text{if } |A| > 1.$$

So the only subsets with commonality values different from zero are possibly the single-element subsets and the empty set. We write here $q(\mathbf{x})$ for $q(\{\mathbf{x}\})$. Since

$$m(\emptyset) \;=\; q(\emptyset) - \sum_{\mathbf{x} \in \Omega_s} q(\mathbf{x}) \;\geq\; 0,$$

we see that

$$q(\emptyset) \;\geq\; \sum_{\mathbf{x} \in \Omega_s} q(\mathbf{x}) \;\geq\; 0.$$

For the corresponding m-function we obtain that

$$m(\{\mathbf{x}\}) \;=\; m(\mathbf{x}) \;=\; q(\mathbf{x}), \qquad m(A) \;=\; 0, \quad \text{if } |A| > 1.$$

Let Φ' denote the subset of those particular set potentials. This set Φ' is closed under combination and marginalization. In fact, if $q_1, q_2 \in \Phi'$, we have that

$q_1 \otimes q_2(A) = q_1(A^{\downarrow s})q_2(A^{\downarrow t}) = 0$ if $|A| > 1$, since in this case either $|A|^{\downarrow s} > 1$ or $|A|^{\downarrow t} > 1$ (or both). On the other hand we have that

$$
\begin{aligned}
q_1 \otimes q_2(\mathbf{x}) &= q_1(\mathbf{x}^{\downarrow s})q_2(\mathbf{x}^{\downarrow t}) \geq 0, \\
q_1 \otimes q_2(\emptyset) &= q_1(\emptyset)q_2(\emptyset) \geq 0.
\end{aligned}
$$

We remark also that $q_1 \otimes q_2$ is by the general theory of set potentials still a commonality function. This shows that Φ' is closed under combination.

To see that Φ' is also closed under marginalization, we look at the corresponding m-functions. And we see that

$$
m^{\downarrow t}(A) = \sum_{B^{\downarrow t} = A} m(B) = 0,
$$

if $|A| > 1$, since in this case $B^{\downarrow t} = A$ implies $|B| > 1$. We obtain also

$$
\begin{aligned}
m^{\downarrow t}(\mathbf{x}) &= \sum_{B^{\downarrow t} = \{\mathbf{x}\}} m(B) = \sum_{\mathbf{y}^{\downarrow t} = \mathbf{x}} m(\mathbf{y}) \geq 0, \\
m^{\downarrow t}(\emptyset) &= m(\emptyset).
\end{aligned}
$$

Note however, that the neutral elements on domains s, represented by the commonality functions $e_s(A) = 1$ for all subsets A of Ω_s do not belong to Φ'. So this shows that (Φ', D) is a weak subalgebra of (Φ, D).

We can associate with each commonality function q in Φ' a potential p_q by defining $p_q(\mathbf{x}) = q(\mathbf{x})$. This is a weak homomorphism between (Φ', D) and the valuation algebra of potentials. Indeed, we have

$$
\begin{aligned}
p_{q_1 \otimes q_2}(\mathbf{x}) &= q_1 \otimes q_2(\mathbf{x}) = q_1(\mathbf{x}^{\downarrow s})q_2(\mathbf{x}^{\downarrow t}) \\
&= p_{q_1}(\mathbf{x}^{\downarrow s})p_{q_2}(\mathbf{x}^{\downarrow t}) = p_{q_1} \otimes p_{q_2}(\mathbf{x}).
\end{aligned}
$$

And we have also, since $q(\mathbf{x}) = m(\mathbf{x})$,

$$
\begin{aligned}
p_{q^{\downarrow t}}(\mathbf{x}) &= q^{\downarrow t}(\mathbf{x}) \\
&= \sum_{\mathbf{y}^{\downarrow t} = \mathbf{x}} q(\mathbf{y}) = \sum_{\mathbf{y}^{\downarrow t} = \mathbf{x}} p_q(\mathbf{y}) = p_q^{\downarrow t}(\mathbf{x}).
\end{aligned}
$$

A weak homomorphism h defines a congruence $\phi \equiv \psi \pmod{\theta_h}$ just like an ordinary homomorphism by $h(\phi) = h(\psi)$. Hence, in the example above, we obtain that $q_1 \equiv q_2 \pmod{\theta_h}$ if, and only if $d(q_1) = d(q_2)$ (say $= s$), and $q_1(\mathbf{x}) = q_2(\mathbf{x})$ for all $\mathbf{x} \in \Omega_s$. The homomorphism theorem too carries over to weak homomorphisms and we therefore obtain a weak embedding of probability potentials into set potentials, defined by $p \rightarrow [q]$ where $[q]$ is the equivalence class of all commonalities with $q(\mathbf{x}) = p(\mathbf{x})$ in Φ'. We saw above (Section 3.3.2) that subsets are embedded in the valuation algebra of set potentials. Here we see now a similar (slightly weaker) embedding of probability potentials into the algebra of set potentials. In this sense set potentials (or belief functions) generalize both subsets and (discrete) probability distributions.

3.4   Null Valuations

Some valuation algebras contain valuations which are incompatible and which, when combined, create something like contradictions. As an example consider disjoint sets. Their combination yields the empty set. This shows that such set constraints are incompatible. When additional valuations are combined, the result remains a contradiction. Such contradictions are represented in valuation algebras by null valuations. They have some importance for theoretical considerations and therefore this section is devoted to their study. Examples of null valuations have already been seen scattered around at different places.

Let (Φ, D) be a labeled valuation algebra. A valuation z_s for s is called the *null valuation* for s, if for all valuations ϕ for s we have

$$\phi \otimes z_s \;=\; z_s. \tag{3.42}$$

If a null valuation exists, then it is *unique*. In fact, suppose z_s' is another null valuation for s besides z_s. Then, since both are null valuations, we must have $z_s = z_s \otimes z_s' = z_s'$.

If (Φ, D) is a labeled valuation algebra with null valuations z_s for s, then two valuations ϕ for s and ψ for t are called *contradictory*, if

$$\phi \otimes \psi \;=\; z_{s \cup t}. \tag{3.43}$$

As we shall see in the examples, the combination of contradictory valuations is something that is to be avoided, since it leads to trivialities.

Let's look at some examples to understand the nature of null valuations. In the valuation algebra of probability potentials, the potential $z_s(\mathbf{x}) = 0$ for all $\mathbf{x} \in \Omega_s$ is the null valuation for s, since $p \otimes z_s(\mathbf{x}) = p(\mathbf{x}) z_s(\mathbf{x}) = 0$ for all $\mathbf{x} \in \Omega_s$. In this example we have clearly $z_s^{\downarrow t} = z_t$ for all $t \subseteq s$, and in particular $z_s^{\downarrow t}(\diamond) = 0$. Similarly we have also $z_s^{\uparrow t} = z_t$ for all $t \supseteq s$. Also, if ϕ is a valuation for s, then $\phi^{\downarrow t} = z_t$ for $t \subseteq s$ implies $\phi = z_s$. In fact,

$$\phi^{\downarrow t}(\mathbf{x}) \;=\; \sum_{\mathbf{y} \in \Omega_{s-t}} \phi(\mathbf{x}, \mathbf{y}) \;=\; 0 \quad \text{for all } \mathbf{x} \in \Omega_t$$

is only possible, if $\phi(\mathbf{x}, \mathbf{y}) = 0$ for all $(\mathbf{x}, \mathbf{y}) \in \Omega_s$, since $\phi(\mathbf{x}, \mathbf{y}) \geq 0$.

Two potentials p_1 for s and p_2 for t are contradictory, if, for all $\mathbf{x} \in \Omega_{s \cup t}$,

$$p_1(\mathbf{x}^{\downarrow s}) p_2(\mathbf{x}^{\downarrow t}) \;=\; 0.$$

This is exactly then the case, if for all $\mathbf{x} \in \Omega_{s \cup t}$ either $p_1(\mathbf{x}^{\downarrow s}) = 0$ or $p_2(\mathbf{x}^{\downarrow t}) = 0$ or both.

In the case of subsets, the null valuation for s is the empty subset of Ω_s. We take here the convention that each frame Ω_s has its own empty subset $\emptyset_s$. Clearly, as with potentials we have $\emptyset_s^{\downarrow t} = \emptyset_t$ for all subsets t of s. And similarly, $\emptyset_s^{\uparrow t} = \emptyset_t$ for all supersets t of s. In terms of indicator functions, we have of course $z_s(\mathbf{x}) = 0$ for all $\mathbf{x} \in \Omega_s$. As in the case of potentials, if C is a subset of Ω_s such that $C^{\downarrow t} = \emptyset_t$, we must have $C = \emptyset_t$. Two subsets are contradictory, if their join is the empty set, $C_1 \bowtie C_2 = \emptyset$. Thus, if the two sets represent constraints on the possible values of the variables, they are *incompatible*.

Possibility potentials have the same null elements $z_s(\mathbf{x}) = 0$. This is true, since for all t-norms T we have $T(p(\mathbf{x}), 0) = 0$. Two possibility potentials p_1 and p_2 are contradictory, if $T(p_1(\mathbf{x}^{\downarrow s}), p_2(\mathbf{x}^{\downarrow t})) = 0$. In the case of the product and the Gödel t-norms, this is the case if either $p_1(\mathbf{x}^{\downarrow s}) = 0$ or $p_2(\mathbf{x}^{\downarrow t}) = 0$ for every configuration $\mathbf{x}$. In the case of the Lukasziewicz t-norm, however, it is sufficient that

$$p_1(\mathbf{x}^{\downarrow s}) + p_2(\mathbf{x}^{\downarrow t}) \leq 1.$$

So it depends on the t-norm whether two potentials are contradictory or not.

In the case of (quasi-) set potentials, the null valuation for s is, in terms of m-, q- and b functions, given by $z_s(A) = 0$ for all subsets A of Ω_s. As in the previous examples, we have that if m is the m-function of a set potential for s, then $m^{\downarrow t} = z_t$ if and only if $m = z_s$. Note that this is *not* true for *quasi-set potentials*! In fact

$$m^{\downarrow t}(A) = \sum_{B^{\downarrow t} = A} m(B) = 0$$

does not imply $m(B) = 0$ for all subsets B, if $m(B)$ is not restricted to be non-negative. We remark that a m-function with $m(A) = 0$ for $A \neq \emptyset$ and $m(\emptyset) > 0$ is not considered as a null valuation. In fact, such set potentials are not unique and they do not satisfy (3.42).

We may also consider the case of normalized set potentials with normalized combination. In terms of m-functions the null elements are then defined by $z_s(A) = 0$ for all $A \neq \emptyset$ and $z_s(\emptyset) = 1$. This definition corresponds to the definition of normalized combination in the case of $K = 0$ (see Section 2.3.6). So, we have for example $m \otimes z_s(A) = 0$, if $A \neq \emptyset$, and $m \otimes z_s(\emptyset) = 1$ according to this definition of combination. This is again the null element. Also, the equation $z_s^{\downarrow t} = z_t$ is valid.

There are also valuation algebras *without* null valuations. Examples are Spohn potentials and Gaussian potentials. These potentials are therefore never contradictory.

As in the examples above, we expect that marginalization of a consistent (non-null) valuation produces a consistent valuation. Such a property is surely desired for a reasonable formalism to represent information or knowledge. However the case of quasi-set potentials shows that this is not necessarily the case. In order to exclude such cases we extend the system of axioms to take care of null elements as follows:

(1) *Semigroup:* Φ is associative and commutative under combination. For all $s \in D$ there are elements e_s and z_s with $d(e_s) = d(z_s) = s$ such that for all $\phi \in \Phi$ with $d(\phi) = s$, $e_s \otimes \phi = \phi$ and $z_s \otimes \phi = z_s$.

Axioms (2) to (6) (labeling, marginalization, transitivity, combination, neutrality and possibly stability) remain as before. But the following axiom is added:

(8) *Nullity:* For $x, y \in D$, $x \subseteq y$, $\phi \in \Phi$ with $d(\phi) = y$,

$$\phi^{\downarrow x} = z_x \quad \text{if, and only if,} \quad \phi = z_y. \tag{3.44}$$

This axiom is satisfied in all examples of valuation algebras considered so far, with the exception of quasi-set potentials (for algebras without null elements, the axiom is trivially satisfied).

Here are two useful results on null valuations:

Lemma 3.10 *Suppose* (Φ, D) *is a stable, labeled valuation algebra which satisfies the nullity axiom. Then*

1. *If* $s \subseteq t$, *then* $z_s^{\uparrow t} = z_t$.

2. *If* $d(\phi) = s$, *then* $\phi \otimes z_t = z_{s \cup t}$.

Proof. (1) By the stability we have $(z_s^{\uparrow t})^{\downarrow s} = z_s$. The nullity axiom implies thus that $(z_s^{\uparrow t}) = z_t$.

(2) We use (1) just proved to obtain $\phi \otimes z_t = (\phi \otimes e_{s \cup t}) \otimes (z_t \otimes e_{s \cup t}) = \phi^{\uparrow s \cup t} \otimes z_t^{\uparrow s \cup t} = \phi^{\uparrow s \cup t} \otimes z_{s \cup t} = z_{s \cup t}$. □

Suppose now, that the valuation algebra (Φ, D) satisfies the nullity axiom and is also stable.

According to Lemma 3.10 (1) and the nullity axiom, $[z_s]_\sigma = \{z_t : z_t \in \Phi\}$. In the quotient valuation algebra $(\Phi/\sigma, D)$ we obtain from Lemma 3.10 (2)

$$
\begin{aligned}
{[\phi]_\sigma} \otimes [z_s]_\sigma &= [\phi \otimes z_s]_\sigma &= [z_{s \cup d(\phi)}]_\sigma, \\
{[z_s]_\sigma^{\Rightarrow x}} &= [z_s^{\rightarrow x}]_\sigma &= [z_x]_\sigma.
\end{aligned} \tag{3.45}
$$

This shows, that $[z_s]_\sigma$ is the *domain-free null valuation* and that the focus of the domain-free null valuation to any domain x is again the domain-free null valuation. Thus, the domain-free null valuation, just as the neutral element, has dimension zero.

This result motivates the introduction of *domain-free null valuations* in domain-free valuation algebras. If (Ψ, D) is a domain-free valuation algebra, then $z \in \Psi$ is a null element, if $\psi \otimes z = z$ for all valuations ψ in Ψ. If there exists a null valuation in a domain-free valuation algebra, then it is unique. The domain-free version of the nullity axiom is as follows: $\psi^{\Rightarrow x} = z$ if and only if $\psi = z$. Therefore, we update the system of axioms for domain-free valuation algebras (Ψ, D) with null element as follows:

(1) *Semigroup:* Ψ is associative and commutative under combination. There is a neutral element $e \in \Psi$ and a null element $z \in \Psi$ such that $e \otimes \psi = \psi$ and $z \otimes \psi = z$ for all $\psi \in \Psi$.

Axioms (2) to (5) (transitivity, combination, neutrality, support) remain as before. But the following new axiom is added:

(6) *Nullity:* For $x \in D$ and $\psi \in \Psi$,

$$\psi^{\Rightarrow x} = z \quad \text{if, and only if,} \quad \psi = z. \tag{3.46}$$

This shows that $\{e, z\}$ is a subalgebra of (Ψ, D), if it satisfies the nullity axiom.

We saw above, that the domain-free algebra associated with a stable, labeled valuation algebra, satisfying the nullity axiom, itself satisfies the nullity axiom. Inversely, if (Ψ, D) is a domain-free valuation algebra, satisfying the nullity axiom, then the associated labeled valuation algebra (Ψ^*, D) satisfies also the nullity axiom, as can easily be verified.

3.5 Regular Valuation Algebras

In two of the most important architectures for computing marginals of probability potentials, *division* of potentials is used (Lauritzen & Spiegelhalter, 1988; Jensen, Lauritzen & Olesen, 1990). Although in valuation algebras in general no division is defined, (Lauritzen & Jensen, 1997) discussed division in semigroups which are union of groups or which are embedded in a union of groups. They showed how the architectures mentioned above apply to such structures. We take up here their approach and develop it in the algebraic framework of valuations algebras. Much of the material discussed here comes from semigroup theory. We refer to (Clifford & Preston, 1967) for semigroups. In semigroup theory necessary and sufficient conditions are stated for a semigroup to be the union of a family of disjoint groups or to be embedded into a semigroup which is such an union. In our context however, the additional question of how this structure interferes with the domain structure and marginalization arises.

In Chapter 4 we shall show how division can be used in computing marginals. But studying division in valuation algebra is also important in its own right. It clarifies the structure of valuation algebras. It is further needed in scaling of valuation algebras and enriches the discussion of conditional independence related to valuation algebras, see Chapter 5.

We start with Croisot's theory of semigroups with inverses (Croisot, 1953) and adapt it to valuation algebras. These are semigroups which can be decomposed into a union of disjoint groups. Let then (Φ, D) be a labeled valuation algebra (not necessarily stable). The basic notion is the one of regularity.

Definition 3.11 Regularity:

1. *An element $\phi \in \Phi$ is called* regular, *if there exists for all $t \subseteq d(\phi)$ an element $\chi \in \Phi$ with $d(\chi) = t$ and such that*

$$\phi = \phi^{\downarrow t} \otimes \chi \otimes \phi. \tag{3.47}$$

2. *The valuation algebra (Φ, D) is called* regular, *if all its elements are regular.*

Note that null valuations are always regular, since we may take for χ any element in the corresponding domain t. In stable valuation algebras, neutral elements are also regular. In this case we take $\chi = e_t$. But neutral elements may also be regular, without being stable. We remark that χ needs not to be unique.

χ depends of course on t. But we do not emphasize this dependency, because it will always be clear from the context to which domain χ refers. In particular, for $t = d(\phi)$, regularity means that

$$\phi \;=\; \phi \otimes \chi \otimes \phi. \tag{3.48}$$

Elements which can be represented like that are called regular in semigroup theory, and that is where we borrowed this concept. In this case we obtain

$$\phi \otimes \chi \otimes \phi \otimes \chi \;=\; \phi \otimes \chi,$$

which shows that $\phi \otimes \chi$ is idempotent. We also have, more generally, for regular elements that

$$\phi^{\downarrow t} \;=\; (\phi^{\downarrow t} \otimes \chi \otimes \phi)^{\downarrow t} \;=\; \phi^{\downarrow t} \otimes \chi \otimes \phi^{\downarrow t}.$$

So, if ϕ is a regular element of a valuation algebra, then its marginals are all regular as elements of the corresponding semigroup.

Furthermore, we see that for $t \subseteq s \subseteq d(\phi)$

$$\phi^{\downarrow s} \;=\; (\phi^{\downarrow t} \otimes \chi \otimes \phi)^{\downarrow s} \;=\; \phi^{\downarrow t} \otimes \chi \otimes \phi^{\downarrow s}$$

(use Lemma 2.1 (3)). So, if ϕ is a regular valuation in a valuation algebra, all its marginals are regular elements of the valuation algebra.

Let's examine some examples:

(1) *Idempotent Valuation Algebras:* In this important case, a solution to the regularity equation (3.47) is either $\chi = e_t$, or $\chi = \phi^{\downarrow t}$. There are many other solutions. So, all valuations are regular, including the null elements. Idempotent valuation algebras are regular, but in a trivial way.

(2) *Probability Potentials:* Let p be a potential on s and $t \subseteq s$. Then

$$p^{\downarrow t} \otimes \chi \otimes p(\mathbf{x}) \;=\; p^{\downarrow t}(\mathbf{x}^{\downarrow t}) \cdot \chi(\mathbf{x}^{\downarrow t}) \cdot p(\mathbf{x}).$$

So we may define

$$\chi(\mathbf{x}) \;=\; \begin{cases} \frac{1}{p^{\downarrow t}(\mathbf{x})} & \text{if } p^{\downarrow t}(\mathbf{x}) > 0, \\ \text{arbitrary} & \text{otherwise} \end{cases} \tag{3.49}$$

Since $p^{\downarrow t}(\mathbf{x}^{\downarrow t}) = 0$ implies that $p(\mathbf{x}) = 0$, this is a solution to the regularity equation. Hence all potentials are regular, including the null potentials. If $p^{\downarrow t}(\mathbf{x}) = 0$ for some configurations $\mathbf{x}$, then χ is not uniquely determined. Probability potentials are the most important model for regular valuation algebras.

(3) *Possibility Potentials:* The situation regarding regularity depends here on the t-norm selected for the valuation algebra. This is a further hint to the

non-uniform nature of possibility theory. For the product t-norm, the situation is exactly as for probability potentials: (3.49) is a solution of the regularity equation (of course, the marginal is defined differently for possibility potentials than for probability potentials). So, with this t-norm the valuation algebra of possibility potentials is regular.

With the Gödel t-norm, the valuation algebra is idempotent, hence trivially regular.

For the Lukasziewicz t-norm, the regularity equation is

$$p(\mathbf{x}, \mathbf{y}) \quad = \quad \max\{0, p(\mathbf{x}, \mathbf{y}) + \max\{0, p^{\downarrow t}(\mathbf{y}) + \chi(\mathbf{y}) - 1\} - 1\}.$$

Hence, if $p(\mathbf{x}, \mathbf{y}) > 0$, we must have

$$\max\{0, p^{\downarrow t}(\mathbf{y}) + \chi(\mathbf{y}) - 1\} \quad = \quad 1.$$

This implies

$$p^{\downarrow t}(\mathbf{y}) + \chi(\mathbf{y}) \quad = \quad 2.$$

But, if $p^{\downarrow t}(\mathbf{y}) < 1$ there is no solution $\chi(\mathbf{y})$ in the interval $[0, 1]$. Thus, there are possibility potentials, which are not regular relative to the Lukasziewicz t-norm. The corresponding valuation algebra of possibility potentials is therefore not regular.

(4) *Spohn Potentials*: The regularity equation in this valuation algebra is

$$p(\mathbf{x}, \mathbf{y}) \quad = \quad p^{\downarrow t}(\mathbf{y}) + \chi(\mathbf{y}) + p(\mathbf{x}, \mathbf{y}).$$

This equation has the solution $\chi(\mathbf{y}) = -p^{\downarrow t}(\mathbf{y})$. But this is no more a Spohn potential, since it takes negative values. Thus, the valuation algebra of Spohn potentials is not regular. If we allow for p also negative integers (quasi-Spohn potentials), then they form still a valuation algebra under addition for combination and minimization for marginalization. This enlarged valuation algebra is regular. So at least the valuation algebra of Spohn potentials is embedded in a larger regular one.

(5) *Set Potentials*: Let $q \in Q^+$ be a commonality function on s and $t \subseteq s$. The regularity equation is in this case

$$q^{\downarrow t} \otimes \chi \otimes q(A) \quad = \quad q^{\downarrow t}(A^{\downarrow t}) \cdot \chi(A^{\downarrow t}) \cdot q(A).$$

$q^{\downarrow t}(A) = 0$ implies $q(B) = 0$ for all $B \subseteq \Omega_s$ such that $B^{\downarrow t} = A$. To see this, use the definition of commonality in terms of m-functions and the marginalization of m-functions:

$$q^{\downarrow t}(A^{\downarrow t}) \quad = \quad \sum_{C: C \supseteq A^{\downarrow t}} m^{\downarrow t}(C) \quad = \quad \sum_{C: C \supseteq A^{\downarrow t}} \sum_{B\ B^{\downarrow t} = C} m(B)$$

$$= \quad \sum_{B: B^{\downarrow t} \supseteq A^{\downarrow t}} m(B) \quad \geq \quad \sum_{B \cdot B \supseteq A} m(B) \quad = \quad q(A) \quad \geq \quad 0.$$

Therefore, the following is a solution to the regularity equation:

$$\chi(A) \;=\; \begin{cases} \frac{1}{q^{\downarrow t}(A)} & \text{if } q^{\downarrow t}(A) > 0, \\ \text{arbitrary} & \text{otherwise.} \end{cases} \tag{3.50}$$

However, in general, this solution is no more a commonality function $\chi \in Q^+$, but represents only a quasi-set potential. Therefore, set potentials are not necessarily regular and the valuation algebra of set potentials is not regular. If, on the other hand, we assume q to represent a quasi-set potential, then $q^{\downarrow t}(A^{\downarrow t}) = 0$ does not necessarily imply that $q(A) = 0$. This means that the regularity equation may have no solution. So the valuation algebra of quasi-set potentials is not regular either (although quasi-set potentials are regular as a semigroup).

Sometimes, non-negative q-functions are considered. They form a semigroup under multiplication. However, if we define marginalization, as usual by going back to corresponding m-functions, then, when we come back to q-functions it is no longer necessarily non-negative. So non-negative q-functions do not form a valuation algebra. So, compared to Spohn potentials, the situation of commonality functions (hence of belief functions) is rather more complicated. It is not clear at this point, whether set potentials are embedded in a regular valuation algebra (see however Section 3.6).

Although, as we have seen, by far not all interesting valuation algebras are regular, the model of probability potentials is sufficiently important to make a study of regular valuation algebras worthwhile and rewarding.

Regularity allows to introduce a notion of an inverse, a notion which is borrowed from semigroup theory.

Definition 3.12 Inverse Valuations: $\phi, \psi \in \Phi$ *are called* inverses, *if*

$$\phi \otimes \psi \otimes \phi \;=\; \phi \quad \text{and} \quad \psi \otimes \phi \otimes \psi \;=\; \psi. \tag{3.51}$$

Note that inverses are not necessarily unique. Here are a few results on inverses:

Lemma 3.13

1. *If ϕ is regular, that is, $\phi = \phi \otimes \chi \otimes \phi$, then ϕ and $\chi \otimes \phi \otimes \chi$ are inverses.*

2. *If ϕ, ψ are inverses, then $f = \phi \otimes \psi$ is idempotent and $f \otimes \phi = \phi$ and $f \otimes \psi = \psi$.*

3. *If ϕ, ψ are inverses, then $d(\phi) = d(\psi)$.*

Proof. (1) We have

$$\phi \otimes (\chi \otimes \phi \otimes \chi) \otimes \phi \;=\; \phi \otimes \chi \otimes (\phi \otimes \chi \otimes \phi) \;=\; \phi \otimes \chi \otimes \phi \;=\; \phi,$$

and

$$(\chi \otimes \phi \otimes \chi) \otimes \phi \otimes (\chi \otimes \phi \otimes \chi) \;=\; \chi \otimes (\phi \otimes \chi \otimes \phi) \otimes \chi \otimes \phi \otimes \chi$$

$$= \chi \otimes (\phi \otimes \chi \otimes \phi) \otimes \chi = \chi \otimes \phi \otimes \chi.$$

(2) Here we obtain

$$f \otimes f = (\phi \otimes \psi) \otimes (\phi \otimes \psi) = (\phi \otimes \psi \otimes \phi) \otimes \psi = \phi \otimes \psi = f.$$

We have also

$$f \otimes \phi = (\phi \otimes \psi) \otimes \phi = \phi,$$

and similarly $f \otimes \psi = \psi$.

(3) $\phi = \phi \otimes \psi \otimes \phi$ implies $d(\psi) \subseteq d(\phi)$ and $\psi = \psi \otimes \phi \otimes \psi$ implies $d(\phi) \subseteq d(\psi)$.
□

The idempotent elements of a regular valuation algebra play an important role. They are inverse to themselves. If f_1, f_2 are two idempotent elements, then $f_1 \otimes f_2$ is idempotent, since $(f_1 \otimes f_2) \otimes (f_1 \otimes f_2) = (f_1 \otimes f_1) \otimes (f_2 \otimes f_2) = f_1 \otimes f_2$. We define $f_1 \leq f_2$ if $f_1 \otimes f_2 = f_2$. This is a partial order between the idempotent elements of a regular valuation algebra, that is,

1. $f \leq f$.

2. $f_1 \leq f_2$ and $f_2 \leq f_1$ implies $f_1 = f_2$.

3. $f_1 \leq f_2$ and $f_2 \leq f_3$ implies $f_1 \leq f_3$.

This order becomes important in our study of regular valuation algebras. Clearly, for any two idempotents, $f_1, f_2 \leq f_1 \otimes f_2$. But, if f is another upper bound of f_1, f_2, then $f \otimes (f_1 \otimes f_2) = f$. So $f_1 \otimes f_2$ is the least upper bound, the supremum of f_1, f_2. Thus, the idempotents of a regular valuation algebra form a *semilattice*.

The neutral elements of the algebra are idempotent, and by the neutrality axiom we have $e_t \leq e_s$ if $t \subseteq s$. Also, since $e_s \otimes f = f$, if $d(f) = s$, we have that $e_s \leq f$ for all idempotents f on s. Furthermore, $f^{\uparrow t} = e_t \otimes f$ for f with $s = d(f) \subseteq t$ are idempotents, and we have that $e_s, f \leq f^{\uparrow t}$. The marginals $f^{\downarrow t}$ however are in general no more idempotent (see examples below).

If S is a subset of Φ and $\phi \in \Phi$, then $\phi \otimes S$ denotes the set of valuations $\{\phi \otimes \psi : \psi \in S\}$. In particular, $\phi \otimes \Phi$ is called a principal ideal generated by ϕ. The following lemma and theorem are well-known in semigroup theory:

Lemma 3.14 *In a regular valuation algebra there exists for all $\phi \in \Phi$ a unique idempotent $f \in \Phi$ such that $\phi \otimes \Phi = f \otimes \Phi$.*

Proof. Since ϕ is regular there is a χ such that $\phi = \phi \otimes \chi \otimes \phi$ and $f = \phi \otimes \chi$ is an idempotent such that $f \otimes \phi = \phi$ (Lemma 3.13 (2)). Therefore $\phi \otimes \psi = f \otimes (\phi \otimes \psi) \in f \otimes \Phi$ and $f \otimes \psi = \phi \otimes (\chi \otimes \psi) \in \phi \otimes \Phi$, hence $\phi \otimes \Phi = f \otimes \Phi$.

Suppose that $f_1 \otimes \Phi = f_2 \otimes \Phi$. Then $f_1 \otimes \chi = f_2$ for some $\chi \in \Phi$. This implies that $f_2 \otimes f_1 = f_1 \otimes \chi \otimes f_1 = f_1 \otimes \chi = f_2$. Similarly we obtain that $f_1 \otimes f_2 = f_2$, hence $f_1 = f_2$.
□

This lemma says that every principal ideal is generated by a unique idempotent valuation.

Theorem 3.15 *In a regular valuation algebra each valuation ϕ has a unique inverse.*

Proof. We know already that regular elements have inverses (Lemma 3.13 (1)). We need only to prove that they are unique. Let then ψ, ψ' be inverses of ϕ so that

$$\phi \otimes \psi \otimes \phi = \phi, \qquad \psi \otimes \phi \otimes \psi = \psi,$$
$$\phi \otimes \psi' \otimes \phi = \phi, \qquad \psi' \otimes \phi \otimes \psi' = \psi'.$$

Then we have $\phi \otimes \psi \otimes \Phi = \phi \otimes \Phi = \phi \otimes \psi' \otimes \Phi$ so that $\phi \otimes \psi = \phi \otimes \psi'$ by the lemma above, since $\phi \otimes \psi$ and $\phi \otimes \psi'$ are idempotents (Lemma 3.13 (2)). But then

$$\psi = \psi \otimes \phi \otimes \psi = \psi' \otimes \phi \otimes \psi = \psi' \otimes \phi \otimes \psi' = \psi'.$$

So, inverses are unique. $\qquad\qquad\qquad\qquad\qquad\qquad\qquad\qquad\qquad\qquad\qquad$ □

Inverses can be illustrated especially for probability potentials, which do form a regular valuation algebra as we have seen above. The unique inverse of a potential p on s is given, according to Lemma 3.13 (1) above, by

$$\chi \otimes p \otimes \chi(\mathbf{x}) = \begin{cases} \frac{1}{p(\mathbf{x})} & \text{if } p(\mathbf{x}) > 0, \\ 0, & \text{otherwise.} \end{cases} \qquad (3.52)$$

if we take χ as defined in (3.49).

The idempotents are potentials f whose values are either 0 or 1. The set $supp(f) = \{\mathbf{x} : f(\mathbf{x}) = 1\}$ is called the support set of f. Clearly, assuming $d(f_1) = d(f_2)$, we have $f_1 \leq f_2$ if, and only if, $supp(f_1) \supseteq supp(f_2)$. For the neutral elements, we have $supp(e_s) = \Omega_s$. Note that the marginal of an idempotent is no more an idempotent in general.

For possibility potentials with the product t-norm we have the same inverse (3.52). But in this example, in contrast to probability potentials, marginals of idempotent valuations remain idempotent.

In idempotent valuation algebras, which are regular too, the unique inverse of a valuation ϕ is ϕ itself. All elements are idempotent. So there is a partial order over the whole set Φ. This case will be explored in detail in Chapter 6.

Next we introduce a so-called Green relation between valuations. We define $\phi \equiv \psi \pmod{\gamma}$ if $\phi \otimes \Phi = \psi \otimes \Phi$. So ϕ and ψ are equivalent modulo γ if they generate the same principal ideal. This is an equivalence relation on Φ as can easily be verified. It is indeed a congruence in the valuation algebra, as will be shown later. As usual we write $[\phi]_\gamma$ for the equivalence class of ϕ. The following lemma is needed to develop the main result below:

Lemma 3.16 *If $\phi \equiv \phi \otimes \psi \pmod{\gamma}$, then $[\phi]_\gamma = \psi \otimes [\phi]_\gamma$*

Proof. We remark that $\phi \in \phi \otimes \Phi$. This is so, since $\phi \otimes \Phi = f \otimes \Phi$ (Lemma 3.14), where $f = \phi \otimes \psi$, ψ inverse to ϕ, is the unique idempotent element generating the principal ideal $\phi \otimes \Phi$. So we have $f \otimes \phi = \phi \in \phi \otimes \Phi$. This fact will be repeatedly used in the following without further reference.

The equivalence postulated in the lemma implies that there are valuations $\epsilon_1, \epsilon_2 \in \Phi$ such that

$$\phi \otimes \epsilon_1 \quad = \quad \phi \otimes \psi, \qquad \phi \otimes \psi \otimes \epsilon_2 \quad = \quad \phi.$$

In fact, we may take $\epsilon_1 = \psi$. We define the following mappings

$$\sigma_1 : \eta \quad \mapsto \quad \eta \otimes \psi,$$
$$\sigma_2 : \eta \quad \mapsto \quad \eta \otimes \epsilon_2.$$

The first map goes from $[\phi]_\gamma$ into $\psi \otimes [\phi]_\gamma$. But $\eta \equiv \phi \pmod{\gamma}$ implies $\eta \otimes \psi \equiv \phi \otimes \psi \equiv \phi \pmod{\gamma}$. This implies that σ_1 maps in fact $[\phi]_\gamma$ into $[\phi]_\gamma$ such that $\psi \otimes [\phi]_\gamma \subseteq [\phi]_\gamma$. Now we see also that $\eta \equiv \phi \pmod{\gamma}$ implies $\eta \otimes \epsilon_2 \equiv \phi \otimes \psi \otimes \epsilon_2 = \phi \pmod{\gamma}$. So σ_2 maps also $[\phi]_\gamma$ into $[\phi]_\gamma$. Finally $\eta \equiv \phi \pmod{\gamma}$ means that there is a $\chi \in \Phi$ such that $\eta = \phi \otimes \chi$. Then we obtain

$$(\eta \otimes \psi) \otimes \epsilon_2 \quad = \quad \chi \otimes (\phi \otimes \psi) \otimes \epsilon_2 \quad = \quad \chi \otimes \phi = \eta.$$

Hence $\sigma_2 \circ \sigma_1$ is the identity mapping from $[\phi]_\gamma$ onto $[\phi]_\gamma$. But this shows that $[\phi]_\gamma = \psi \otimes [\phi]_\gamma$. □

Of much more direct interest is the next lemma. It leads to the core of the present discussion.

Lemma 3.17 *Let* (Φ, D) *be a regular valuation algebra. Then, for all* $\phi \in \Phi$, *the equivalence class* $[\phi]_\gamma$ *is a commutative group.*

Proof. We note first that by Lemma 3.14 above, that there is a unique idempotent f in the class $[\phi]_\gamma$. It follows, that $\phi \otimes \phi \in [\phi]_\gamma$. In fact, there is a χ such that $\phi = \phi \otimes \chi \otimes \phi$ and $\phi \otimes \chi$ is the idempotent f in $[\phi]_\gamma$. For each $\psi \in \Phi$ we have then $\phi \otimes \psi = \phi \otimes \chi \otimes \phi \otimes \psi$. So $\phi \otimes \Phi = \phi \otimes \phi \otimes \Phi$.

Now we apply Lemma 3.16 repeatedly to prove that $[\phi]_\gamma$ is closed under combination. So let $\psi, \eta \in [\phi]_\gamma$. $\phi \equiv \phi \otimes \phi \pmod{\gamma}$ implies then $[\phi]_\gamma = \phi \otimes [\phi]_\gamma$. Therefore we see that $\phi \otimes \psi \in [\phi]_\gamma$ or $\phi \equiv \phi \otimes \psi \pmod{\gamma}$. This in turn implies that $[\phi]_\gamma = \psi \otimes [\phi]_\gamma$. Therefore we conclude that $\psi \otimes \eta \in [\phi]_\gamma$. This shows that $[\phi]_\gamma$ is closed under combination. The class inherits commutativity and associativity from the valuation algebra it is contained in.

We have already seen that $f \otimes \phi = \phi$. So the idempotent in $[\phi]_\gamma$ is the unit element in $[\phi]_\gamma$. We have to show that there is an inverse of every element ϕ in the class. Let χ be as above and define

$$\phi^{-1} \quad = \quad \chi \otimes \phi \otimes \chi.$$

to be the inverse in the regular semigroup. Then we obtain $\phi \otimes \phi^{-1} = (\phi \otimes \chi) \otimes (\phi \otimes \chi) = f \otimes f = f$. Also $\phi \equiv \phi^{-1} \pmod{\gamma}$. In fact, for $\psi \in \Phi$, we have that

$\phi \otimes \psi = \phi \otimes \phi^{-1} \otimes \phi \otimes \psi \in \phi^{-1} \otimes \Phi$. Conversely, $\phi^{-1} \otimes \psi = \chi \otimes \phi \otimes \chi \otimes \psi \in \phi \otimes \Phi$. So ϕ^{-1} is also the group inverse. This completes the proof. $\square$

Altogether we have proved with these results the following fundamental theorem:

Theorem 3.18 *If* (Φ, D) *is a regular valuation algebra, then* Φ *is the union of disjoint groups* $[\phi]_\gamma$,

$$\Phi = \bigcup_{\phi \in \Phi} [\phi]_\gamma. \tag{3.53}$$

This is essentially a theorem of semigroup theory. We are now going to link it with the domain structure of labeled valuation algebras. But before, we note, that the partial order of idempotent elements carries over to the family of groups $[\phi]_\gamma$, since each group contains exactly one idempotent element, and each idempotent valuation f belongs to exactly one group (the group $[f]_\gamma$). These groups form therefore even a semilattice.

We denote the uniquely determined group a valuation $\phi \in \Phi$ belongs to by $\gamma(\phi)$, that is, $\gamma(\phi) = [\phi]_\gamma$. We call $\gamma(\phi)$ also the *support* of ϕ (this is motivated by the example of potentials, see above). Its idempotent element, that is, the group unit, is denoted accordingly by $f_{\gamma(\phi)}$. So we define

$$\gamma(\phi) \leq \gamma(\psi) \quad \text{if, and only if,} \quad f_{\gamma(\phi)} \leq f_{\gamma(\psi)}. \tag{3.54}$$

Since

$$f_{\gamma(\phi)} \otimes f_{\gamma(\psi)} = f_{\gamma(\phi)} \vee f_{\gamma(\psi)} = f_{\gamma(\phi \otimes \psi)}$$

we see that $\gamma(\phi \otimes \psi)$ is the supremum of $\gamma(\phi)$ and $\gamma(\psi)$ and we write $\gamma(\phi) \vee \gamma(\psi)$ for it. Thus

$$\gamma(\phi \otimes \psi) = \gamma(\phi) \vee \gamma(\psi). \tag{3.55}$$

Hence, Φ is the union of a *semilattice* of groups. If $\gamma(\phi) \geq \gamma(\psi)$, then

$$\phi \otimes f_{\gamma(\psi)} = \phi \otimes f_{\gamma(\phi)} \otimes f_{\gamma(\psi)} = \phi \otimes f_{\gamma(\phi)} = \phi. \tag{3.56}$$

So an idempotent is a neutral element for some valuation ϕ, if its support is smaller than that of ϕ.

The inverse of a valuation ϕ in the group $[\phi]_\gamma$ will in what follows be denoted as in the proof of the lemma above by ϕ^{-1}. So we have

$$\phi \otimes \phi^{-1} = f_{\gamma(\phi)}. \tag{3.57}$$

Remark that

$$(\phi \otimes \psi)^{-1} = \phi^{-1} \otimes \psi^{-1}.$$

We claimed above that the Green relation γ is a congruence in the regular valuation algebra. This is formally stated in the next theorem.

Theorem 3.19 *If (Φ, D) is a regular valuation algebra, then the Green relation γ is a congruence in this algebra.*

Proof. We have already seen that γ is an equivalence relation.

We show first, that $\phi \equiv \psi \pmod{\gamma}$, or $\gamma(\phi) = \gamma(\psi)$, implies that $d(\phi) = d(\psi)$. $\gamma(\phi) = \gamma(\psi)$ means that $\phi \otimes \Phi = \psi \otimes \Phi$. Thus, there are valuations ϵ, ϵ' such that

$$\phi = \psi \otimes \epsilon, \qquad \psi = \phi \otimes \epsilon'.$$

By the labeling axiom we have then that $d(\psi) \subseteq d(\phi)$ and $d(\phi) \subseteq d(\psi)$, hence $d(\phi) = d(\psi)$. This result shows that each sub-semigroup Φ_s of valuations with domains s decompose themselves already into a semigroup of groups,

$$\Phi_s = \bigcup_{\phi \in \Phi_s} [\phi]_\gamma.$$

Next, if $\gamma(\phi_1) = \gamma(\psi_1)$ and $\gamma(\phi_2) = \gamma(\psi_2)$ imply (by (3.55)) that $\gamma(\phi_1 \otimes \phi_2) = \gamma(\psi_1 \otimes \psi_2)$, hence $\phi_1 \otimes \phi_2 \equiv \psi_1 \otimes \psi_2 \pmod{\gamma}$. So γ is compatible with combination.

Finally, we prove that $\gamma(\phi) = \gamma(\psi)$ implies $\gamma(\phi^{\downarrow t}) = \gamma(\psi^{\downarrow t})$. Let $\eta \in \gamma(\phi^{\downarrow t})$. Then there is a valuation $\epsilon \in \Phi_t$ such that

$$\eta = \phi^{\downarrow t} \otimes \epsilon = (\phi \otimes \epsilon)^{\downarrow t}.$$

But $\phi \otimes \epsilon \in \phi \otimes \Phi = \psi \otimes \Phi$. There is thus a ϵ' such that $\phi \otimes \epsilon = \psi \otimes \epsilon'$. Hence we find that

$$\eta = (\psi \otimes \epsilon')^{\downarrow t}.$$

Thanks to the regularity there is a χ with $d(\chi) = d(\psi)$ such that $\psi = \chi \otimes \psi^{\downarrow t}$. Thus, we see finally that

$$\eta = (\psi^{\downarrow t} \otimes \chi \otimes \epsilon')^{\downarrow t} = \psi^{\downarrow t} \otimes (\chi \otimes \epsilon')^{\downarrow t}.$$

This implies $\gamma(\phi^{\downarrow t}) = \gamma(\eta) \geq \gamma(\psi^{\downarrow t})$. By symmetry, we have also $\gamma(\psi^{\downarrow t}) \geq \gamma(\phi^{\downarrow t})$, hence $\gamma(\phi^{\downarrow t}) = \gamma(\psi^{\downarrow t})$. $\square$

There are some more relations between the domain structure of a regular valuation algebra and the Green relation γ.

Lemma 3.20

1. Assume $\phi, \psi \in \Phi_s$, $t \subseteq s$. Then $\gamma(\psi) \leq \gamma(\phi)$ implies $\gamma(\psi^{\downarrow t}) \leq \gamma(\phi^{\downarrow t})$

2. $\gamma(\phi^{\downarrow t}) \leq \gamma(\phi)$ for all $t \subseteq d(\phi)$.

Proof. (1) We have to show that $\gamma(\phi^{\downarrow t}) = \gamma(\phi^{\downarrow t} \otimes \psi^{\downarrow t})$ or that

$$\phi^{\downarrow t} \otimes \Phi_t \;=\; \phi^{\downarrow t} \otimes \psi^{\downarrow t} \otimes \Phi_t.$$

Assume therefore, that $\eta \in \gamma(\phi^{\downarrow t})$. Then $d(\eta) = t$ and there is a valuation $\epsilon \in \Phi_t$ such that

$$\eta \;=\; \phi^{\downarrow t} \otimes \epsilon \;=\; (\phi \otimes \epsilon)^{\downarrow t}.$$

But $\phi \otimes \epsilon \in \phi \otimes \Phi = \phi \otimes \psi \otimes \Phi$. Therefore, there is a valuation $\epsilon' \in \Phi$ such that

$$\phi \otimes \epsilon \;=\; \phi \otimes \psi \otimes \epsilon'.$$

By the regularity of the algebra,

$$\phi \;=\; \phi^{\downarrow t} \otimes \chi \otimes \phi, \qquad \psi \;=\; \psi^{\downarrow t} \otimes \chi' \otimes \psi$$

for some $\chi, \chi' \in \Omega_t$. Hence we obtain

$$\begin{aligned}
\eta \;&=\; (\phi \otimes \psi \otimes \epsilon')^{\downarrow t} \\
&=\; ((\phi^{\downarrow t} \otimes \psi^{\downarrow t}) \otimes (\chi \otimes \chi' \otimes \phi \otimes \psi \otimes \epsilon')^{\downarrow t}.
\end{aligned}$$

This shows that $\gamma(\eta) = \gamma(\phi^{\downarrow t}) \geq \gamma(\phi^{\downarrow t} \otimes \psi^{\downarrow t})$. But, of course, $\gamma(\phi^{\downarrow t}) \leq \gamma(\phi^{\downarrow t} \otimes \psi^{\downarrow t})$, hence $\gamma(\phi^{\downarrow t}) = \gamma(\phi^{\downarrow t} \otimes \psi^{\downarrow t})$.

(2) We have $\gamma(\phi^{\downarrow t}) = \gamma((\phi^{\downarrow t})^{-1})$. By regularity, we have

$$\phi \;=\; \phi \otimes \chi \otimes \phi^{\downarrow t}, \qquad d(\chi) \;=\; t, \tag{3.58}$$

which implies

$$\phi^{\downarrow t} \;=\; \phi^{\downarrow t} \otimes \chi \otimes \phi^{\downarrow t},$$

From

$$\phi^{\downarrow t} \;=\; \phi^{\downarrow t} \otimes (\phi^{\downarrow t})^{-1} \otimes \phi^{\downarrow t}$$

we deduce that

$$(\phi^{\downarrow t})^{-1} \;=\; f_{\gamma(\phi^{\downarrow t})} \otimes \chi.$$

This shows that $\gamma(\phi^{\downarrow t}) \geq \gamma(\chi)$. Therefore, we obtain from (3.58)

$$\gamma(\phi) \;=\; \gamma(\phi) \vee \gamma(\chi) \vee \gamma(\phi^{\downarrow t}) \;=\; \gamma(\phi) \vee \gamma(\phi^{\downarrow t}).$$

This means that $\gamma(\phi^{\downarrow t}) \leq \gamma(\phi)$. $\square$

Let

$$\chi \;=\; (\phi^{\downarrow t})^{-1} \otimes \phi. \tag{3.59}$$

Then, by point (2) of the lemma we obtain

$$\chi \otimes \phi^{\downarrow t} \;=\; \phi^{\downarrow t} \otimes (\phi^{\downarrow t})^{-1} \otimes \phi \;=\; f_{\gamma(\phi^{\downarrow t})} \otimes \phi \;=\; \phi. \tag{3.60}$$

In the case of probability potentials for example, we see that

$$\chi(\mathbf{x}, \mathbf{y}) \;=\; \frac{p(\mathbf{x}, \mathbf{y})}{p^{\downarrow t}(\mathbf{y})}, \tag{3.61}$$

is essentially a family of conditional distributions, $\chi(\mathbf{x}, \mathbf{y}) = p(\mathbf{x}|\mathbf{y})$, if $p^{\downarrow t}(\mathbf{y}) > 0$. And then we have also

$$p(\mathbf{x}, \mathbf{y}) \;=\; p(\mathbf{x}|\mathbf{y}) p^{\downarrow t}(\mathbf{y}). \tag{3.62}$$

Such "conditional" elements will be studied more in detail in Chapter 5.

As a consequence of Theorem 3.19 above, we see that each sub-semigroup Φ_s of all valuations with domain s, decomposes itself already into a semilattice of groups. The group $\gamma(e_s)$ is the smallest element in this semilattice, if the valuation algebra has neutral elements. The elements of this group are called *positive*. The set

$$\Phi_p \;=\; \bigcup_{s \in D} \gamma(e_s)$$

of positive elements is closed under combination: assume $\phi, \psi \in \Phi_p$ with domains x and y. Then $\gamma(\phi) = \gamma(e_x)$ and $\gamma(\psi) = \gamma(e_y)$, hence $\gamma(\phi \otimes \psi) = \gamma(e_x \otimes e_y) = \gamma(e_{x \cup y})$. Thus $\phi \otimes \psi \in \Phi_p$. Further, from $\gamma(e_s) = \gamma(\phi)$ it follows that $\gamma(e_s^{\downarrow t}) = \gamma(\phi^{\downarrow t})$. If the valuation algebra is stable, then $\gamma(e_s^{\downarrow t}) = \gamma(e_t) = \gamma(\phi^{\downarrow t})$. This shows that Φ_p is closed under marginalization in stable algebras. Even if the algebra is not stable, this holds, if the congruence γ is stable.

Note that the positive valuations with domain s form already a group, since $\Phi_{p,s} = \gamma(e_s)$.

A particularity of positive valuations is, that the equation $\psi = \phi \otimes \chi$ for $\phi \in \gamma(e_s)$, $d(\psi) = s$ has exactly one solution χ with $d(\chi) = s$. In fact, $\psi \otimes \phi^{-1}$ is a solution. If χ, χ' are solutions with $d(\chi) = d(\chi') = s$, then

$$\phi \otimes \chi \;=\; \phi \otimes \chi'$$

then positivity of ϕ leads to

$$e_s \otimes \chi \;=\; e_s \otimes \chi'$$

if both sides of the equation above are multiplied by ϕ^{-1}. So, since $d(\chi) = d(\chi') = s$ we obtain indeed $\chi = \chi'$.

We illustrate these results once more using probability potentials. All potentials with the same support $supp(p) = \{\mathbf{x} : p(\mathbf{x}) > 0\}$ form a group. Its unit element is the potential f whose value is 1 for all configurations of the support, and 0 outside the support. The inverse of a potential p is the potential defined by $1/p(\mathbf{x})$ on the support and 0 otherwise. The semilattice of the groups corresponds to the semilattice of the supports, looked at as subsets of Ω_r, with intersection as the supremum. The whole regular valuation algebra of potentials is the union of these groups. The positive elements are just what they ought to be: potentials p with $p(\mathbf{x}) > 0$ for all $\mathbf{x} \in \Omega_s$. Since the congruence based on support sets is stable, the positive elements form a subalgebra of the algebra of all probability potentials. The same holds for possibility potentials with the product t-norm.

In idempotent valuation algebras each element forms a one-element group by itself. That is why the decomposition of the valuation algebra into groups is of no great interest in this case. The neutral elements e_s are the only "positive" elements.

3.6   Separative Valuation Algebras

Many valuation algebras are not regular, and fall thus not under the theory developed in the previous section. Examples are provided by set potentials, Spohn potentials and Gaussian potentials. Still, some notion of division can be introduced also in more general circumstances as has been noted in (Lauritzen & Jensen, 1997). This is in particular the case, if a valuation algebra, although not being itself regular, can be embedded into a regular valuation algebra, which, as we have seen, is a union of disjoint groups. Such embeddings have been studied in semigroup theory and we base our discussion on this theory (see (Clifford & Preston, 1967) for more details about semigroup theory).

Let (Φ, D) be a labeled valuation algebra. A sufficient condition to embed a semigroup Φ into a group G, is that for all $\phi \in \Phi$

$$\phi \otimes \psi \;=\; \phi \otimes \psi'.$$

implies $\psi = \psi'$. If the semigroup Φ satisfies this condition it is called *cancellative* (Clifford & Preston, 1967). We cannot expect this condition to be fulfilled in Φ, since for example $\phi \otimes \psi = \phi \otimes \psi \otimes e_s$ if $d(\phi) = s$ and $d(\psi) < s$. However, the subsemigroups Φ_x may be cancellative, as the examples below will show. Therefore, we call the valuation algebra (Φ, D) *cancellative*, if for all $x \in D$ the semigroups Φ_x are cancellative.

The classical construction of the group G_x associated with the semigroup Φ_x goes as follows: We consider the pairs (ϕ, ψ) of elements of Φ_x and define

$$(\phi, \psi) \;=\; (\phi', \psi') \quad \text{if} \quad \phi \otimes \psi' \;=\; \phi' \otimes \psi.$$

Multiplication between such pairs is then, for $a = (\phi, \psi), b = (\phi', \psi')$ defined by

$$ab \;=\; (\phi \otimes \phi', \psi \otimes \psi').$$

This is well defined, since, if $(\phi_1, \psi_1) = (\phi_2, \psi_2)$ and $(\phi_1', \psi_1') = (\phi_2', \psi_2')$, then $(\phi_1 \otimes \phi_1', \psi_1 \otimes \psi_1') = (\phi_2 \otimes \phi_2', \psi_2 \otimes \psi_2')$. Multiplication is clearly commutative and associative. The unit of multiplication is defined by pairs $u = (\gamma, \gamma)$, which are all equal and for which $ua = (\gamma \otimes \phi, \gamma \otimes \psi) = (\phi, \psi)$. For $a = (\phi, \psi)$ define $b = (\psi, \phi)$. Then $ab = (\phi \otimes \psi, \psi \otimes \phi) = u$. So b is the inverse to a, for which we write a^{-1}. This shows that G_x is a group.

Let

$$\Phi^* = \bigcup_{x \in D} G_x.$$

We are now going to give Φ^* the structure of a semigroup by defining a combination operation in it. Let $\phi_x, \psi_x \in \Phi_x$ and $\phi_y, \psi_y \in \Phi_y$, so that $a = (\phi_x, \psi_x) \in G_x$ and $b = (\phi_y, \psi_y) \in G_y$. Then we define

$$a \otimes b = (\phi_x \otimes \phi_y, \psi_x \otimes \psi_y). \tag{3.63}$$

Since $\phi_x \otimes \phi_y, \psi_x \otimes \psi_y \in \Phi_{x \vee y}$, we have that $a \otimes b \in G_{x \vee y}$.

First, we need to verify that this operation is well defined, that it does not depend on the particular definition of a and b. Suppose that $(\phi_x, \psi_x) = (\phi'_x, \psi'_x)$ and $(\phi_y, \psi_y) = (\phi'_y, \psi'_y)$. Then

$$\phi_x \otimes \psi'_x = \phi'_x \otimes \psi_x, \qquad \phi_y \otimes \psi'_y = \phi'_y \otimes \psi_y,$$

hence

$$(\phi_x \otimes \phi_y) \otimes (\psi'_x \otimes \psi'_y) = (\phi'_x \otimes \phi'_y) \otimes (\psi_x \otimes \psi_y),$$

such that

$$(\phi_x \otimes \phi_y, \psi_x \otimes \psi_y) = (\phi'_x \otimes \phi'_y, \psi'_x \otimes \psi'_y).$$

Next, associativity is proved for $a = (\phi_x, \psi_x)$, $b = (\phi_y, \psi_y)$ and $c = (\phi_z, \psi_z)$. We have

$$
\begin{aligned}
a \otimes (b \otimes c) &= (\phi_x, \psi_x) \otimes (\phi_y \otimes \phi_z, \psi_y \otimes \psi_z) \\
&= (\phi_x \otimes (\phi_y \otimes \phi_z), \psi_x \otimes (\psi_y \otimes \psi_z)) \\
&= ((\phi_x \otimes \phi_y) \otimes \phi_z, (\psi_x \otimes \psi_y) \otimes \psi_z) \\
&= (\phi_x \otimes \phi_y, \psi_x \otimes \psi_y) \otimes (\phi_z, \psi_z) \\
&= (a \otimes b) \otimes c.
\end{aligned}
$$

Commutativity is evident. So Φ^* is a commutative semigroup, which is a union of disjoint groups.

It remains to show that Φ, as a semigroup, is embedded in Φ^*. We claim that the mapping $\phi \mapsto (\phi \otimes \phi, \phi)$ is an embedding. It is one-to-one since $(\phi \otimes \phi, \phi) = (\psi \otimes \psi, \psi)$ for $\phi, \psi \in \Phi_x$ implies $\phi = \psi$ because the semigroups Φ_x are cancellative. Furthermore, we have

$$\phi \otimes \psi \mapsto ((\phi \otimes \psi) \otimes (\phi \otimes \psi), \phi \otimes \psi) = ((\phi \otimes \phi), \phi) \otimes ((\psi \otimes \psi), \psi).$$

So it is a semigroup homomorphism, hence an embedding.

Every group G_x has a unit element, which we denote by f_x. If the valuation algebra (Φ, D) has neutral elements, then we have $f_x = e_x$ for all domains x. More precisely, we have $f_x = (e_x \otimes e_x, e_x) = (e_x, e_x)$; but from now on we identify elements of Φ with their image in Φ^*. We note that in the general case

$$f_x \otimes f_y \;=\; (f_x \otimes f_x) \otimes (f_y \otimes f_y) \;=\; (f_x \otimes f_y) \otimes (f_x \otimes f_y)$$

So $f_x \otimes f_y$ is an idempotent belonging to $G_{x \cup y}$, hence the unit element of this group. So, even in the general case a kind of neutrality axiom is valid, but possibly for neutral elements outside Φ. This implies, if $x \subseteq d(\phi) = y$,

$$\phi \otimes f_x \;=\; \phi \otimes f_y \otimes f_x \;=\; \phi \otimes f_y \;=\; \phi.$$

Hence, the unit elements in lower domains are neutral elements for valuations in upper domains.

Now, every valuation ϕ with $d(\phi) = x$ has an inverse ϕ^{-1} in G_x, which in general is no more a valuation of Φ. In particular, we have that, for $t \subseteq d(\phi)$,

$$\phi^{\downarrow t} \otimes (\phi^{\downarrow t})^{-1} \otimes \phi \;=\; f_t \otimes \phi \;=\; \phi.$$

As a first example we consider set potentials. Let $q, q_1, q_2 \in Q^+$ be commonality functions on the same domain. An equation

$$q(A) q_1(A) \;=\; q(A) q_2(A)$$

implies $q_1(A) = q_2(A)$ if, and only if, $q(A) > 0$ for all subsets A of the domain. So the valuation algebra of set potentials is not cancellative, but the subalgebra of positive commonalities, $q(A) > 0$ for all subsets A of the domain, is cancellative. So it is embedded into the union of groups of the quotients q_1/q_2 of commonality functions with group multiplication defined as ordinary multiplication of quotients. If $d(q_1) = d(q_2) = s$, then q_1/q_2 is defined for subsets A of Ω_s by

$$q_1/q_2(A) \;=\; \frac{q_1(A)}{q_2(A)}$$

For $a = q_1/q_1' \in G_s$ and $b = q_2/q_2' \in G_t$, the combination in Φ^* is defined by

$$a \otimes b(A) \;=\; \frac{q_1(A^{\downarrow s}) q_2(A^{\downarrow t})}{q_1'(A^{\downarrow s}) q_2'(A^{\downarrow t})}.$$

Set potentials have neutral elements, which are thus also the unities of the groups G_x.

For Spohn potentials in Φ_x,

$$p(\mathbf{x}) + p_1(\mathbf{x}) \;=\; p(\mathbf{x}) + p_2(\mathbf{x})$$

implies that $p_1 = p_2$. So the valuation algebra of Spohn potentials is cancellative. If p_1 and p_2 are two Spohn potentials on a domain x, then the elements of the group G_x can be identified with the differences $p_1(\mathbf{x}) - p_2(\mathbf{x})$. These functions are in general no more Spohn potentials, since they may take negative value. Clearly, Φ^* is in this case the semigroup of mappings of Ω_s for $s \subseteq r$ into integers.

Let's turn to Gaussian potentials. They are not regular. But assume that $(\mu, \mathbf{K}), (\mu_1, \mathbf{K}_1), (\mu_2, \mathbf{K}_2)$ are Gaussian potentials on the same domain s. Then

$$\mathbf{K} + \mathbf{K}_1 \;=\; \mathbf{K} + \mathbf{K}_2$$

implies $\mathbf{K}_1 = \mathbf{K}_2$. And

$$(\mathbf{K} + \mathbf{K}_1)^{-1} \cdot (\mathbf{K} \cdot \mu + \mathbf{K}_1 \cdot \mu_1) \;=\; (\mathbf{K} + \mathbf{K}_2)^{-1} \cdot (\mathbf{K} \cdot \mu + \mathbf{K}_2 \cdot \mu_2)$$

implies, in view of the above, that $\mu_1 = \mu_2$. So, the valuation algebra of Gaussian potentials is cancellative. The embedding semigroup Φ^* essentially consists of quotients of Gaussian densities, which are themselves no more Gaussian potentials. The quotient of a Gaussian potential $g(\mathbf{x})/g^{\downarrow t}(\mathbf{x}^{\downarrow t})$ is a family of conditional Gaussian distributions. This example shows, that embedding a valuation algebra into something larger is not always just an abstract construction, but may well have a significant meaning.

Possibility potentials with Lukasziewicz t-norm are not cancellative. In fact, assume that in

$$\max\{0, p(\mathbf{x}) + p_1(\mathbf{x}) - 1\} \;=\; \max\{0, p(\mathbf{x}) + p_2(\mathbf{x}) - 1\}$$

we have $p(\mathbf{x}) + p_1(\mathbf{x}) - 1 \le 0$. Then $p_2(\mathbf{x})$ can be arbitrary as long as $p(\mathbf{x}) + p_2(\mathbf{x}) - 1 \le 0$. So we need not have $p_1 = p_2$.

We remark that positive probability potentials are also cancellative. As we have seen in the previous section they already form by themselves a group, since quotients of positive potentials are again positive potentials. That is, what makes this example particularly simple. This holds also in a similar way for positive densities in general. Consider variables with real values. Then the domains Ω_s are simply s-dimensional real spaces $\mathbf{R}^{|s|}$. We call *densities* on s non-negative real-valued functions $f(\mathbf{x}) \ge 0$, whose integral

$$\int\limits_{-\infty}^{+\infty} f(\mathbf{x}) d\mathbf{x}$$

exists and is finite. We may either consider measurable functions and Lebesgue integrals or continuous functions and Riemann integrals. In the former case densities f and g are considered as equal, if $f(\mathbf{x}) = g(\mathbf{x})$ almost everywhere (see the end of Section 2.4). Suppose that $s = x \cup y$, x, y disjoint. Then

$$f^{\downarrow x}(\mathbf{x}) \;=\; \int\limits_{-\infty}^{+\infty} f(\mathbf{x}, \mathbf{y}) d\mathbf{y}$$

is still a density, the marginal of f with respect to x (of course, we assume here that $\mathbf{x} \in \Omega_x$ and $\mathbf{y} \in \Omega_y$).

Assume f to be a density on s. Then we may consider the corresponding normalized (proper) density

$$\frac{f(\mathbf{x})}{\int f(\mathbf{x})d\mathbf{x}}.$$

In the following we consider only normalized (proper) densities, such that

$$\int f(\mathbf{x})d\mathbf{x} = 1.$$

We add to this set of densities the null function $f(\mathbf{x}) = 0$ (almost everywhere). We remark, that the marginal of a proper density is still a proper density, and the marginal of the null function is still a null function. So, the family of densities is closed under marginalization. If f and g are two densities on s and t respectively, then we define a normalized combination as follows,

$$f \otimes g(\mathbf{x}) = \frac{f(\mathbf{x}^{\downarrow s})g(\mathbf{x}^{\downarrow t})}{\int f(\mathbf{x}^{\downarrow s})g(\mathbf{x}^{\downarrow t})d\mathbf{x}},$$

provided the integral in the denominator exists. Otherwise, or if f or g is a null function, we define $f \otimes g(\mathbf{x}) = 0$ everywhere. It is not difficult to verify that this combination is commutative and associative. We claim that the combination axiom is also satisfied. Indeed, let f and g be densities on s and t respectively, and let $\mathbf{x}, \mathbf{y}, \mathbf{z}$ be configurations for $s - t$, $s \cap t$ and $t - s$. Then,

$$
\begin{aligned}
(f \otimes g)^{\downarrow s}(\mathbf{x}, \mathbf{y}) &= \int \frac{f(\mathbf{x}, \mathbf{y})g(\mathbf{y}, \mathbf{z})}{\int f(\mathbf{x}, \mathbf{y})g(\mathbf{y}, \mathbf{z})d\mathbf{x}d\mathbf{y}d\mathbf{z}} d\mathbf{z} \\
&= \frac{f(\mathbf{x}, \mathbf{y}) \int g(\mathbf{y}, \mathbf{z})d\mathbf{z}}{\int f(\mathbf{x}, \mathbf{y})g(\mathbf{y}, \mathbf{z})d\mathbf{x}d\mathbf{y}d\mathbf{z}} \\
&= \frac{f(\mathbf{x}, \mathbf{y})g^{\downarrow s \cap t}(\mathbf{y})}{\int f(\mathbf{x}, \mathbf{y})g^{\downarrow s \cap t}(\mathbf{y})d\mathbf{x}d\mathbf{y}} \\
&= f \otimes g^{\downarrow s \cap t}(\mathbf{x}, \mathbf{y}).
\end{aligned}
$$

This holds, provided the integral in the denominator exists. Otherwise, it can be verified, that both $(f \otimes g)^{\downarrow s}$ and $f \otimes g^{\downarrow s \cap t}$ become the null function. This proves that the combination axiom holds. Densities, together with null function, form therefore a valuation algebra (without neutral elements).

But this valuation algebra is not cancellative. Positive densities $f(\mathbf{x}) > 0$ for all $\mathbf{x} \in \Omega_s$ however form a subalgebra which is cancellative. Like for Gaussian potentials, quotients of a density with its marginal to some domain t gives a family of conditional density functions

$$f(\mathbf{y}|\mathbf{x}) = \frac{f(\mathbf{x}, \mathbf{y})}{f^{\downarrow t}(\mathbf{x})}.$$

These conditional densities are no more densities, since they are not integrable. If however $x \subseteq s - t$, then the integrals

$$\int_{-\infty}^{+\infty} f(\mathbf{y}, \mathbf{z}|\mathbf{x})dy = \frac{f^{\downarrow z}(\mathbf{x}, \mathbf{z})}{f(^{\downarrow t}(\mathbf{x}))} \qquad (3.64)$$

exist. So, in the embedding group of conditional densities, partial marginalization is possible (see Section 2.4).

These examples show that the condition of cancellativity of valuation algebras is too restrictive to capture all interesting cases such as densities or set potentials. We introduce a weaker condition which is sufficient to embed a valuation algebra into a union of groups. We are going to consider valuation algebras (Φ, D) in which a *congruence* γ exists, which has the property that for all $\phi \in \Phi$ and $t \subseteq d(\phi)$,

$$\phi^{\downarrow t} \otimes \phi \equiv \phi \pmod{\gamma}. \qquad (3.65)$$

The equivalence classes $[\phi]_\gamma$ of such a congruence γ are semigroups, since $\phi \equiv \psi$ (mod γ) implies, using (3.65), that

$$\phi \otimes \psi \equiv \phi \otimes \phi \equiv \phi \pmod{\gamma}.$$

So Φ decomposes into a family of disjoint semigroups

$$\Phi = \bigcup_{\phi \in \Phi} [\phi]_\gamma.$$

Since γ is a congruence, this holds also for every Φ_x,

$$\Phi_x = \bigcup_{\phi \in \Phi_x} [\phi]_\gamma.$$

We note further that the fact that γ is a congruence implies also

$$[\phi]_\gamma \otimes [\psi]_\gamma = \{\phi' \otimes \psi' : \phi' \in [\phi]_\gamma, \psi' \in [\psi]_\gamma\} = [\phi \otimes \psi]_\gamma.$$

This allows us to define a partial order between the semigroups $[\phi]_\gamma$: $[\psi]_\gamma \leq [\phi]_\gamma$ if $[\psi]_\gamma \otimes [\phi]_\gamma = [\phi]_\gamma$. This is indeed a partial order:

1. *Reflexivity:* $[\phi]_\gamma \leq [\phi]_\gamma$, since $[\phi]_\gamma \otimes [\phi]_\gamma = [\phi \otimes \phi]_\gamma = [\phi]_\gamma$ by (3.65).

2. *Antisymmetry:* $[\psi]_\gamma \leq [\phi]_\gamma$ and $[\phi]_\gamma \leq [\psi]_\gamma$ imply $[\psi]_\gamma = [\psi]_\gamma$. Indeed, the two conditions imply $[\phi]_\gamma \otimes [\psi]_\gamma = [\phi]_\gamma$ and $[\phi]_\gamma \otimes [\psi]_\gamma = [\psi]_\gamma$, hence $[\phi]_\gamma = [\psi]_\gamma$.

3. *Transitivity:* $[\psi]_\gamma \leq [\phi]_\gamma$ and $[\phi]_\gamma \leq [\eta]_\gamma$ imply $[\psi]_\gamma \leq [\eta]_\gamma$. In fact, $[\psi]_\gamma \otimes [\phi]_\gamma = [\phi]_\gamma$ and $[\phi]_\gamma \otimes [\eta]_\gamma = [\eta]_\gamma$ imply $[\psi]_\gamma \otimes [\eta]_\gamma = [\psi]_\gamma \otimes [\phi]_\gamma \otimes [\eta]_\gamma = [\psi]_\gamma \otimes [\eta]_\gamma.$

Again using (3.65) we see that $[\phi]_\gamma, [\psi]_\gamma \leq [\phi \otimes \psi]_\gamma$. On the other hand, assume $[\phi]_\gamma, [\psi]_\gamma \leq [\eta]_\gamma$, then this means that

$$[\phi]_\gamma \otimes [\eta]_\gamma \;\; = \;\; [\psi]_\gamma \otimes [\eta]_\gamma \;\; = \;\; [\eta]_\gamma$$

and (3.65) allows us to deduce that

$$\begin{aligned}[\phi \otimes \psi]_\gamma \otimes [\eta]_\gamma &= [\phi]_\gamma \otimes [\psi]_\gamma \otimes [\eta]_\gamma = [\phi]_\gamma \otimes [\psi]_\gamma \otimes [\eta]_\gamma \otimes [\eta]_\gamma \\ &= [\eta]_\gamma \otimes [\eta]_\gamma = [\eta]_\gamma.\end{aligned}$$

This shows that $[\phi \otimes \psi]_\gamma$ is the supremum of $[\phi]_\gamma$ and $[\psi]_\gamma$,

$$[\phi]_\gamma \vee [\psi]_\gamma \;\; = \;\; [\phi \otimes \psi]_\gamma. \tag{3.66}$$

The semigroups $[\phi]_\gamma$ form a *semilattice*.

After these preparative remarks, we define now a class of valuation algebras, which can be embedded into a semigroup, which is a union of groups.

Definition 3.21 Separative Valuation Algebras: *A labeled valuation algebra* (Φ, D) *is called* separative, *if*

1. *There is a congruence γ in (Φ, D), which satisfies (3.65).*

2. *The semigroups $[\phi]_\gamma$ are all cancellative.*

Regular valuation algebras are separative valuation algebras. In this case, the particular congruence γ, introduced in the previous section, produces not only sub-semigroups of Φ, but already groups. The former case in the present section, where the semigroups Φ_x were cancellative, is a special case of separative valuation algebras. In this particular case the congruence γ is the relation of having the same domain and the semigroups $[\phi]_\gamma$ are the semigroups Φ_x and the semilattice Y is the lattice of subsets of D. As in this particular case, in the more general case of a separative valuation algebra, each semigroup $[\phi]_\gamma$ can be embedded into the group of pairs (ϕ, ψ) of elements of $[\phi]_\gamma$. We call this group $\gamma(\phi)$. These groups can be assumed as disjoint. The partial order between semigroups $[\phi]_\gamma$ carries over to the groups $\gamma(\phi)$: $\gamma(\psi) \leq \gamma(\phi)$ if $[\psi]_\gamma \leq [\phi]_\gamma$. The unit element of group $\gamma(\phi)$ is denoted by $f_{\gamma(\phi)}$.

Let

$$\Phi^* \;\; = \;\; \bigcup_{\phi \in \Phi} \gamma(\phi). \tag{3.67}$$

Within Φ^* we define an operation of combination as above: If $a = (\phi_a, \psi_a)$ and $b = (\phi_b, \psi_b)$ are elements of Φ^*, then

$$a \otimes b \;\; = \;\; (\phi_a \otimes \phi_b, \psi_a \otimes \psi_b).$$

As before, it can be shown that this operation is well defined, commutative and associative. Φ^* is thus a commutative semigroup.

The mapping from Φ into Φ^* defined by

$$\phi \;\longmapsto\; (\phi \otimes \phi, \phi)$$

is, again as before, a semigroup-homomorphism, and because of the cancellativity of the semigroup $[\phi]_\gamma$ one-to-one. So Φ is embedded, as a semigroup, into Φ^*. In what follows we tacitly assume that $\Phi \subseteq \Phi^*$.

We have proved the following theorem:

Theorem 3.22 *The semigroup Φ of a separative valuation algebra (Φ, D) is embedded into a semigroup, which is the union of a semilattice of groups.*

Note that $f_{\gamma(\phi)} \otimes f_{\gamma(\psi)}$ is idempotent, since

$$(f_{\gamma(\phi)} \otimes f_{\gamma(\psi)}) \otimes (f_{\gamma(\phi)} \otimes f_{\gamma(\psi)}) \;=\; (f_{\gamma(\phi)} \otimes f_{\gamma(\phi)}) \otimes (f_{\gamma(\psi)} \otimes f_{\gamma(\psi)})$$
$$=\; f_{\gamma(\phi)} \otimes f_{\gamma(\psi)}.$$

This implies that $f_{\gamma(\phi)} \otimes f_{\gamma(\psi)}$ is the unity of the group $\gamma(\phi) \otimes \gamma(\psi) = \gamma(\phi \otimes \psi)$,

$$f_{\gamma(\phi)} \otimes f_{\gamma(\psi)} \;=\; f_{\gamma(\phi \otimes \psi)}.$$

Also, if $\gamma(\psi) \leq \gamma(\phi)$, then, for all $\phi' \in \gamma(\phi)$,

$$\phi' \otimes f_{\gamma(\psi)} \;=\; \phi' \otimes f_{\gamma(\phi)} \otimes f_{\gamma(\psi)} \;=\; \phi' \otimes f_{\gamma(\phi)} \;=\; \phi'.$$

So far we have used (3.65) only in the special case of $\phi \otimes \phi \equiv \phi \pmod \gamma$. In this sense, up to now, our development belongs to semigroup theory. But now we shall exploit (3.65) to examine the domain structure of the valuation algebra and marginalization. First, we note that, for $\phi, \psi \in \Phi$, $\gamma(\phi) = \gamma(\psi)$ entails that $d(\phi) = d(\psi)$ and $\gamma(\phi^{\downarrow t}) = \gamma(\psi^{\downarrow t})$. This follows from the assumption that γ is a congruence. Furthermore, Lemma 3.20 is still valid in separative valuation algebras.

Lemma 3.23

1. *Assume $\phi, \psi \in \Phi_s$, $t \subseteq s$. Then $\gamma(\phi) \leq \gamma(\psi)$ implies $\gamma(\phi^{\downarrow t}) \leq \gamma(\psi^{\downarrow t})$.*

2. *$\gamma(\phi^{\downarrow t}) \leq \gamma(\phi)$ for all $t \subseteq d(\phi)$.*

Proof. We note first, that (2) follows from (3.65).

In order to prove (1), assume $\gamma(\phi) \leq \gamma(\psi)$ and $t \subseteq d(\phi) = d(\psi)$. Then $\gamma(\psi^{\downarrow t}) \leq \gamma(\psi) = \gamma(\phi)$ or $\gamma(\psi^{\downarrow t} \otimes \phi) = \gamma(\phi)$. But this implies that $\gamma((\psi^{\downarrow t} \otimes \phi)^{\downarrow t}) = \gamma(\psi^{\downarrow t} \otimes \phi^{\downarrow t}) = \gamma(\phi^{\downarrow t})$, hence $\gamma(\psi^{\downarrow t}) \leq \gamma(\phi^{\downarrow t})$. $\quad\square$

We are now going to extend the operations of labeling and marginalization from Φ, at least partially, to Φ^*. To an element $\eta \in \Phi^*$ we associate a label $d(\eta) = d(\phi)$, if $\eta \in \gamma(\phi)$ for some $\phi \in \Phi$. This label is well defined, and depends not on the particular representant ϕ of the group $\gamma(\phi)$, because γ is a congruence. Also, if $\eta \in \Phi$, then the label just defined, corresponds to the label defined in the valuation algebra (Φ, D). Note also that $d(\eta^{-1}) = d(\eta)$.

Consider $\phi, \psi \in \Phi^*$. Then, if $\phi \in \gamma(\phi')$ and $\psi \in \gamma(\psi')$, for some $\phi', \psi' \in \Phi$, hence $d(\phi) = d(\phi')$ and $d(\psi) = d(\psi')$, we obtain $\phi \otimes \psi \in \gamma(\phi' \otimes \psi')$. Therefore,

$$d(\phi \otimes \psi) \;=\; d(\phi' \otimes \psi') \;=\; d(\phi') \cup d(\psi') \;=\; d(\phi) \cup d(\psi).$$

Hence, the *labeling axiom* is still valid in Φ^*. This proves also that the set of all elements of Φ^* with domain x,

$$\Phi_x^* \;=\; \bigcup_{\phi \in \Phi_x} \gamma(\phi), \tag{3.68}$$

is itself a semigroup.

Suppose that every semilattice of groups $\gamma(\phi)$ with domain x has a *minimal* element, say γ_x, for every domain $x \in D$. As before (see Section 3.5), the elements of γ_x are called positive. Denote the unity of group γ_x by e_x. Then,

$$e_x \otimes \phi \;=\; \phi$$

for all $\phi \in \Phi_x^*$. This tells us that every semigroup Φ_x^* has a neutral element. And, for $x, y \in D$, the element $e_x \otimes e_y$ is the unity element of the group $\gamma_x \otimes \gamma_y$ with domain $x \cup y$. Indeed, let $\gamma(\phi)$ be any group with domain $x \cup y$. Then,

$$\gamma_x \;\leq\; \gamma(\phi^{\downarrow x}) \;\leq\; \gamma(\phi),$$
$$\gamma_y \;\leq\; \gamma(\phi^{\downarrow y}) \;\leq\; \gamma(\phi).$$

Therefore, $\gamma_x \otimes \gamma_y \leq \gamma(\phi)$. Hence, $\gamma_x \otimes \gamma_y$ is the smallest element in the semilattice of groups with domain $x \cup y$. That is, $\gamma_x \otimes \gamma_y = \gamma_{x \cup y}$ or

$$e_x \otimes e_y \;=\; e_{x \cup y}.$$

Hence, the *neutrality axiom* is satisfied in Φ^*, if the semigroups Φ_x^* have neutral elements.

Next, we introduce a partial operation of marginalization into Φ^*. Any element $\eta \in \Phi^*$ is a pair (a quotient) (ϕ, ψ), where $\phi, \psi \in \Phi$ and $\gamma(\eta) = \gamma(\phi) = \gamma(\psi)$. Assume $d(\eta) = t$, such that $d(\phi) = d(\psi) = t$ too. Now, $\phi = (\phi \otimes \phi, \phi)$ and $\psi = (\psi \otimes \psi, \psi)$ or $\psi^{-1} = (\psi, \psi \otimes \psi)$ (here we consider Φ as a subset of Φ^*). Thus, we obtain

$$\eta \;=\; (\phi, \psi) \;=\; (\phi \otimes \phi, \phi) \otimes (\psi, \psi \otimes \psi) \;=\; \phi \otimes \psi^{-1}.$$

Suppose now that $\psi = e_t \otimes \psi'$ for some $\psi' \in \Phi$ with $d(\psi') \subseteq t$. Then $\gamma(\psi) = \gamma(e_t) \vee \gamma(\psi')$ such that $\gamma(\psi') \leq \gamma(\psi) = \gamma(\phi)$. Therefore we conclude,

$$\eta \;=\; \phi \otimes (e_t \otimes \psi')^{-1} \;=\; \phi \otimes e_t \otimes \psi'^{-1} \;=\; \phi \otimes \psi'^{-1}.$$

If $d(\psi') \subseteq s \subseteq t$, then we define

$$\eta^{\downarrow s} \;=\; \phi^{\downarrow s} \otimes \psi'^{-1}. \tag{3.69}$$

We have to show that this definition is unambiguous. So, assume that

$$\phi \otimes \psi^{-1} = \phi' \otimes \psi'^{-1},$$

where $\phi, \psi, \phi', \psi' \in \Phi$, $d(\psi) \subseteq d(\phi)$, $d(\psi') \subseteq d(\phi')$ and $\gamma(\psi) \leq \gamma(\phi)$, $\gamma(\psi') \leq \gamma(\phi')$. This equality leads to

$$\phi \otimes \psi' = \phi' \otimes \psi.$$

From the first equality, we deduce also that

$$\gamma(\phi) = \gamma(\phi \otimes \psi^{-1}) = \gamma(\phi' \otimes \psi'^{-1}) = \gamma(\phi'),$$

hence $d(\phi) = d(\phi')$. Assume now $d(\psi), d(\psi') \subseteq s \subseteq d(\phi) = d(\phi')$. Then,

$$\phi^{\downarrow s} \otimes \psi' = (\phi \otimes \psi')^{\downarrow s} = (\phi' \otimes \psi)^{\downarrow s} = \phi'^{\downarrow s} \otimes \psi. \qquad (3.70)$$

Now, $\gamma(\psi) \leq \gamma(\phi)$ implies that

$$\gamma(\phi^{\downarrow s}) = \gamma((\phi \otimes \psi)^{\downarrow s}) = \gamma(\phi^{\downarrow s} \otimes \psi).$$

That is, we have $\gamma(\phi^{\downarrow s}) = \gamma(\phi'^{\downarrow s}) \geq \gamma(\psi)$ and, in the same way, $\gamma(\phi^{\downarrow s}) = \gamma(\phi'^{\downarrow s}) \geq \gamma(\psi')$. Therefore, we obtain from (3.70)

$$\phi^{\downarrow s} \otimes \psi^{-1} = \phi'^{\downarrow s} \otimes \psi'^{-1},$$

hence $(\phi \otimes \psi^{-1})^{\downarrow s} = (\phi' \otimes \psi'^{-1})^{\downarrow s}$. So, marginalization is well defined.

Let $\eta \in \Phi$. Then $\eta = \eta \otimes \eta^{\downarrow \emptyset} \otimes (\eta^{\downarrow \emptyset})^{-1}$. Here we have $d(\eta \otimes \eta^{\downarrow \emptyset}) \supseteq d(\eta^{\downarrow \emptyset}) = \emptyset$ and $\gamma(\eta \otimes \eta^{\downarrow \emptyset}) \geq \gamma(\eta^{\downarrow \emptyset})$. So, for the valuations $\eta \in \Phi$, the new marginalization introduced above is defined, and

$$\begin{aligned}(\eta \otimes \eta^{\downarrow \emptyset} \otimes (\eta^{\downarrow \emptyset})^{-1})^{\downarrow s} &= (\eta \otimes \eta^{\downarrow \emptyset})^{\downarrow s} \otimes (\eta^{\downarrow \emptyset})^{-1} \\ &= \eta^{\downarrow s} \otimes \eta^{\downarrow \emptyset} \otimes (\eta^{\downarrow \emptyset})^{-1} = \eta^{\downarrow s}.\end{aligned}$$

Hence, the new marginalization introduced into Φ^* is indeed an extension of the marginalization defined in Φ.

Consider $\eta = \phi \otimes \psi^{-1}$, with $\phi, \psi \in \Phi$, $\gamma(\psi) \leq \gamma(\phi)$, and $d(\psi) \subseteq t \subseteq s \subseteq d(\phi)$. Then $\eta^{\downarrow t}$, $\eta^{\downarrow s}$, as well as $(\eta^{\downarrow s})^{\downarrow t}$ are defined. The last marginal is defined since $\gamma(\phi) = \gamma(\phi \otimes \psi)$, hence $\gamma(\phi^{\downarrow s}) = \gamma((\phi \otimes \psi)^{\downarrow s}) = \gamma(\phi^{\downarrow s} \otimes \psi)$, thus $\gamma(\psi) \leq \gamma(\phi^{\downarrow s})$. Hence, we obtain

$$\begin{aligned}\eta^{\downarrow t} = \phi^{\downarrow t} \otimes \psi^{-1} &= (\phi^{\downarrow s})^{\downarrow t} \otimes \psi^{-1} \\ &= (\phi^{\downarrow s} \otimes \psi^{-1})^{\downarrow t} = (\eta^{\downarrow s})^{\downarrow t}.\end{aligned}$$

This proves that the *transitivity axiom* is valid in Φ^*.

Finally, consider $\eta_1 = \phi_1 \otimes \psi_1^{-1}$ and $\eta_2 = \phi_2 \otimes \psi_2^{-1}$ with $d(\eta_1) = x$ and $d(\eta_2) = y$. As usual we assume $\gamma(\psi_1) \leq \gamma(\phi_1)$ and $\gamma(\psi_2) \leq \gamma(\phi_2)$. Let z be a domain such that $x \subseteq z \subseteq x \cup y$ and $d(\psi_2) \subseteq y \cap z$. Then $\eta_2^{\downarrow y \cap z}$ exists. And so does $(\eta_1 \otimes \eta_2)^{\downarrow z}$, since

$$\eta_1 \otimes \eta_2 = (\phi_1 \otimes \phi_2) \otimes (\psi_1 \otimes \psi_2)^{-1}$$

and

$$d(\psi_1 \otimes \psi_2) \quad = \quad d(\psi_1) \cup d(\psi_2) \quad \subseteq \quad z \quad \subseteq \quad x \cup y \quad = \quad d(\phi_1 \otimes \phi_2)$$
$$\gamma(\psi_1 \otimes \psi_2) \quad = \quad \gamma(\psi_1) \vee \gamma(\psi_2) \quad \leq \quad \gamma(\phi_1) \vee \gamma(\phi_2) \quad = \quad \gamma(\phi_1 \otimes \phi_2).$$

We obtain therefore

$$\eta_1 \otimes \eta_2^{\downarrow y \cap z} \quad = \quad \phi_1 \otimes \phi_2^{\downarrow y \cap z} \otimes \psi_1^{-1} \otimes \psi_2^{-1}$$
$$= \quad (\phi_1 \otimes \phi_2)^{\downarrow z} \otimes (\psi_1 \otimes \psi_2)^{-1} \quad = \quad (\eta_1 \otimes \eta_2)^{\downarrow z}. \quad (3.71)$$

Therefore, the *combination axiom* holds in Φ^* too.

We have proved the following theorem:

Theorem 3.24 *If (Φ, D) is a separative valuation algebra, Φ^* as defined above, then (Φ^*, D), with the operation of labeling, combination and partial marginalization as defined above, is a valuation algebra with partial marginalization.*

Note that the existence of neutral elements is not necessarily assumed in this theorem. But if they exist, then they satisfy the neutrality axiom, as shown above.

We turn now to some examples. If we consider *densities*, then the relation

$$f \equiv g \pmod{\gamma} \quad \text{if} \quad d(f) = d(g) \text{ and } f(\mathbf{x}) = 0 \Leftrightarrow g(\mathbf{x}) = 0$$

is clearly a congruence in the labeled valuation algebra of densities. Let

$$supp(f) \quad = \quad \{\mathbf{x} : f(\mathbf{x}) > 0\}$$

be the support set of a density f. Then, two densities f and g are equivalent, if they have the same support sets, $supp(f) = supp(g)$. We have that $f \equiv f \otimes f^{\downarrow t}$ $(\bmod~\gamma)$ for all densities. This holds because for continuous densities, we have as for discrete potentials, that $f^{\downarrow t}(\mathbf{x}^{\downarrow t}) = 0$ implies that $f(\mathbf{x}) = 0$. For measurable densities this holds almost everywhere, and we may select f such that $f(\mathbf{x}) = 0$. The semigroup of densities on the same support set is also clearly cancellative. It is therefore embedded into the group of quotients of densities. The valuation algebra of densities is therefore separative. It is embedded into a semigroup Φ^* which is the union of the semilattice of groups of quotients of densities with identical support. The functions $e_x(\mathbf{x}) = 1$ for all $\mathbf{x} \in \Omega_x$ are the neutral elements of the valuation algebra (Φ^*, D). Marginalization is defined for quotients $f(\mathbf{x}, \mathbf{y})/g(\mathbf{y})$, where $g(\mathbf{y}) = 0$ implies $f(\mathbf{x}, \mathbf{y}) = 0$ for all $\mathbf{x}$. Then marginalization corresponds to integration over components of $\mathbf{x}$. In particular, conditional densities $f(\mathbf{x}|\mathbf{y}) = f(\mathbf{x}, \mathbf{y})/f^{\downarrow t}(\mathbf{y})$ (where we assume $f(\mathbf{x}|\mathbf{y}) = 0$ when $f^{\downarrow t}(\mathbf{y}) = 0$), are partially integrable, namely over components of $\mathbf{x}$.

In the example of set potentials we have a similar situation. We define a congruence between commonalities by

$$q_1 \equiv q_2 \pmod{\gamma} \quad \text{if} \quad q_1(A) = 0 \Leftrightarrow q_2(A) = 0. \quad (3.72)$$

Again we have that $q \equiv q \otimes q^{\downarrow t}$ (mod γ). The semilattice of semigroups induced by this congruence corresponds still to support families $supp(q) = \{A \subseteq \Omega_s : q(A) > 0\}$. The elements of the groups G_α are quotients $q_1(A)/q_2(A)$ of commonalities, defined to vanish, if $A \notin supp(q_1) = supp(q_2)$. The unities of these groups are commonalities defined by $e(A) = 1$ for $A \in supp(e)$, $e(A) = 0$ otherwise.

To conclude, we remark that (Hewitt & Zuckerman, 1956) have introduced a necessary and sufficient condition for a commutative semigroup to be embedded into a semigroup which is the union of disjoint groups. This condition, which is called separativity in semigroup theory is that

$$\phi \otimes \psi = \phi \otimes \phi = \psi \otimes \psi \qquad (3.73)$$

imply always that $\phi = \psi$. In a commutative semigroup there is always a particular congruence, which causes the decomposition of the semigroup into a semilattice of so-called archimedean semigroups (Tamura & Kimura, 1954). Separativity is then necessary and sufficient for these archimedean components to be cancellative. Of course, separative valuation algebras satisfy the semigroup separativity condition, as do more particularly, regular algebras. Unfortunately, it is in general very difficult to see what these archimedean components are exactly in examples and to check their compatibility with marginalization as expressed by condition (3.65). There are also valuation algebras, whose semigroup is not even separative. An example is provided by possibility potentials with Lukasziewicz t-norm. So there is no way to introduce a division in this example.

3.7 Scaled Valuation Algebras

The examples of probability potentials and of set potentials have shown the importance of normalization from a semantical point of view. Only normalized potentials are probability distributions; and only normalized set potentials represent true belief functions. In view of the importance of this scaling operation we study it here in the abstract framework of valuation algebras.

The framework to discuss scaling is given by *separative*, labeled valuation algebras such as probability potentials, densities and set potentials. We are going to examine in this framework marginalization to the empty set of variables. Remember that the frame of the empty set, $\Omega_\emptyset$, has by convention one single configuration, denoted by $\diamond$. That is, we have $\Omega_\emptyset = \{\diamond\}$. Thus, for any $\mathbf{x} \in \Omega_s$ we have $\mathbf{x}^{\downarrow \emptyset} = \diamond$. Furthermore, if $A \subseteq \Omega_s$ is not empty, it follows that $A^{\downarrow \emptyset} = \Omega_\emptyset$. The power set of $\Omega_\emptyset$ consists only of the empty set and $\Omega_\emptyset$ itself.

Null valuations play an important role in scaling. We therefore assume that the valuation algebra satisfies the nullity axiom. We remark that each null element z_s forms by itself a group. Since $\phi \otimes z_s = z_s$ if $d(\phi) \leq s$, it follows that $\gamma(\phi) \leq \gamma(z_s)$ whenever $d(\phi) \leq s$. In particular, we have that $\gamma(z_t) \leq \gamma(z_s)$ if $t \subseteq s$.

We define

$$\phi^{\downarrow} = \phi \otimes (\phi^{\downarrow \emptyset})^{-1}. \qquad (3.74)$$

$\phi^{\downarrow}$ is called the *normalization* of ϕ. Note that

$$\phi = \phi^{\downarrow} \otimes \phi^{\downarrow \emptyset}. \qquad (3.75)$$

In particular, we have that $z_s^{\downarrow} = z_s$. In a stable valuation algebra, we have also that $e_s^{\downarrow} = e_s$, since $e_{\emptyset} = e_s^{\downarrow \emptyset}$ is its own inverse. From (3.74) we see that $\gamma(\phi) \leq \gamma(\phi^{\downarrow})$ and from (3.75) we conclude that $\gamma(\phi^{\downarrow}) \leq \gamma(\phi)$. Therefore, we have $\gamma(\phi^{\downarrow}) = \gamma(\phi)$.

Let's look at some examples. First consider *probability potentials*. Since this algebra is regular, the normalization of each potential belongs to the algebra itself. If p is a non-null potential, then

$$p^{\downarrow}(\mathbf{x}) = \frac{p(\mathbf{x})}{p^{\downarrow \emptyset}(\diamond)} = \frac{p(\mathbf{x})}{\sum_{\mathbf{x} \in \Omega_s} p(\mathbf{x})}.$$

The null potential is already normalized. We see that $(p^{\downarrow})^{\downarrow \emptyset}(\diamond) = 1$.

Gaussian potentials are normalized, since $(\mu, \mathbf{K}) \otimes (\mu, \mathbf{K})^{\downarrow \emptyset} = (\mu, \mathbf{K}) \otimes (\diamond, \diamond) = (\mu, \mathbf{K})$.

Densities are scaled, $f = f^{\downarrow}$, since $f^{\downarrow \emptyset}(\diamond) = 1$.

For *possibility potentials* p we have

$$p^{\downarrow \emptyset}(\diamond) = \max_{\mathbf{x} \in \Omega_s} p(\mathbf{x}).$$

With the product t-norm we obtain, if p is not the null potential,

$$p^{\downarrow}(\mathbf{x}) = \frac{p(\mathbf{x})}{\max_{\mathbf{x} \in \Omega_s} p(\mathbf{x})}.$$

In this case, similar to probability potentials, we have

$$(p^{\downarrow})^{\downarrow \emptyset}(\diamond) = 1.$$

With the Gödel t-norm the valuation algebra is idempotent. In an idempotent valuation algebra, each valuation is trivially normalized. The algebra with Lukasziewicz t-norm is not separative, hence no scaling is possible.

In the case of *Spohn potentials*

$$p^{\downarrow \emptyset}(\diamond) = \min_{\mathbf{x} \in \Omega_s} p(\mathbf{x})$$

and

$$p^{\downarrow}(\mathbf{x}) = p(\mathbf{x}) - p^{\downarrow \emptyset}(\diamond).$$

Normalized Spohn potentials have the property that

$$(p^{\downarrow})^{\downarrow \emptyset}(\diamond) = 0.$$

Spohn (Spohn, 1988) originally considered for semantical reasons only such normalized potentials.

Next we consider *set potentials*. Here, we have to consider two cases. The normalization of a commonality function q is defined by

$$q^{\downarrow}(A) = \frac{q(A)}{q^{\downarrow \emptyset}(\diamond)},$$

if $A \neq \emptyset$ and $q^{\downarrow \emptyset}(\diamond) \neq 0$. The denominator is

$$q^{\downarrow \emptyset}(\diamond) = m^{\downarrow \emptyset}(\diamond) = \sum_{A \neq \emptyset} m(A).$$

This is different from zero, if $m(A) \neq 0$ for at least one non-empty set A. This is the first case to be considered. Thus we obtain for the m-function of the normalized set potential, for $A \neq \emptyset$,

$$m^{\downarrow}(A) = \sum_{B \supseteq A} (-1)^{|B-A|} q^{\downarrow}(A) = \frac{\sum_{B \supseteq A} (-1)^{|B-A|} q(A)}{\sum_{A \neq \emptyset} m(A)} \qquad (3.76)$$
$$= \frac{m(A)}{\sum_{A \neq \emptyset} m(A)}.$$

For the empty set we obtain

$$q^{\downarrow}(\emptyset) = \frac{q(\emptyset)}{q^{\downarrow \emptyset}(\emptyset)} > 0,$$

and

$$q^{\downarrow \emptyset}(\emptyset) = m^{\downarrow \emptyset}(\emptyset) + m^{\downarrow \emptyset}(\diamond) = \sum_{A} m(A) = q(\emptyset).$$

This shows that $q^{\downarrow}(\emptyset) = 1$. Hence

$$m^{\downarrow}(\emptyset) = \sum_{A} (-1)^{|A|} q^{\downarrow}(A) = \frac{\sum_{A \neq \emptyset} (-1)^{|A|} q(A)}{\sum_{A \neq \emptyset} m(A)} + 1$$
$$= \frac{\sum_{A} (-1)^{|A|} q(A) - q(\emptyset)}{\sum_{A \neq \emptyset} m(A)} + 1$$
$$= \frac{m(\emptyset)}{\sum_{A \neq \emptyset} m(A)} - \frac{m(\emptyset) + \sum_{A \neq \emptyset} m(A)}{\sum_{A \neq \emptyset} m(A)} + 1 = 0.$$

So, we obtain the normalized m-function just as defined in Section 2.3.6. There is however also the case of $m(A) = 0$ for all $A \neq \emptyset$. Then we have $q(\emptyset) = m(\emptyset)$ and $q(A) = 0$ for $A \neq \emptyset$ and also $q^{\downarrow \emptyset}(\emptyset) = m^{\downarrow \emptyset}(\emptyset)$ and $q^{\downarrow \emptyset}(\diamond) = 0$. If $m(\emptyset) > 0$, then we obtain $q^{\downarrow}(\emptyset) = 1$ and $q^{\downarrow}(A) = 0$ for $A \neq \emptyset$. This corresponds to the normalized m-function $m^{\downarrow}(\emptyset) = 1$ and $m^{\downarrow}(A) = 0$ for $A \neq \emptyset$. All the mass goes in this case to the empty set. If $m(\emptyset) = 0$, then we have $q^{\downarrow}(A) = m^{\downarrow}(A) = 0$ for all subsets A.

If the valuation algebra (Φ, D) is not regular, we have no guarantee in general that $\phi^{\downarrow}$ belongs to Φ. However this is the case for all examples considered above. Even if $\phi^{\downarrow}$ does not belong to Φ, the marginal of $\phi^{\downarrow}$ is defined for all domains $t \subseteq d(\phi)$. Therefore, for simplicity, we assume for the following that $\phi^{\downarrow} \in \Phi$ for all $\phi \in \Phi$. This allows then to state a number of important properties of normalized valuations.

Lemma 3.25 *Assume $\phi^{\downarrow} \in \Phi$ for all $\phi \in \Phi$. Then*

1. $d(\phi^{\downarrow}) = d(\phi)$.

2. $(\phi^{\downarrow})^{\downarrow} = \phi^{\downarrow}$.

3. $(\phi \otimes \psi)^{\downarrow} = (\phi^{\downarrow} \otimes \psi^{\downarrow})^{\downarrow}$.

4. $(\phi^{\downarrow})^{\downarrow t} = (\phi^{\downarrow t})^{\downarrow}$ *for all $t \subseteq d(\phi)$.*

Proof. (1) From $\phi = \phi^{\downarrow} \otimes \phi^{\downarrow\emptyset}$ follows $d(\phi) = d(\phi^{\downarrow}) \cup \emptyset = d(\phi^{\downarrow})$.

(2) From $\phi = \phi^{\downarrow} \otimes \phi^{\downarrow\emptyset}$ we obtain $\phi^{\downarrow\emptyset} = (\phi^{\downarrow})^{\downarrow\emptyset} \otimes \phi^{\downarrow\emptyset}$, hence $(\phi^{\downarrow})^{\downarrow\emptyset} = f_{\gamma(\phi^{\downarrow\emptyset})}$. From $\phi^{\downarrow} = \phi \otimes (\phi^{\downarrow\emptyset})^{-1}$ we conclude that $\gamma(\phi^{\downarrow\emptyset}) \leq \gamma(\phi) = \gamma(\phi^{\downarrow})$. Thus,

$$(\phi^{\downarrow})^{\downarrow} \;=\; \phi^{\downarrow} \otimes ((\phi^{\downarrow})^{\downarrow\emptyset})^{-1} \;=\; \phi^{\downarrow} \otimes f_{\gamma(\phi^{\downarrow\emptyset})} \;=\; \phi^{\downarrow}.$$

(3) We have

$$\begin{aligned}
\phi \otimes \psi \;&=\; (\phi^{\downarrow} \otimes \phi^{\downarrow\emptyset}) \otimes (\psi^{\downarrow} \otimes \psi^{\downarrow\emptyset}) = (\phi^{\downarrow} \otimes \psi^{\downarrow}) \otimes (\phi^{\downarrow\emptyset} \otimes \psi^{\downarrow\emptyset}) \\
&=\; (\phi^{\downarrow} \otimes \psi^{\downarrow})^{\downarrow} \otimes (\phi^{\downarrow} \otimes \psi^{\downarrow})^{\downarrow\emptyset} \otimes (\phi^{\downarrow\emptyset} \otimes \psi^{\downarrow\emptyset}) \\
&=\; (\phi^{\downarrow} \otimes \psi^{\downarrow})^{\downarrow} \otimes ((\phi^{\downarrow} \otimes \phi^{\downarrow\emptyset}) \otimes (\psi^{\downarrow} \otimes \psi^{\downarrow\emptyset}))^{\downarrow\emptyset} \\
&=\; (\phi^{\downarrow} \otimes \psi^{\downarrow})^{\downarrow} \otimes (\phi \otimes \psi)^{\downarrow\emptyset}.
\end{aligned}$$

But this implies $(\phi^{\downarrow} \otimes \psi^{\downarrow})^{\downarrow} = (\phi \otimes \psi)^{\downarrow}$ since $\gamma((\phi \otimes \psi)^{\downarrow\emptyset}) \leq \gamma(\phi \otimes \psi) = \gamma(\phi^{\downarrow} \otimes \psi^{\downarrow}) = \gamma((\phi^{\downarrow} \otimes \psi^{\downarrow})^{\downarrow})$.

(4) We have on the one hand

$$\phi^{\downarrow t} \;=\; (\phi^{\downarrow t})^{\downarrow} \otimes \phi^{\downarrow\emptyset}$$

and on the other hand

$$\phi^{\downarrow t} = (\phi^{\downarrow} \otimes \phi^{\downarrow\emptyset})^{\downarrow t} = (\phi^{\downarrow})^{\downarrow t} \otimes \phi^{\downarrow\emptyset} \tag{3.77}$$

From this we conclude that $(\phi^{\downarrow})^{\downarrow t} = (\phi^{\downarrow t})^{\downarrow}$ since $\gamma(\phi^{\downarrow\emptyset}) \leq \gamma((\phi^{\downarrow})^{\downarrow t}), \gamma((\phi^{\downarrow t})^{\downarrow})$.
□

We introduce the relation $\phi \equiv \psi \pmod{\theta}$ if $\phi^{\downarrow} = \psi^{\downarrow}$. This is an equivalence relation on Φ as is quickly verified. It is more as the next lemma shows.

Lemma 3.26 *Assume that $\phi^\downarrow \in \Phi$ for all $\phi \in \Phi$. Then θ is a congruence in* (Φ, D)

Proof. (1) $\phi \equiv \psi \pmod \theta$ and Lemma 3.25 (1) imply

$$d(\phi) \;=\; d(\phi^\downarrow) \;=\; d(\psi^\downarrow) \;=\; d(\psi)$$

(2) Assume $\phi \equiv \psi \pmod \theta$ and $\chi \in \Phi$. Then we have by Lemma 3.25 (3)

$$(\phi \otimes \chi)^\downarrow \;=\; (\phi^\downarrow \otimes \chi^\downarrow)^\downarrow \;=\; (\psi^\downarrow \otimes \chi^\downarrow)^\downarrow \;=\; (\psi \otimes \chi)^\downarrow$$

So we have $\phi \otimes \chi \equiv \psi \otimes \chi \pmod \theta$.

(3) Assume again $\phi \equiv \psi \pmod \theta$ and $t \subseteq d(\phi) = d(\psi)$. Then, by Lemma 3.25 (4) we obtain

$$(\phi^{\downarrow t})^\downarrow \;=\; (\phi^\downarrow)^{\downarrow t} \;=\; (\psi^\downarrow)^{\downarrow t} \;=\; (\psi^{\downarrow t})^\downarrow$$

and so, $\phi^{\downarrow t} \equiv \psi^{\downarrow t} \pmod \theta$. □

Let now $\Phi^\downarrow$ be the set of all normalized valuations,

$$\Phi^\downarrow \;=\; \{\phi^\downarrow : \phi \in \Phi\}. \tag{3.78}$$

This set is closed under marginalization, as Lemma 3.25 (4) above shows. Combination of normalized valuations however generally does not yield normalized valuations. We are going to define a new operation of normalized combination by

$$\phi^\downarrow \oplus \psi^\downarrow = (\phi^\downarrow \otimes \psi^\downarrow)^\downarrow. \tag{3.79}$$

Note that labeling is well defined in $(\Phi^\downarrow, D)$.

Theorem 3.27 $(\Phi^\downarrow, D)$ *is a labeled valuation algebra with combination defined by (3.79), labeling and marginalization as in (Φ, D).*

Proof. (1) *Semigroup.* Clearly, the normalized combination is commutative. That it is associative follows from Lemma 3.25 (3)

$$
\begin{aligned}
(\phi^\downarrow \oplus \psi^\downarrow) \oplus \gamma^\downarrow \;&=\; ((\phi^\downarrow \otimes \psi^\downarrow)^\downarrow \otimes \gamma^\downarrow)^\downarrow \\
&=\; ((\phi^\downarrow \otimes \psi^\downarrow) \otimes \gamma^\downarrow)^\downarrow \;=\; (\phi^\downarrow \otimes (\psi^\downarrow \otimes \gamma^\downarrow))^\downarrow \\
&=\; (\phi^\downarrow \otimes (\psi^\downarrow \otimes \gamma^\downarrow)^\downarrow)^\downarrow \;=\; \phi^\downarrow \oplus (\psi^\downarrow \oplus \gamma^\downarrow).
\end{aligned}
$$

Φ does not necessarily have neutral elements (see examples below). But if Φ has neutral elements e_s, then, for ϕ with $d(\phi) = s$ we have, by Lemma 3.25 (3)

$$\phi^\downarrow \oplus e_s^\downarrow \;=\; (\phi \otimes e_s)^\downarrow \;=\; \phi^\downarrow.$$

So $e_s^\downarrow$ are neutral elements in $\Phi^\downarrow$.

(2) *Labeling.* This is inherited from (Φ, D): $d(\phi^\downarrow \oplus \psi^\downarrow) = d((\phi^\downarrow \otimes \psi^\downarrow)^\downarrow) = d(\phi^\downarrow \otimes \psi^\downarrow) = d(\phi^\downarrow) \cup d(\psi^\downarrow)$.

(3) *Marginalization.* Follows from $d((\phi^\downarrow)^{\downarrow x}) = d((\phi^{\downarrow x})^\downarrow) = d(\phi^{\downarrow x}) = x$.

(4) *Transitivity.* We have, by Lemma 3.25 (4)

$$((\phi^\downarrow)^{\downarrow x})^{\downarrow y} \;=\; ((\phi^{\downarrow x})^{\downarrow y})^\downarrow \;=\; (\phi^{\downarrow y})^\downarrow \;=\; (\phi^\downarrow)^{\downarrow y}.$$

(5) *Combination.* Assume $d(\phi^\downarrow) = x$, $d(\psi^\downarrow) = y$ and $x \subseteq z \subseteq x \cup y$. Using Lemma 3.25 we obtain

$$(\phi^\downarrow \oplus \psi^\downarrow)^{\downarrow z} \;=\; ((\phi \otimes \psi)^{\downarrow z})^\downarrow \;=\; (\phi \otimes \psi^{\downarrow y \cap z})^\downarrow \;=\; \phi^\downarrow \oplus (\psi^\downarrow)^{\downarrow y \cap z}.$$

(6) *Neutrality.* If there are neutral elements e_s in Φ, then the neutral elements $e_s^\downarrow$ inherit the neutrality axiom from e_s by Lemma 3.25. $\square$

The mapping $\phi \mapsto \phi^\downarrow$ is a homomorphism from (Φ, D) onto $(\Phi^\downarrow, D)$. The later valuation algebra is called the *scaled valuation algebra* associated with the algebra (Φ, D).

For *probability potentials, Spohn potentials* and *set potentials*, the scaled algebras use normalized combination as introduced in Chapter 2.

We remark that we may also introduce scaling in a valuation algebra (Φ, D) with partial marginalization. Elements $\phi \in \Phi$, for which marginals are defined for all $t \subseteq d(\phi)$, are called *(generalized) densities*. As an example consider non-negative Lebesgue-integrable functions (see Section 2.4). $\phi^\downarrow$ is then only defined for densities. Marginalization is defined for densities ϕ, hence for $\phi^\downarrow$. Combination is defined as follows:

$$\phi^\downarrow \oplus \psi^\downarrow \;=\; \begin{cases} (\phi^\downarrow \otimes \psi^\downarrow)^\downarrow, & \text{if } \phi^\downarrow \otimes \psi^\downarrow \text{ is a density,} \\ z_{d(\phi^\downarrow) \cup d(\psi^\downarrow)}, & \text{otherwise.} \end{cases}$$

It can be verified that normalized (proper) densities form a valuation algebra. The normalized densities obtained from Lebesgue-integrable non-negative functions are proper densities.

We note also that an idempotent valuation algebra is itself scaled, since in this case $\phi^\downarrow = \phi$. Another example of scaled valuations is provided by *Gaussian potentials*.

4 Local Computation

4.1 Fusion Algorithm

In this chapter we discuss different ways to compute marginals of a valuation which factors into a combination of valuations. We have seen in Chapter 2 that the complexity of the operations of combination and marginalization tends to increase exponentially with the size of the domains of the valuations involved. A crude measure of the size of a domain s is its cardinality $|s|$, the number of variables in the domain. A better measure would be the cardinality of the frame $|\Omega_s|$. So we are interested in methods where the operations needed to compute a marginal can be limited to small domains. This basic problem has been recognized early in the development of Bayesian networks. (Lauritzen & Spiegelhalter, 1988) was the pioneering work which showed how join trees (also called junction trees, or Markov trees) can be used to compute marginals of large multidimensional discrete probability distributions, if they factor into factors with small domains. Shenoy and Shafer (Shenoy & Shafer 1990) introduced the axioms of valuation algebras needed to generalize computation on join trees from probability to other formalisms, especially belief functions. In the sequel many refinements and several different architectures for computing on join trees have been proposed. We refer to (Shafer, 1996) and (Cowell et. al., 1999; Dawid, 1992) for a discussion of this subject relative to probability theory. Relatively few contributions appeared concerning computation on join trees for other formalisms than probability theory. One notable exception is the paper (Lauritzen & Jensen, 1997) which discusses the most important architectures in the context of abstract valuation algebras. We refer also to (Dechter, 1999) for related algorithms, which can be applied to a number of different formalisms.

The common feature of all these architectures is that they assume a factorization of the valuation to be considered. They organize the computations in a series of combinations and marginalizations. And the important point is that they do it in such a way, that each of these operations takes place in a *domain of one of the factors*. Never a larger domain is needed. Such schemes of computation are called *local computations*. If the domains of the factors are all sufficiently small, then local computations become feasible. The subject of this chapter is precisely to study various ways to arrange for local computations in valuation algebras.

This chapter takes up the discussion in (Shafer, 1991) and (Lauritzen & Jensen, 1997) and places the subject into the framework developed in the previous Chapter 3. We hope thereby to provide a firmer foundation for the different architectures for local computation in join trees. Furthermore, we hope to lay the basis for genuinely generic architectures for inference with various formalisms, both for numerical approaches to uncertainty like probability, possibility, Spohn disbelief potentials, belief functions, etc. as well as for symbolic, logical approaches. The second kind of formalism has been much less considered from the point of view of valuation algebras, than the numerical methods, see however (Dechter, 1999). Therefore, the following Chapters 6 and 7 will be devoted to a more detailed discussion of the related idempotent valuation algebras.

The most basic approach to compute a marginal is by successive elimination of variables. We will see that this leads in a natural way to graphical structures, called join trees and computation based on join trees. Join trees are the essential ingredient for local computation. Computation on join trees can be arranged in different ways, which lead to different architectures. The most important ones are the so-called Shenoy-Shafer architecture, the Lauritzen-Spiegelhalter architecture and the HUGIN architecture. The first one is the most general one, in the sense that it requires only a valuation algebra without particular special properties. The latter two architectures require division, as for example provided by a regular or a separative valuation algebra. Idempotent valuation algebras can be treated by a simplified architecture derived from the Lauritzen-Spiegelhalter architecture.

We start in this section by examining variable elimination. Let (Φ, D) be a valuation algebra, with or without neutral elements. Assume that we have a factorization of some $\phi \in \Phi$,

$$\phi = \phi_1 \otimes \phi_2 \otimes \cdots \otimes \phi_m, \tag{4.1}$$

and let $d(\phi_i) = s_i$. Then we have

$$d(\phi) = s = s_1 \cup s_2 \cup \cdots \cup s_m.$$

Arrange the variables of s into some sequence $X_1, X_2, \ldots, X_n$, where $n = |s|$. In order to compute $\phi^{\downarrow \{X_n\}}$ we eliminate the other variables in this sequence. Now, to eliminate a variable, say X_1, from ϕ, which is a valuation on a potentially large domain s, may be very expensive. The following simple lemma however shows that this is not really necessary.

Lemma 4.1 *In a valuation algebra, if $\phi \in \Phi$ is factored as (4.1), then*

$$\phi^{-X_1} = \psi^{-X_1} \otimes \left(\bigotimes_{i:X_1 \notin s_i} \phi_i \right). \tag{4.2}$$

where

$$\psi = \left(\bigotimes_{\imath \, X_1 \in s_\imath} \phi_\imath \right)$$

Proof. Define

$$\psi' = \left(\bigotimes_{\imath . X_1 \notin s_\imath} \phi_\imath \right),$$

so that $X_1 \notin d(\psi')$. Then, according to the combination axiom (written for variable elimination, see Section 2.2), we have $\phi^{-X_1} = (\psi \otimes \psi')^{-X_1} = \psi^{-X_1} \otimes \psi'$. This proves (4.2). $\square$

So we have to eliminate variable X_1 only in a domain

$$\bigcup_{\imath \cdot X_1 \in s_\imath} s_\imath,$$

which, in many cases, will be much smaller than s. And the same consideration will apply to the elimination of the further variables $X_2, X_3, \ldots$.

This algorithm of successive variable eliminations using local variable elimination has been put into a form called *fusion algorithm* by Shenoy (Shenoy, 1992): if we compare the factors of ϕ in (4.1) with those of ϕ^{-X_1} in (4.2), then we observe that in deleting X_1, the factors of ϕ which do not contain X_1 in their domains remain unchanged, and the factors that do contain X_1 in their domain are first combined and then X_1 is eliminated from their combination. This operation is called *fusion*. A formal definition is as follows: let $Fus_Y(\{\phi_1, \phi_2, \ldots, \phi_m\})$ denote the set of valuations after fusing the valuation in the set $\{\phi_1, \phi_2, \ldots, \phi_m\}$ with respect to a variable Y,

$$Fus_Y(\{\phi_1, \phi_2, \ldots, \phi_m\}) = \{\psi^{-Y}\} \cup \{\phi_\imath : Y \notin d(\phi_\imath)\}, \qquad (4.3)$$

where

$$\psi = \left(\bigotimes_{\imath \, Y \in d(\phi_\imath)} \phi_\imath \right).$$

Using this notation, the result of Lemma 4.1 can be expressed as

$$\phi^{-X_1} = \bigotimes Fus_{X_1}(\{\phi_1, \phi_2, \ldots, \phi_m\}).$$

It follows then, by repeatedly applying Lemma 4.1 and the axiom of commutativity of elimination of variables (see Section 2.2), that

$$\phi^{\downarrow\{X_n\}} = \bigotimes Fus_{X_{n-1}}(\cdots (Fus_{X_2}(Fus_{X_1}(\{\phi_1, \phi_2, \ldots, \phi_m\}))) \cdots).$$

This algorithm to compute a marginal applies to any valuation algebra, with or without neutral elements.

We construct now a graphical representation of this process of fusion, following (Shafer, 1996): start with the list $l = \{s_1, s_2, \ldots, s_m\}$ of domains and form a first node with

$$s' = \bigcup_{\imath : X_1 \in s_\imath} s'_\imath.$$

This is the beginning of the graph G we are going to construct. Change then the list l to

$$l \cup \{s' - \{X_1\}\} - \{s_\imath : X_1 \in s_\imath\}.$$

In words: remove all domains containing the variable X_1 which is eliminated, but add the domain of the combined valuations whose domains contain X_1, but without X_1. Clearly this corresponds to the fusion operation $Fus_{X_1}(\{\phi_1, \phi_2, \ldots, \phi_m\})$. Annotate finally in the new list l the entry $s' - \{X_1\}$ with a link to the new node s'.

Before step i, when variable $X_\imath$ is eliminated, we have some graph G and a list $l = \{s'_1, \ldots, s'_{m'}\}$. The entries of l are domains which do not contain the eliminated variables $X_1, \ldots, X_{\imath-1}$. We have before this step also

$$\phi^{-\{X_1, \cdot, X_{\imath-1}\}} = \phi'_1 \otimes \cdots \otimes \phi'_{m'}, \tag{4.4}$$

with $d(\phi'_\imath) = s'_\imath$. Now, to the graph G is added a new node

$$s' = \bigcup_{\jmath : X_\imath \in s'_\jmath} s'_\jmath.$$

Then, all links found in the annotated domains absorbed by s', that is, in $\{s'_\jmath : X_\imath \in s'_\jmath\}$, are transformed into edges leading to the new node s'. Finally the list l is modified into

$$l \cup \{s' - \{X_\imath\}\} - \{s'_\jmath : X_\imath \in s'_\jmath\}. \tag{4.5}$$

This corresponds again to the fusion operation: all domains which are absorbed in s' are removed, and the new domain $s' - \{X_\imath\}$ is added. Finally, the new domain $s' - \{X_\imath\}$ is annotated with a link to the new node s' in the graph G. This step is repeated for $i = 1$ to n. After step n, the list l is empty.

In Fig. 4.1 this procedure is illustrated for the example in Section 2.1 (see (2.6)).

We remark that each node in the graph can be given the number of the variable eliminated when the node is introduced. In this numbering $i < j$ means that the node j is introduced after node i. We say that variable $X_\imath$ is eliminated in node i. It appears in no node introduced later. When $X_\imath$ belongs to some node introduced earlier than i, then $X_\imath$ belongs to all nodes on the path from this node to $\imath$, the latter included.

Initial list:

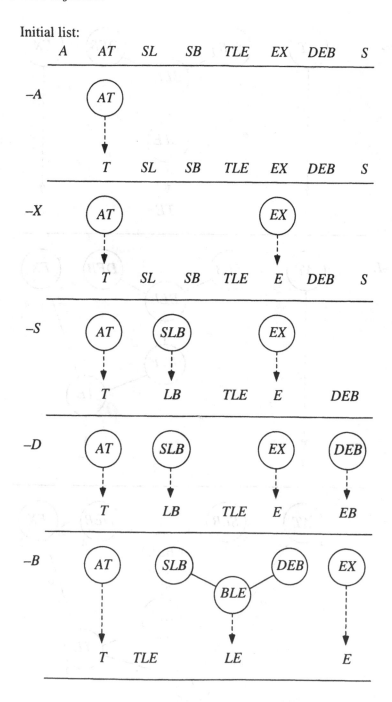

Figure 4.1: Construction of the graph representing the fusion algorithm for example (2.6), Section 2.1 with the sequence A, X, S, D, B, L, E, T for variable elimination (continued in Fig. 4.2).

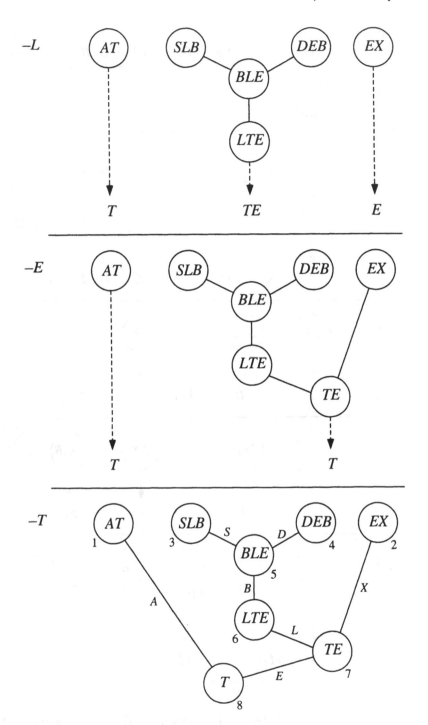

Figure 4.2: Continuation of Fig. 4.1.

Lemma 4.2 *At the end of the fusion algorithm, the graph G is a tree.*

Proof. Note that each node is connected to some node which is later introduced into G. The only exception is the last node. This is so, because when a node is newly introduced, a link to an annotated domain in the list l is created. At some later step, this domain will be absorbed, since at the end the list l is empty. In this moment, the node is linked by an edge to the new node. Hence, there is a path from each node to the last node introduced, therefore there is a path between any pair of nodes. This means that the graph G is connected.

By the remark above, each node is linked to only *one* later node by an edge. Therefore, there can be no cycle in G. Because, assume there is a cycle, then there is a node with least number i on the cycle. If then j and k are the two neighbors of i on the cycle, then $i < j, k$. But this means that both j and k are later introduced than i and both are linked to i. But this is not possible. □

The tree G has many more properties. The most important one is, that it is a join tree.

Definition 4.3 *A tree, whose nodes are domains s, is called a join tree, if for any pair of nodes s', s'', if $X \in s' \cap s''$, then $X \in s$ for all nodes on the path between s' and s''.*

Join trees are also sometimes called junction trees or Markov trees.

Lemma 4.4 *At the end of the fusion algorithm, the graph G is a join tree.*

Proof. We know already that it is a tree. Select two nodes s' and s'' and $X \in s' \cap s''$. We go down from s' along to later nodes, until we arrive at the node where X is eliminated. Let k' be the number of this node. Similarly, we go down from s'' to later nodes until the node where X is eliminated. Assume the number is k''. But X is eliminated in exactly one node. So $k' = k''$ and the path from s' to s'' goes from s' to $k' = k''$ and from there to s''. X belongs to all nodes of the paths from s' to $k' = k''$ and s'' to $k' = k''$, hence to all nodes of the path from s' to s''. □

For later reference, we state two further properties of G. For a node s, we call the neighbor nodes s', which are introduced *before* s its *parents*, and the unique neighbor, which is introduced later, its *child*. Denote the set of parents of the node i by $pa(i)$ and the child by $ch(i)$. Note that the set of parents may be empty. Then we call the node a *leaf*. On the other hand, only the latest node has no child, it is also called the *root*. We denote by t_i the domain in node i of the graph G

Lemma 4.5 *At the end of the fusion algorithm the following holds in the graph G:*

1. *For all i we have $t_i \cap t_{ch(i)} = t_i - \{X_i\}$ and for all $j \in pa(i)$ we have $t_j \cap t_i = t_j - \{X_j\}$.*

2. *For all i we have*

$$t_i = \left(\bigcup_{j \in pa(i)} t_j - \{X_j\} \right) \cup \left(\bigcup_{j: X_1, .., X_{i-1} \notin s_j, X_i \in s_j} s_j \right). \tag{4.6}$$

Proof. (1) When a node i is created, then $t_i - X_i$ is annotated in the list l. When node i is linked to its unique child $ch(i)$, then $t_i - \{X_i\} \subseteq t_{ch(i)}$, but $X_i \notin t_{ch(i)}$, since the variable has already been eliminated. Hence we see that $t_i \cap t_{ch(i)} = t_i - \{X_i\}$. The second part follows from this.

(2) Note that

$$t_i = \bigcup_{j: X_i \in s'_j} s'_j, \tag{4.7}$$

where the s'_j are domains in the actual list l. Some of these domains are annotated. They give the first part on the right hand side of (4.6). Further there may be domains s'_i in the union (4.7), which are not annotated. This means that they correspond to original domains s_i of factors ϕ_i, which have not been eliminated in the list l so far. But this means that they do not contain variables already eliminated, that is X_1 to X_{i-1}. This gives the second part in (4.6). □

The join tree constructed by a fusion algorithm can be used to give an alternative picture of the fusion algorithm. We observe, that for all domains s_i of the factors ϕ_i of ϕ there is a node s in the join tree G which covers s_i, that means, $s_i \subseteq s$. In fact, let X_j be the first variable of s_i in the elimination sequence. Then we have $s_i \subseteq t_j$. We assume for the moment that the valuation algebra has neutral elements. Then, we put every factor ϕ_i on the node t_j, if X_j is the first variable of the elimination sequence in s_i. On nodes j, where there are several factors ϕ_i we combine them and obtain ψ_j. On nodes j where there are no factors, we put the neutral element $\psi_j = e_{t_j}$. Fig. 4.3 illustrates this for the example of Fig. 4.1.

Assume that node i contains the valuation ψ_i when variable i is eliminated, and its successor $ch(i)$ contains $\psi_{ch(i)}$. Then, in the fusion algorithm, node i computes $\psi_i^{-X_i}$ and sends this valuation as a message to node $ch(i)$. The receiving node combines the incoming message with its own valuation, $\psi_{ch(i)} \otimes \psi_i^{-X_i}$, and stores the result. Here we visualize nodes as (virtual) processors, which store valuations, compute and send messages and process them. This is the picture used also in the following discussion throughout this chapter.

On a leaf j of the join tree we have a valuation ψ_j with the property that $d(\psi_j) = t_j$. This follows from Lemma 4.5 (2). Also by Lemma 4.5 (2) the same holds for any node j, once all incoming messages have been processed, that is before the elimination of variable X_j.

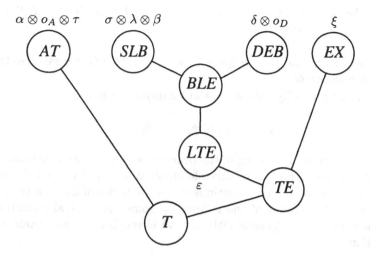

Figure 4.3: Putting factors on the join tree constructed from the fusion algorithm, illustrated by the example of Fig. 4.1.

To conclude the discussion of the fusion algorithm, we remark that the message passing view of the algorithm remains valid, even if the valuation algebra has no neutral elements (for example for densities or Gaussian potentials). In this case we leave the nodes j where there are no factors simply empty. Or, if the valuation algebra has positive elements, we put neutral elements e_{t_j} on them. If a node is empty, then upon receiving a message, it simply stores the message without any previous combination. Lemma 4.5 (2) guarantees then that before elimination of variable X_j we have a valuation $\psi_j \in \Phi$ on node j with $X_j \in d(\psi_j)$ and variable elimination or marginalization is therefore always well defined. So, the fusion algorithm works indeed for the most general kind of valuation algebras. All we need are essentially the transitivity and especially the combination axiom. But this basic scheme can be refined in several ways. And some of these refinements require more structure, like for example a notion of division.

4.2 Collect Algorithm

A message passing algorithm similar to the fusion algorithm can be defined on general join trees, which are not necessarily defined by a sequence of variable eliminations. All what we require is that a valuation ϕ factors according to a join tree. What we mean by this is described in the following definition.

Definition 4.6 Property (J): *A valuation ϕ has property (J), if there is a factorization of ϕ,*

$$\phi = \psi_1 \otimes \psi_2 \otimes \cdots \otimes \psi_m \tag{4.8}$$

such that the domains $d(\psi_i) = s_i$ are the nodes of a join tree. (4.8) is also called a join tree factorization.

If ϕ has property (J) with respect to some join tree G, we say also that ϕ factors according to G.

In practice we usually have an initial factorization of ϕ

$$\phi \;=\; \phi_1 \otimes \phi_2 \otimes \cdots \otimes \phi_n$$

like for example at the beginning of the fusion algorithm. If we construct a join tree G by a certain sequence of variable eliminations, and put the factors ϕ_i on the nodes, then we obtain valuations ψ_j on the nodes of the join tree, some of which are combinations of the factors ϕ_i, some are neutral elements (we assume here valuation algebras with neutral elements). So we obtain a new factorization

$$\phi \;=\; \psi_1 \otimes \psi_2 \otimes \cdots \otimes \psi_m,$$

since the domains of the added neutral elements are smaller than $d(\phi)$. We have $d(\psi_j) \subseteq t_j$ on all nodes of the join tree. If we replace ψ_j by $\psi_j \otimes e_{t_j}$, we obtain finally a factorization of ϕ according to the join tree G constructed by the fusion algorithm.

So any initial factorization of a valuation ϕ may be turned into a factorization according to a join tree. The fusion algorithm provides a method for doing that. Note that different sequences of variable eliminations may lead to different join trees, with different cardinalities of the nodes. Loosely speaking, we want join trees whose nodes have small cardinalities. So the problem of the construction of an optimal join tree arises. We do not want to enter into a discussion of this optimization problem. We remark only that the problem has been shown to be NP-hard (Arnborg, et. al., 1987), but that there exist good heuristics. Fig. 4.4 shows as an example another join tree for the same example as in Fig. 4.3.

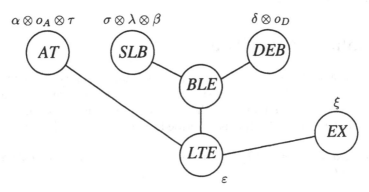

Figure 4.4: Another join tree for a factorization of the example of Fig. 4.3.

Assume that we have a factorization (4.8) according to a join tree G. We want to compute the marginal of ϕ relative to one of the domains s_i, say relative to s_m, to fix ideas. We take node s_m as a *root* in the graph G and direct all edges towards this root. It is then always possible to number the nodes of this directed graph (and hence the factors of ϕ) in such a way, that if j is a node on the directed path from i to the root m, then $j > i$. For example, if G is the join tree constructed by a fusion algorithm, this numbering may correspond to the sequence of variable eliminations. As in this case we denote by $pa(i)$ the set of parents of i (neighboring nodes j of i with $j < i$) and by $ch(i)$ the unique child of i (the neighbor j of i with $i < j$).

We use now again the picture of nodes as virtual processors, passing messages between them. The messages will be scheduled in the sequence of the numbering just introduced. Each node i transmits its message "inwards" towards its child. Therefore, a node emits its message only when it has received the messages from all its parents. Let $\psi_j^{(i)}$ denote the content of the storage at node j before step i. At the beginning we have $\psi_j^{(1)} = \psi_j$.

At step i, node i computes the message

$$\mu_{i \to ch(i)} = \psi_i^{(i) \downarrow s_i \cap s_{ch(i)}}. \tag{4.9}$$

This message is sent to the child $ch(i)$ which updates its storage

$$\psi_{ch(i)}^{(i+1)} = \psi_{ch(i)}^{(i)} \otimes \mu_{i \to ch(i)}. \tag{4.10}$$

The storage at all other nodes do not change at step i, $\psi_j^{(i+1)} = \psi_j^{(i)}$, for $j \neq ch(i)$.

This algorithm is called the *collect algorithm*. Note that it is a local algorithm: All marginalizations and combinations take place within the domains s_i of the factors of ϕ. It is a basic scheme for all architectures based on computations on join trees. The collect algorithm also can be formulated in a more informal way, independent of any numbering, by the following rules:

- *Rule 1*: Each node waits to send its message to its child until it has received the messages from all its parents. This means that leaves can send their messages right away.

- *Rule 2*: When a node is ready to send its message, it marginalizes its current content to the intersection of its domain and those of its child.

- *Rule 3*: When a node receives a message, it replaces its current content with the combination of its current content and the message.

These rules summarize also the fusion algorithm viewed as a message passing process. The specialty of the fusion algorithm is that in rule 2 marginalization actually consists only in the elimination of exactly one variable, whereas in the general case possibly more than one variable must be removed in rule

2. The rule-based formulation above indicates also that the collect algorithm can be executed partially in parallel.

We have to prove that this procedure really produces the marginal of ϕ relative to domain s_m. For this purpose we prove first the following lemma.

Lemma 4.7 *Define*

$$y_i \;=\; \bigcup_{j=i}^{m} s_j, \qquad i = 1, 2, \dots, m. \tag{4.11}$$

Then, for $i = 1, 2, \dots, m$

$$\left(\bigotimes_{j=i}^{m} \psi_j^{(i)}\right)^{\downarrow y_{i+1}} \;=\; \bigotimes_{j=i+1}^{m} \psi_j^{(i+1)} \;=\; \phi^{\downarrow y_{i+1}}. \tag{4.12}$$

Proof. We show first that $d(\psi_j^{(i)}) = s_j$ for all j and i. And we do this by induction on i. For $i = 1$ we have indeed $d(\psi_j^{(1)}) = d(\psi_j) = s_j$ for $j = 1$ to m. Assume now that $d(\psi_j^{(i)}) = s_j$ holds for i. Then according to (4.10) and (4.9)

$$d(\psi_{ch(i)}^{(i+1)}) \;=\; d(\psi_{ch(i)}^{(i)}) \cup (s_i \cap s_{ch(i)}) \;=\; s_{ch(i)}.$$

Furthermore, for all $j \neq ch(i)$, we have $d(\psi_j^{(i+1)}) = d(\psi_j^{(i)}) = s_j$. So indeed $d(\psi_j^{(i)}) = s_j$ for all j and i.

Since G is a join tree we conclude that

$$s_i \cap s_{ch(i)} \;=\; s_i \cap y_{i+1},$$

because each variable X which belongs to s_i and some s_j with $j > i$ must belong to $s_{ch(i)}$, since each path from i to $j > i$ passes through $s_{ch(i)}$. Therefore, using the combination axiom we obtain

$$\left(\bigotimes_{j=i}^{m} \psi_j^{(i)}\right)^{\downarrow y_{i+1}} \;=\; \left(\psi_i^{(i)} \otimes \left(\psi_{ch(i)}^{(i)} \otimes \bigotimes_{j=i+1, j \neq ch(i)}^{m} \psi_j^{(i)}\right)\right)^{\downarrow y_{i+1}}$$

$$=\; \psi_i^{(i)\downarrow s_i \cap y_{i+1}} \otimes \psi_{ch(i)}^{(i)} \otimes \bigotimes_{j=i+1, j \neq ch(i)}^{m} \psi_j^{(i)}$$

$$=\; \psi_i^{(i)\downarrow s_i \cap s_{ch(i)}} \otimes \psi_{ch(i)}^{(i)} \otimes \bigotimes_{j=i+1, j \neq ch(i)}^{m} \psi_j^{(i)}$$

$$=\; \psi_{ch(i)}^{(i+1)} \otimes \bigotimes_{j=i+1, j \neq ch(i)}^{m} \psi_j^{(i+1)}$$

$$=\; \bigotimes_{j=i+1}^{m} \psi_j^{(i+1)}.$$

The rightmost equality in (4.12) is again proved by induction on i. For $i = 1$ we have indeed

$$\left(\bigotimes_{j=i}^{m} \psi_j^{(1)} \right)^{\downarrow y_2} = \left(\bigotimes_{j=i}^{m} \psi_j \right)^{\downarrow y_2} = \phi^{\downarrow y_2}.$$

Assume now, that it holds for i, that is,

$$\bigotimes_{j=i}^{m} \psi_j^{(i)} = \phi^{\downarrow y_i}.$$

Then, by transitivity of marginalization,

$$\bigotimes_{j=i+1}^{m} \psi_j^{(i+1)} = \left(\bigotimes_{j=i}^{m} \psi_j^{(i)} \right)^{\downarrow y_{i+1}} = (\phi^{\downarrow y_i})^{\downarrow y_{i+1}} = \phi^{\downarrow y_{i+1}}.$$

This proves (4.12) for all i. □

The soundness of the collect algorithm is a corollary of this lemma.

Theorem 4.8 *At the end of the collect algorithm we have in node m the marginal of ϕ relative to s_m,*

$$\psi_m^{(m)} = \phi^{\downarrow s_m}. \tag{4.13}$$

Proof. (4.13) follows from (4.12) for $i = m$, since $y_m = s_m$. □

This result can be slightly generalized. The set of nodes j which are linked by a directed path to node i in the join tree G form by themselves a join tree, which we call G_i. Node i is included in G_i.

Lemma 4.9 *At the end of the collect algorithm, each node i contains the marginal relative to s_i of the factors associated to the nodes of the tree G_i,*

$$\psi_i^{(i)} = \left(\bigotimes_{j \in G_i} \psi_j \right)^{\downarrow s_i}. \tag{4.14}$$

Proof. The node i is the root of the join tree G_i. So Theorem 4.8 applies to this reduced join tree. □

We remark that the collect algorithm applies also to valuation algebras with partial marginalization (like for example the extension of separative algebras), provided

$$\bigotimes_{j \in G_i} \psi_j$$

has a marginal relative to the domain $s_i \cap s_{ch(i)}$ for all subtrees G_i of the join tree G. This guarantees that all marginals occurring in the collect algorithm exist. In particular, the messages $\psi_i^{(i)\downarrow s_i \cap s_{ch(i)}}$ are defined (see Lemma 4.9). For example, if a factorization (4.1) of a valuation ϕ of a separative algebra (Φ, D) with positive elements is transformed by the fusion algorithm into a factorization according to a join tree, then some of the factors can be in the extension Φ^* of Φ. But according to Section 4.1 this factorization satisfies the condition above. We shall return to this important case in Section 4.5.

4.3 Computing Multiple Marginals

When the marginal of a valuation ϕ is to be computed for several domains s_i of a factorization, then the collect algorithm could be repeated with the different domains as roots. This means each time redirecting some edges, whereas other edges keep their direction. Fig. 4.5 illustrates this. More precisely, the edges on the path between the old and new root change the direction, whereas all other edges keep their direction. It is clear, that the messages on the edges which keep their direction do not change. Hence, repeating the collect algorithm for different roots involves a lot of redundant computations.

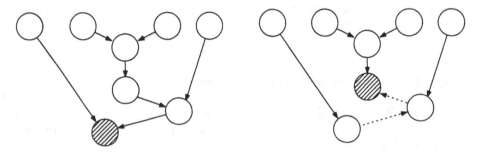

Figure 4.5: Redirecting edges in a join tree, when the root changes.

Rather we want to reuse the messages computed in the first collect algorithm as far as possible. Therefore, instead of redirecting edges, we will leave the edges *undirected* and associate directions with the *messages*. Also, if each node sends a message to each of its neighbors in the undirected join tree, instead of just to its child in the rooted join tree, then we can compute marginals for every domain s_i of the factorization of ϕ according to the join tree. For this purpose, we introduce *mailboxes* or *storages* on each edge between two nodes of the join tree. These mailboxes will serve to store the two messages passing through the edge, one in each direction. See Fig. 4.6 for a schematic representation of the situation.

The messages any node emits are scheduled as follows: A node i emits a message to its neighbor j, when it has received all the message from its other neighbors.

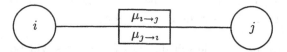

Figure 4.6: The message mailboxes on an edge between two neighboring nodes of a join tree.

Let $ne(i)$ denote the set of all the neighbors of node i in the join tree corresponding to the factorization of (4.8) of ϕ. Then, the message sent to neighbor j is

$$\mu_{i \to j} = \left(\psi_i \otimes \bigotimes_{k \in ne(i), k \neq j} \mu_{k \to i} \right)^{\downarrow s_i \cap s_j}. \tag{4.15}$$

The leaves can emit messages right at the beginning to their unique neighbors. There will always be a node which can send a message until all nodes have emitted messages to all their neighbors. Then the algorithm stops. Fig. 4.7 shows a possible sequence of messages send in this scheme. This computation scheme is called the *Shenoy-Shafer architecture*.

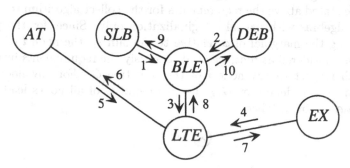

Figure 4.7: A possible sequence of messages sent in a join tree.

We claim that at the end of this algorithm the marginal of ϕ relative to all domains s_i can be easily computed.

Theorem 4.10 *At the end of the message passing in the Shenoy-Shafer architecture, we obtain at node i*

$$\phi^{\downarrow s_i} = \psi_i \otimes \left(\bigotimes_{j \in ne(i)} \mu_{j \to i} \right). \tag{4.16}$$

Proof. The messages $\mu_{k \to j}$ do not depend on the actual schedule used to compute them. So we may always select i as a root, direct the edges to this root and number the nodes as in the corresponding collect algorithm. Then

the message passing to this root corresponds exactly to the collect algorithm
and the proposition follows from Theorem 4.8.   ☐

In fact, whatever possible scheduling of messages will be selected, there
will always be a determined node r which is the first to receive all its incoming
messages. In Fig. 4.7 this is node LTE which after step 5 has received the
messages from its three neighbors. If we direct the edges of the join tree
according to the messages sent so far, we obtain a rooted join tree with node
r as root. And the computation corresponds exactly to the collect algorithm
with this root. The only difference is that the incoming messages are not
combined with the content of the nodes but stored in the mailboxes. The
nodes keep ψ_i in their registers. Therefore, the first phase of the computations
according to the Shenoy-Shafer architecture is called the *collect phase* . The
algorithm does not stop at this point, but the messages propagate "outward"
from the root. In Fig. 4.7 these messages are no. 6 to 10. This second phase is
called *distribute phase* . So the computations in the Shenoy-Shafer architecture
can always be decomposed into these two phases.

The computations according to Shenoy-Shafer architecture are, like the
collect algorithm, possible for any valuation algebra. In order to obtain a factorization of ϕ according to a join tree, for example by variable elimination
in the fusion algorithm, we may need to assume the existence of neutral elements. We stated above the requirements for the collect algorithm to work for
valuation algebras with partial marginalization only. Since, for the purpose
of computing the marginal of ϕ relative to domain s_i, the node i of the join
tree can be considered as root, we need to satisfy the requirements introduced
above with respect to any node i of the join tree G. For any node i, and
neighbor k of of i, denote by $G_{i,k}$ the subtree of G of all nodes leading to i,
without passing by k. Then

$$\bigotimes_{j \in G_{i,k}} \psi_j$$

must have a marginal with respect to $s_i \cap s_k$. This allows then to compute the
message

$$\mu_{i \to k} \;=\; \left(\bigotimes_{j \in G_{i,k}} \psi_j \right)^{\downarrow s_i \cap s_k}$$

(see Lemma 4.9). The problem of local computations in valuation algebras
with partial marginalization will be examined more in detail in Section 4.5
below.

The architecture contains however some inefficiencies. When a node has
more than three neighbors, then in computing the messages and also in deter-
mining the marginal, several combinations of incoming messages are repeated.
Also some combinations may take place on larger domains than necessary, see
for example (Kohlas & Shenoy, 2000). In order to remove these inefficiencies

Shenoy (Shenoy, 1997 b) proposed to work with *binary join trees* in which every node has at most three neighbors. Then the inefficiencies mentioned above disappear. The prize is a somewhat larger join tree and correspondingly more mail boxes. Any factorization according to a join tree may be extended to a factorization according to a binary join tree (Shenoy, 1997 b). But there are also other architectures, besides this refinement of Shenoy-Shafer architecture, which avoid the inefficiencies of the crude Shenoy-Shafer architecture. The price to be paid for these improvements is however the assumption of more structure in the underlying valuation algebra. We shall see in the following sections that some notion of division is needed. So these architectures require a regular or separative valuation algebra.

4.4 Architectures with Division

In the Shenoy-Shafer architecture, contrary to the collect algorithm, the incoming messages in a node during the collect phase are not combined with the content of the node. This is done only at the end of the distribute phase, when the node marginal is computed. The reason is, that when messages are combined during the collect phase, then there is one message too many in the distribute phase, namely the one coming from the neighbor to which a message is computed in the distribute phase. If the valuation algebra allows division, then we can combine messages in the collect phase and divide out the messages which are superfluous in the distribute phase. This idea has been realized for probability networks in (Lauritzen & Spiegelhalter, 1988) and in an improved form in (Jensen, Lauritzen & Olesen, 1990). That it can be generalized to valuation algebras has been shown in (Lauritzen & Jensen, 1997), and also in a somewhat different, and more general, way in (Shafer, 1991). Here we show first how these methods work for regular algebras . The more general case of separative algebras will be examined in Section 4.5.

Assume then that (Φ, D) is a regular valuation algebra. We suppose that $\phi \in \Phi$ is factored according to a join tree G, that is

$$\phi = \psi_1 \otimes \psi_2 \otimes \cdots \otimes \psi_m, \qquad (4.17)$$

where $d(\psi_i) = s_i$ and the domains s_i are the nodes of the join tree G. As in the collect algorithm we select an (arbitrary) node as the *root* and direct the edges of G towards the root. We may again number the nodes such that $i < j$ if node j is on the directed path from i to the root. As in the collect algorithm we can without loss of generality assume that s_m is the root and the indices of the nodes s_i correspond already to this numbering. $pa(i)$ and $ch(i)$ denote again the set of parents and the unique child of node i in the directed version of the join tree. Note that $j < i$ for all $j \in pa(i)$ and $i < ch(i)$.

The first algorithm to be presented consists of a collect and a distribute phase. The *collect phase* corresponds to the collect algorithm of Section 4.2. The messages in this phase correspond therefore to messages in the Shenoy-

Shafer architecture. So, when node i is ready to send its message to its child $ch(i)$, then in its store we have, just as in the collect algorithm,

$$\eta_i = \psi_i \otimes \bigotimes_{j \in pa(i)} \mu_{j \to i}. \tag{4.18}$$

The message sent is (see (4.15))

$$\mu_{i \to j} = \eta_i^{\downarrow s_i \cap s_j} = \left(\psi_i \otimes \bigotimes_{k \in ne(i), k \neq j} \mu_{k \to i} \right)^{\downarrow s_i \cap s_j}.$$

The node $ch(i)$ combines the incoming message with its current content. In contrast to the collect algorithm however, in node i we divide the message sent to $ch(i)$ out, so that we store there

$$\eta_i \otimes (\eta_i^{\downarrow s_i \cap s_{ch(i)}})^{-1} = \psi_i \otimes \bigotimes_{j \in pa(i)} \mu_{j \to i} \otimes (\mu_{i \to j})^{-1}. \tag{4.19}$$

instead of $\eta_i = \psi_i^{(i)}$. At the end of the collect phase, we have in node m the marginal $\phi^{\downarrow s_m}$, since the collect phase corresponds to the collect algorithm, up to different storages in the nodes. But this last remark does not concern the root m, because it has no child.

In the following distribute phase, each node, once it has received the message from its child, sends messages to all its parents. The phase starts with the root, which has no child, and which can therefore immediately begin to compute messages. Note that we may proceed in the inverse sequence of the node numbering. The form of the messages is exactly the same as in the collect phase: the marginal of the content of node i relative to the domains $s_i \cap s_j$ for $j \in pa(i)$. The receiving node j will, as in the collect phase, combine the incoming message with its current content (4.19). The algorithm stops, when, at the end of the distribute phase, all leaves have processed their incoming message. This organization of the message passing scheme in a join tree is called *Lauritzen-Spiegelhalter architecture*.

We claim that then all nodes contain the marginals of ϕ relative to their domain.

Theorem 4.11 *At the end of the computations in the Lauritzen-Spiegelhalter architecture, each node i of the join tree G contains the marginal $\phi^{\downarrow s_i}$.*

Proof. We prove the theorem by induction on the node numbers i. The theorem holds for node m by Theorem 4.8. Assume now that it holds for all $j > i$. We are going to prove that it holds then also for i. Note that the content of i in the distribute phase, before processing a message is (see (4.19))

$$\psi_i \otimes \bigotimes_{j \in pa(i)} \mu_{j \to i} \otimes (\mu_{i \to ch(i)})^{-1}.$$

By the assumption of induction, since $ch(i) > i$, and by Theorem 4.10, at node $ch(i)$ we have, when it is ready to send its messages in the distribute phase,

$$\phi^{\downarrow s_{ch(i)}} = \psi_{ch(i)} \otimes \bigotimes_{j \in ne(ch(i))} \mu_{j \to ch(i)}.$$

So the message sent to node i is

$$\phi^{\downarrow s_i \cap s_{ch(i)}} = \left(\psi_{ch(i)} \otimes \bigotimes_{j \in ne(ch(i))} \mu_{j \to ch(i)} \right)^{\downarrow s_i \cap s_{ch(i)}}$$

$$= \left(\psi_{ch(i)} \otimes \bigotimes_{j \in ne(ch((i))), j \neq i} \mu_{j \to ch(i)} \right)^{\downarrow s_i \cap s_{ch(i)}} \otimes \mu_{i \to ch(i)}$$

$$= \mu_{ch(i) \to i} \otimes \mu_{i \to ch(i)}. \tag{4.20}$$

So, we obtain at node i, when we combine the incoming message with the actual content, using Theorem 4.10,

$$\psi_i \otimes \left(\bigotimes_{j \in pa(i)} \mu_{j \to i} \right) \otimes (\mu_{i \to ch(i)})^{-1} \otimes \mu_{ch(i) \to i} \otimes \mu_{i \to ch(i)}$$

$$= \phi^{\downarrow s_i} \otimes f_{\gamma(\mu_{i \to ch(i)})} = \phi^{\downarrow s_i}.$$

The last equality follows since, by Lemma 4.9,

$$\mu_{i \to ch(i)} = \left(\bigotimes_{j \in G_i} \psi_j \right)^{\downarrow s_i \cap s_{ch(i)}},$$

which implies that

$$\gamma(\mu_{i \to ch(i)}) = \gamma \left(\left(\bigotimes_{j \in G_i} \psi_j \right)^{\downarrow s_i \cap s_{ch(i)}} \right) \leq \gamma(\phi^{\downarrow s_i})$$

(see Lemma 3.20). This shows that node i contains $\phi^{\downarrow s_i}$ at the end of the distribute phase. $\square$

A closer look at the computations in the Lauritzen-Spiegelhalter architecture reveals two additional properties: note that in the collect phase, when a message is sent from i to $ch(i)$, then at node i the content is divided by the message $\mu_{i \to ch(i)}$, whereas at node $ch(i)$ the content is combined by the same message. Since at the beginning of the collect phase the contents of the nodes combine to ϕ, this remains true during the collect phase. In particular, at the end of the collect phase we have thus

$$\phi = \phi^{\downarrow s_m} \otimes \bigotimes_{i=1}^{m-1} \eta_i. \tag{4.21}$$

During the distribute phase on the other hand, we combine at each step the content of node i with $\phi^{\downarrow s_i \cap s_{ch(i)}}$ to get $\phi^{\downarrow s_i}$. The domains $s_i \cap s_{ch(i)}$ are called separators in the join tree. Let $\mathcal{S}$ denote the set of separators in the join tree and $\mathcal{V}$ the set of domains or nodes in the join tree. Then by the remark above, since we start the distribute phase with contents on the nodes which combine to ϕ and we multiply at each step with a marginal of ϕ relative to a separator, we have at the end

$$\phi \otimes \bigotimes_{s \in \mathcal{S}} \phi^{\downarrow s} \;=\; \bigotimes_{v \in \mathcal{V}} \phi^{\downarrow v}. \tag{4.22}$$

This can also be written as

$$\phi \;=\; \bigotimes_{v \in \mathcal{V}} \phi^{\downarrow v} \otimes \bigotimes_{s \in \mathcal{S}} (\phi^{\downarrow s})^{-1}. \tag{4.23}$$

Another architecture, also based on division has been proposed by (Jensen, Lauritzen & Olesen, 1990) for probability potentials. It is again divided into a collect and distribute phase. The collect phase is like in the collect algorithm. The only change is that between any node i and its child $ch(i)$ a new node, the separator, with domain $s_i \cap s_{ch(i)}$ is introduced. And the message

$$\mu_{i \to ch(i)} \;=\; \eta_i^{\downarrow s_i \cap s_{ch(i)}}$$

sent in the collect phase is stored there. Note that this is indeed the same message as in the Shenoy-Shafer architecture. As in the Lauritzen-Spiegelhalter architecture, we have the marginal $\phi^{\downarrow s_m}$ stored in the root m at the end of the collect phase.

In the distribute phase, which follows the collect phase, as in the Lauritzen-Spiegelhalter architecture, each node i sends a message to all of its parents $pa(i)$, once it has received the message from its child $ch(i)$. The root m can start immediately to send messages. The message $\mu_{ch(i) \to i}$ from $ch(i)$ to i is as in the Lauritzen-Spiegelhalter architecture the marginal of the content of the node $ch(i)$ relative to the domain $s_i \cap s_{ch(i)}$. But in contrast to the later architecture, the message goes first to the separator, where it is divided by the storage there,

$$\mu_{ch(i) \to i} \otimes \left(\mu_{i \to ch(i)} \right)^{-1}. \tag{4.24}$$

This message is then sent to node i, where it is combined with the content of this node, to give

$$\eta_i \otimes \mu_{ch(i) \to i} \otimes \left(\mu_{i \to ch(i)} \right)^{-1}. \tag{4.25}$$

The separator stores $\mu_{ch(i) \to i}$. This way of organizing the computations on a join tree is called the *HUGIN architecture* , because it is the architecture implemented in the HUGIN software. The difference to the Lauritzen-Spiegelhalter architecture is that the division by the collect message is not done on node

i during the collect phase, but on the separator, during the distribute phase. The advantage is that the separator has a smaller domain than node i, so that division there tends to be less costly.

We claim that the HUGIN architecture produces the correct results at the end of the computations. More precisely the following holds:

Theorem 4.12 *At the end of the computations in the HUGIN architecture, each node $v \in \mathcal{V}$ contains the marginal $\phi^{\downarrow v}$ and each separator $s \in \mathcal{S}$ the marginal $\phi^{\downarrow s}$.*

Proof. The theorem is proved by induction over the nodes $v \in \mathcal{V}$. In fact, at the end of the collect phase we have already stated that the root m contains $\phi^{\downarrow s_m}$. In the distribute phase we go backwards in the inverse sequence as in the collect phase. So suppose that the theorem holds for all nodes from m down to some $i+1$. Let $ch(i) = j > i$. Then j sends its message to i passing by the separator $s = s_i \cap s_{ch(i)}$. The new value in the separator becomes then $\mu_{ch(i) \to i} = \phi^{\downarrow s}$. The old content of this separator, dating from the collect phase is the message $\mu_{i \to ch(i)}$. Hence, the message passed to node i is

$$\phi^{\downarrow s} \otimes (\mu_{i \to ch(i)})^{-1}.$$

And node i, also dating from the collect phase, contains the valuation

$$\eta_i = \psi_i \otimes \bigotimes_{k \in pa(i)} \mu_{k \to i}.$$

So, the new value in i will be

$$\psi_i \otimes \bigotimes_{k \in pa(i)} \mu_{k \to i} \otimes \phi^{\downarrow s} \otimes (\mu_{i \to ch(i)})^{-1}.$$

But we have seen in the proof of Theorem 4.11, (4.20) that

$$\phi^{\downarrow s} = \phi^{\downarrow s_i \cap s_{ch(i)}} = \mu_{ch(i) \to i} \otimes \mu_{i \to ch(i)}.$$

Therefore, we obtain for the new content of node v

$$\psi_i \otimes \bigotimes_{k \in ne(i)} \mu_{k \to i} \otimes f_{\gamma(\mu_{i \to ch(i)})} = \phi^{\downarrow s_i} \otimes f_{\gamma(\mu_{i \to ch(i)})} = \phi^{\downarrow s_i}.$$

The last equality follows in the same way as in the proof of Theorem 4.11. This completes the proof by induction. $\square$

The computations in the HUGIN architecture may be described in a more uniform way. At the beginning of the algorithm, put not only the factors ψ_i on the nodes i of the join tree, but also $e_{s_i \cap s_{ch(i)}}$ on the separators between nodes i and $ch(i)$. If the algebra has no neutral elements, then put a group unity $f_{s_i \cap s_{ch(i)}}$ on the separator, such that $\gamma(f) \leq \gamma(\psi_i), \gamma(\psi_{ch(i)})$ (assuming that such a lower bound exists).

As before, let S denote the set of separators and $\mathcal{V}$ the set of original nodes in the join tree G. Then we have at the beginning of the computations

$$\phi \otimes \bigotimes_{s \in S} \psi_s = \bigotimes_{v \in \mathcal{V}} \psi_v,$$

where $\psi_s = e_s$ (or $\psi_s = f_s$) for the separators. This implies also that

$$\phi = \bigotimes_{v \in \mathcal{V}} \psi_v \otimes \bigotimes_{s \in S} \psi_s^{-1}. \tag{4.26}$$

Now, we can describe the message passing during both the collect and the distribute phase in a uniform way: let v, w be two neighboring nodes in the join tree G with separator s between them. If η_v, η_w, η_s are the contents of nodes v, w, s before v sends a message to w, and $\eta_v^*, \eta_w^*, \eta_s^*$ the contents of the nodes after the processing of the message, then

$$\begin{aligned}
\eta_v^* &= \eta_v, \\
\eta_s^* &= \eta_v^{\downarrow s}, \\
\eta_w^* &= \eta_w \otimes \eta_s^* \otimes \eta_s^{-1}.
\end{aligned} \tag{4.27}$$

This is valid during the collect as well as during the distribute phase.

At the end of the computations with Lauritzen-Spiegelhalter architecture, we have seen that formula (4.22) holds. It says essentially, that at the end of computations, ϕ equals the combination of the contents of the nodes $v \in \mathcal{V}$ divided by the contents of the separators $s \in S$. We claim that in the HUGIN architecture this property holds all the time, from the start to the end.

Theorem 4.13 *In the HUGIN architecture, we have always*

$$\phi \otimes \bigotimes_{s \in S} \eta_s = \bigotimes_{v \in \mathcal{V}} \eta_v. \tag{4.28}$$

Proof. The proof goes by induction over the steps of the algorithm. We have seen above that the formula (4.28) holds at the beginning of the computations. Let v', w', s' be the nodes involved in a message passing step and assume that the formula holds before the step. Note that in the collect phase, before the message passing step, $\eta_{s'} = e_{s'}$ (or $\eta_{s'} = f_{s'}$), hence $\gamma(\eta_{s'}) \leq \gamma(\phi)$ and in the distribute phase, before the message passing step, $\gamma(\eta_{s'}) \leq \gamma(\phi^{\downarrow w'}) \leq \gamma(\phi)$ (see (4.25)). Therefore, after this message passing step we have (using the notation introduced above)

$$\phi \otimes \left(\bigotimes_{s \in S} \eta_s^* \right) = \phi \otimes \left(\bigotimes_{s \in S, s \neq s'} \eta_s \right) \otimes \eta_{v'}^{\downarrow s'}$$

$$= \phi \otimes \left(\bigotimes_{s \in S, s \neq s'} \eta_s \right) \otimes \eta_{v'}^{\downarrow s'} \otimes f_{\gamma(\eta_{s'})}$$

$$= \phi \otimes \left(\bigotimes_{s \in S, s \neq s'} \eta_s \right) \otimes \eta_{v'}^{\downarrow s'} \otimes (\eta_{s'} \otimes (\eta_{s'})^{-1})$$

$$= \phi \otimes \left(\bigotimes_{s \in S} \eta_s \right) \otimes \eta_{v'}^{\downarrow s'} \otimes (\eta_{s'})^{-1}.$$

Therefore, using the assumption of induction we obtain

$$\phi \otimes \left(\bigotimes_{s \in S} \eta_s^* \right) = \left(\bigotimes_{v \in V, v \neq w'} \eta_v \right) \otimes \eta_{w'} \otimes \eta_{v'}^{\downarrow s'} \otimes (\eta_{s'})^{-1}$$

$$= \left(\bigotimes_{v \in V} \eta_v^* \right).$$

So (4.28) holds during the whole computation process in the HUGIN architecture. □

In the HUGIN architecture we started the computations with the factors ψ_i of ϕ in the nodes $v \in V$ of the join tree and with neutral elements e_s in the separators. We shall now show that we may start the computations also more generally with other contents in the nodes. Let, as before, G be a join tree with node set V and separators S.

Theorem 4.14 *Assume*

$$\phi = \bigotimes_{v \in V} \psi_v \otimes \bigotimes_{s \in S} \psi_s^{-1}, \tag{4.29}$$

where we assume $d(\psi_v) = v$, $d(\psi_s) = s$, *and moreover that*

$$\gamma(\psi_s) \leq \gamma(\psi_v), \gamma(\psi_w) \tag{4.30}$$

for any separator s between two nodes v and w. Then, if we put ψ_v into node $v \in V$ and ψ_s into separator $s \in S$ at the start of the computation in the HUGIN architecture, then at the end we have $\phi^{\downarrow v}$ for $v \in V$ and $\phi^{\downarrow s}$ for $s \in S$.

Proof. (Shafer, 1996). The join tree G is assumed to be rooted with a root r. Then $s(v)$ designates the separator between v and its child $ch(v)$ in the directed join tree. From (4.29) it follows that

$$\phi = \psi_r \otimes \bigotimes_{v \in V - \{r\}} (\psi_v \otimes \psi_{s(v)}^{-1}).$$

Put $\psi'_r = \psi_r$, $\psi'_v = \psi_v \otimes \psi^{-1}_{s(v)}$ for all non-root nodes and $\psi'_s = e_s$ (or $\psi'_s = f_{\gamma(\psi_s)}$) in all separators. We apply the HUGIN computations both with starting valuations ψ_v, ψ_s and ψ'_v, ψ'_s and denote by η_v, η_s and η'_v, η'_s the contents of the nodes and separators during the HUGIN computations in the two cases. Note that in the second case we have the ordinary HUGIN computations as presented above.

We remark that at the start of the HUGIN computations we have the relations

$$
\begin{aligned}
\eta_r &= \eta'_r, \\
\eta_v &= \eta'_v \otimes \psi_{s(v)}, \qquad \text{for all } v \neq r, \\
\eta_s &= \eta'_s \otimes \psi_s, \qquad \text{for all } s.
\end{aligned}
\tag{4.31}
$$

We remark further that $\gamma(\psi_s) \leq \gamma(\psi_v) = \gamma(\eta_v) = \gamma(\eta'_v)$ at the beginning, if s is a separator between v and some other node. These inequalities remain valid during the collect phase, since the support of η_v and η'_v can only increase.

We claim that the relations (4.31) are invariant during the collect phase of the HUGIN computations. We prove this by induction on the nodes v. In fact, consider the step, when node v sends its message to node $w = ch(v)$ and assume that the relations hold up to node v. Then only $\eta_{s(v)}$ and η_w change:

$$
\eta^*_{s(v)} = \eta^{\downarrow s(v)}_v = \eta'^{\downarrow s(v)}_v \otimes \psi_{s(v)} = \eta'^*_{s(v)} \otimes \psi_{s(v)}.
$$

And also

$$
\begin{aligned}
\eta^*_w &= \eta_w \otimes \eta^*_{s(v)} \otimes \eta^{-1}_{s(v)} \\
&= \eta'_w \otimes \psi_{s(w)} \otimes \eta^*_{s(v)} \otimes \eta^{-1}_{s(v)} \\
&= \eta'_w \otimes \psi_{s(w)} \otimes \eta'^*_{s(v)} \otimes \psi_{s(v)} \otimes (\eta'_{s(v)} \otimes \psi_{s(v)})^{-1} \\
&= \eta'_w \otimes \eta'^*_{s(v)} \otimes \eta'^{-1}_{s(v)} \otimes \psi_{s(w)} \\
&= \eta'^*_w \otimes \psi_{s(w)}.
\end{aligned}
$$

This holds also for the root r, where the term ψ_s does not appear. In particular, at the end of the collect phase we have that $\eta_r = \eta'_r = \phi^{\downarrow r}$.

At the beginning of the distribute phase, the root r sends the same messages to its neighbor in the two HUGIN computations. We show now, that, if a node emits the same messages in the two computations, then the receiving node will have the same content after processing the message. Hence assume that node w emits a message to v. Then

$$
\eta^*_{s(v)} = \eta^{\downarrow s(v)}_w = \eta'^{\downarrow s(v)}_w = \eta'^*_{s(v)}
\tag{4.32}
$$

and the separator obtains the same new content in both cases. Further

$$
\begin{aligned}
\eta^*_v &= \eta_v \otimes \eta^*_{s(v)} \otimes \eta^{-1}_{s(v)} \\
&= \eta'_v \otimes \psi_{s(v)} \otimes \eta'^*_{s(v)} \otimes (\eta'_{s(v)} \otimes \psi_{s(v)})^{-1} \\
&= \eta'_v \otimes \eta'^*_{s(v)} \otimes \eta'^{-1}_{s(v)} \otimes f_{\gamma(\psi_{s(v)})} \\
&= \eta'^*_v.
\end{aligned}
$$

Here, we used the fact that $\gamma(\psi_{s(v)}) \leq \gamma(\eta'_v)$. So, we conclude by induction, from Theorem 4.12, that at the end of the distribute phase $\eta_v = \eta'_v = \phi^{\downarrow v}$ for all $v \in \mathcal{V}$ and $\eta_s = \eta'_s = \phi^{\downarrow s}$ for all $s \in \mathcal{S}$. □

We remark that idempotent valuation algebras are regular algebras, where $\phi^{-1} = \phi$. Therefore, both the Lauritzen-Spiegelhalter architecture and the HUGIN architecture apply to this case. But they become much simpler. In the collect phase of the Lauritzen-Spiegelhalter architecture, the division in the nodes v has no effect. So the collect phase corresponds exactly to the collect algorithm. The distribute phase does not change. In this way, collect and distribute phase become fully identical. In both phases the following rules apply:

- *Rule 1*: The message a node v sends to w is the marginal of its content relative to $v \cap w$,

- *Rule 2*: The receiving node combines the message with its content.

According to (4.22) we have at the end of the computation, thanks to idempotency,

$$\phi = \bigotimes_{v \in \mathcal{V}} \phi^{\downarrow v}. \tag{4.33}$$

The HUGIN architecture becomes identical to the Lauritzen-Spiegelhalter architecture, if we consider the separators as ordinary nodes in the latter case (the join tree augmented by the separators remains a join tree).

4.5 Computations in Valuation Algebras with Partial Marginalization

We start by considering a factorization of an element ϕ of a *separative* valuation algebra,

$$\phi = \phi_1 \otimes \phi_2 \otimes \cdots \otimes \phi_m, \tag{4.34}$$

where $\phi, \phi_1, \ldots \phi_m \in \Phi$. We noted already that variable elimination in the fusion algorithm works for any valuation algebra, hence in particular for a separative one. However, if we construct the join tree associated with a sequence of variable eliminations in the fusion algorithm, see Section 4.1, then we may need to introduce neutral elements on some nodes which do not belong to Φ, but only to the embedding valuation algebra (Φ^*, D) (see Theorem 3.24). This leads then to a new factorization

$$\phi = \psi_1 \otimes \psi_2 \otimes \cdots \otimes \psi_n, \tag{4.35}$$

where this time, $\phi \in \Phi$, but $\psi_1, \ldots, \psi_n \in \Phi^*$ and $s_i = d(\psi_i)$ form a join tree for $i = 1, \ldots, n$. This is an indication that it is important to study local computation of marginals of valuations of valuation algebras with partial marginalization only. We shall give immediately another motivation.

An important valuation algebra with partial marginalization only is the algebra of nonnegative real-valued functions on frames $\mathbf{R}^n$ (see Section 2.4). We take this example as a model for the following discussion. Let (Φ^*, D) a valuation algebra with partial marginalization. We define a special class of elements in Φ^* as follows:

Definition 4.15 Kernels: *A valuation $\phi \in \Phi^*$ is called a* kernel *for h given t, if*

1. $d(\phi) = h \cup t$,

2. *For $t \subseteq s \subseteq h \cup t$, $\phi^{\downarrow s}$ exists, and $\phi^{\downarrow t} = e_t$.*

Then h is called the head *and t the* tail *of the kernel.*

A kernel ϕ with $t = \emptyset$ is also called a *density*. In this case ϕ is scaled, $\phi = \phi^{\downarrow}$ or $\phi^{\downarrow \emptyset} = e_\emptyset$. These concepts are clearly motivated by the example mentioned above. If $f(\mathbf{x}, \mathbf{y})$ is a nonnegative function with $\mathbf{x}$ and $\mathbf{y}$ configurations for sets of variables t and h, then f is a kernel for h given t, if

$$\int f(\mathbf{x}, \mathbf{y}) dy = 1 \qquad \text{for (almost) all } \mathbf{x}.$$

And $f(\mathbf{x})$ is a density, if

$$\int f(\mathbf{x}) dx = 1.$$

More precisely, these are kernels and densities relative to the Lebesgue measure. In general, if (Φ, D) is a *separative* valuation algebra with *positive elements*, then

$$\phi \otimes (\phi^{\downarrow t})^{-1}$$

is a kernel, if ϕ is positive. Examples are provided by Gaussian potentials, which are all positive. The kernels are in this case conditional Gaussian densities.

We state a few elementary results about kernels.

Lemma 4.16 *Let ϕ and ψ be kernels with heads $h(\phi)$, $h(\psi)$ and tails $t(\phi)$, $t(\psi)$, such that the head of ψ is disjoint to the domain of ϕ, $d(\phi) \cap h(\psi) = \emptyset$, then $\phi \otimes \psi$ is a kernel with head $h(\phi) \cup h(\psi)$ and tail $t(\phi) \cup (t(\psi) - h(\phi))$.*

Proof. Let $h_1 = h(\phi)$, $h_2 = h(\psi)$, $t_1 = t(\phi)$ and $t_2 = t(\psi)$. Then, since $\psi^{\downarrow t_2}$ exists and equals e_{t_2}, the combination axiom gives us

$$(\phi \otimes \psi)^{\downarrow h_1 \cup t_1 \cup t_2} = \phi \otimes \psi^{\downarrow t_2} = \phi \otimes e_{t_2}.$$

Note that

$$\phi \otimes e_{t_2} = \phi \otimes e_{t_2 - h_1}.$$

Now, since $\psi^{\downarrow t_2}$ exists and equals e_{t_2}, the transitivity and the combination axioms imply that

$$
\begin{aligned}
(\phi \otimes \psi)^{\downarrow t_1 \cup (t_2 - h_1)} &= ((\phi \otimes \psi)^{\downarrow h_1 \cup t_1 \cup t_2})^{\downarrow t_1 \cup (t_2 - h_1)} \\
&= (e_{t_2 - h_1} \otimes \phi)^{\downarrow t_1 \cup (t_2 - h_1)} = e_{t_2 - h_1} \otimes \phi^{\downarrow t_1} \\
&= e_{t_2 - h_1} \otimes e_{t_1} = e_{t_1 \cup (t_2 - h_1)}.
\end{aligned}
$$

This proves that $\phi \otimes \psi$ is indeed a kernel with head and tail as indicated in the lemma. $\qquad\Box$

The important special case of kernels is the case of densities, which are valuations whose marginals exist for all subsets of their domain. It is well known from probability theory, that such densities are often defined as products of kernels. This is especially the case with models defined on the base of probability networks such as Bayesian networks, see for example (Cowell et. al., 1999; Shafer, 1996). Usually only discrete probability potentials are considered in these frameworks. But such graphical models apply also to more general valuation algebras with kernels. Therefore, we study factorizations of densities into combinations of kernels, following (Shafer, 1996). The basic result is the following lemma:

Lemma 4.17 *Let ϕ_1 and ϕ_2 be kernels with tails $t_1 = \emptyset$, t_2 and heads h_1, h_2, such that h_2 is disjoint of h_1 and $t_2 \subseteq h_1$. Then*

1. $\phi_1 \otimes \phi_2$ is a density.

2. $(\phi_1 \otimes \phi_2)^{\downarrow h_1} = \phi_1$.

Proof. (1) follows from Lemma 4.16.

(2) Note that $\phi_2^{\downarrow t_2}$ exists and equals e_{t_2}. Hence, by the combination axiom, we obtain

$$(\phi_1 \otimes \phi_2)^{\downarrow h_1} = \phi_1 \otimes \phi_2^{\downarrow t_2} = \phi_1 \otimes e_{t_2} = \phi_1.$$

$\qquad\Box$

This lemma generalizes immediately to the next lemma:

Lemma 4.18 *Let $\phi_1, \phi_2, \ldots, \phi_n$ be kernels with heads h_i and tails t_i, such that $t_1 = \emptyset$, $t_i \subseteq d(\phi_1) \cup \cdots \cup d(\phi_{i-1})$, and h_i disjoint to $d(\phi_1) \cup \cdots \cup d(\phi_{i-1})$. Then*

1. $\phi_1 \otimes \phi_2 \otimes \cdots \otimes \phi_n$ is a density.

2. For $i = 1, \ldots, n-1$,

$$\phi_1 \otimes \phi_2 \otimes \cdots \otimes \phi_i = (\phi_1 \otimes \phi_2 \otimes \cdots \otimes \phi_n)^{\downarrow d(\phi_1) \cup \cdots \cup d(\phi_i)}.$$

Proof. This follows easily from the previous Lemma 4.17 by induction on i.

$\square$

A sequence of kernels $\phi_1, \phi_2, \ldots, \phi_n$ satisfying the conditions of the lemma above is called a *construction sequence* (Shafer, 1996). Such construction sequences factor densities into kernels. So here we have another example of a factorization of a valuation whose marginals exist for all subsets of its domain, but whose factors have only partially defined marginals. In such a construction sequence generally other valuations representing evidence are added as factors in such a way that the combination remains a valuation whose marginals exist, and which especially can be scaled. Consider for example the situation of Lemma 4.17 above. Assume that ψ is a valuation on h_2 representing evidence, such that $\phi_1 \otimes \phi_2 \otimes \psi$ still possesses marginals for all subsets of $h_1 \cup h_2$. Then the marginal $(\phi_1 \otimes \phi_2 \otimes \psi)^{\downarrow h_1}$ exists and is scalable. The formula

$$((\phi_1 \otimes \phi_2 \otimes \psi)^{\downarrow h_1})^{\downarrow}$$

represents then a generalized version of *Bayes' theorem*. The question arises whether the different architectures for computing marginals discussed so far, apply to factorizations in separative valuation algebras, where some factors are only in the extension Φ^*, not in Φ. The answer is affirmative in all cases, provided the marginals needed exist.

Consider a factorization

$$\phi \quad = \quad \psi_1 \otimes \psi_2 \otimes \cdots \otimes \psi_m,$$

where $\phi \in \Phi$, $\psi_i \in \Phi^*$ and the domains $d(\psi_i) = s_i$ form a join tree. Take m as the root node and suppose that the numbering of ψ_i is such that $i < j$ if node j is on the path from i to m. Assume that the marginals

$$\mu_{i \to ch(i)} \quad = \quad \psi_i^{(i) \downarrow s_i \cap s_{ch(i)}}$$

(see (4.9)) exist all. This is for example the case if $\psi_1, \ldots, \psi_m$ form a construction sequence and we take the reverse numbering. Then the collect algorithm works in this scheduling of messages, since all marginals needed exist by assumption. The subsequent distribute phase works then too, since the message there are marginals of $\phi \in \Phi$.

We claim that in this case all messages

$$\mu_{i \to j} \quad = \quad \left(\psi_i \otimes \bigotimes_{k \in ne(i), k \neq j} \mu_{k \to i} \right)^{\downarrow s_i \cap s_j} \tag{4.36}$$

in the join tree exist. In fact, we have for $j = m$,

$$\phi^{\downarrow s_m} \quad = \quad \psi_m \otimes \bigotimes_{k \in ne(m)} \mu_{k \to m}$$

(see (4.16). Here all messages $\mu_{k \to m}$ come from the collect phase and exist therefore. By the combination axiom we obtain

$$
\begin{aligned}
\phi^{\downarrow s_m \cap s_j} \otimes \mu_{j \to m}^{-1} &= \left(\phi^{\downarrow s_m} \otimes \mu_{j \to m}^{-1} \right)^{\downarrow s_m \cap s_j} \\
&= \left(\psi_i \otimes \bigotimes_{k \in ne(i), k \neq j} \mu_{k \to i} \right)^{\downarrow s_m \cap s_j} = \mu_{m \to j}.
\end{aligned}
$$

So $\mu_{m \to j}$ exists. By induction on i, we conclude by the same argument that $\mu_{ch(i) \to i}$ exists for all i. As a consequence we note that we may carry out the collect algorithm with any node i as root in the join tree. In particular, we see that the Shenoy-Shafer architecture applies in this case.

When we turn to *Lauritzen-Spiegelhalter architecture* we note immediately that the collect phase works, since it uses the same messages as the ordinary collect algorithm. According to the proof of Theorem 4.11, a node i contains $\phi^{\downarrow s_i}$ when it emits its messages to its outward neighbors during the distribute phase. The messages sent from i to node j is $\phi^{\downarrow s_i \cap s_j}$. Since all marginals of ϕ exist, this shows that the messages in the distribute phase are well defined too. Hence the Lauritzen-Spiegelhalter architecture applies also to the present case of a join tree factorization in a valuation algebra with partial marginalization. The representations (4.22) and (4.23) remain valid.

Regarding the HUGIN architecture, we remark first, that the messages are the same as in the Lauritzen-Spiegelhalter architecture. The difference lies only in the storage in the separators. Therefore, if we start with the valuations ψ_i in the nodes i and the valuations $e_{s_i \cap s_{ch(i)}}$ such that $\gamma(e_{s_i \cap s_{ch(i)}}) \leq \gamma(\psi_i), \gamma(\psi_{ch(i)})$ in the separators (compare Section 4.4), then all marginals needed exist and are well defined. Furthermore, Theorems 4.12 and 4.13 carry over to the present case.

In the more general case we start with valuations $\psi_v \in \Phi^*$ in the nodes $v \in \mathcal{V}$ of the join tree and $\psi_s \in \Phi^*$ in the separators $s \in \mathcal{S}$, such that

$$
\phi = \bigotimes_{v \in \mathcal{V}} \psi_v \otimes \bigotimes_{s \in \mathcal{S}} (\psi_s)^{-1} \tag{4.37}
$$

and moreover

$$
\gamma(\psi_s) \leq \gamma(\psi_v), \gamma(\psi_w)
$$

for any separator s between nodes v and w. In order to verify that the HUGIN architecture works also in this general case, we recall from the proof of Theorem 4.14 that we may transform equation (4.37) into

$$
\phi = \psi_r \otimes \bigotimes_{v \in \mathcal{V} - \{r\}} (\psi_v \otimes \psi_{s(v)}^{-1}), \tag{4.38}
$$

if r is the node chosen as root in the join tree and $s(v)$ the separator between node v and its neighbor towards the root. With this form we are back to the original HUGIN computations, which we recognized as feasible. If η'_v, η'_s denote the contents of nodes v and separators s during the HUGIN computations, then we know that $\eta'^{\downarrow s}_v$ exists. But in the proof of Theorem 4.14 we proved also that during the collect phase we have $\eta_v = \eta'_v \otimes \psi_s$, if η_v denotes the content of node v in the HUGIN computations with the general form (4.37). So, by induction over the steps of the collect phase, we see, using the combination axiom, that the marginal $\eta^{\downarrow s}_v$ exists,

$$\eta^{\downarrow s}_v = (\eta'_v \otimes \psi_s)^{\downarrow s} = \eta'^{\downarrow s}_v \otimes \psi_s.$$

Therefore, the HUGIN computations are feasible in the collect phase in this more general case too. According to the proof of Theorem 4.14, in the distribute phase, the messages of the computations based on the general form and on the reduced form (4.38) are the same. Hence the HUGIN architecture works for the general form also.

We conclude therefore this section by stating that all the architectures discussed in the previous Section 4.4 for regular algebras work in separative algebras (Φ, D) too, even if the join tree factorization of a valuation $\phi \in \Phi$ contains factors $\psi \in \Phi^*$ of the embedding valuation algebra (Φ^*, D) with partial marginalization. It is essentially the combination axiom of valuation algebras with partial marginalization, which guarantees this. More, generally, the Lauritzen-Spiegelhalter and the HUGIN architectures work also in any valuation algebra with partial marginalization, provided ϕ possesses all its marginals for $t \subseteq d(\phi)$.

4.6   Scaling and Updating

In Section 3.7 we noted that in many applications, we are interested in scaled valuations for semantic reasons. This is surely the case for probability potentials. But also, if we want to interpret set potentials as belief functions, then scaling is necessary (Dempster, 1967; Shafer, 1976; Kohlas & Monney, 1995). Spohn potentials too get their real semantic flavor only in the scaled form (Spohn, 1988). So, if we have a valuation $\phi \in \Phi$, given in a join tree factorization

$$\phi = \psi_1 \otimes \psi_2 \otimes \cdots \otimes \psi_m, \tag{4.39}$$

with domains $d(\psi_i) = s_i$, then we often want to compute the marginals of the scaled version of ϕ, that is $(\phi^{\downarrow})^{\downarrow s_i}$. Note that we have to assume a separative (or regular) valuation algebra, in order that scaling is defined.

Of course, we can invoke Lemma 3.25 (3) and write

$$\phi^{\downarrow} = (\psi_1 \otimes \psi_2 \otimes \cdots \otimes \psi_m)^{\downarrow} = (\psi_1^{\downarrow} \otimes \psi_2^{\downarrow} \otimes \cdots \otimes \psi_m^{\downarrow})^{\downarrow}$$
$$= \psi_1^{\downarrow} \oplus \psi_2^{\downarrow} \oplus \cdots \oplus \psi_m^{\downarrow}$$

This is still a join tree factorization, but in the valuation algebra $(\Phi^{\downarrow}, D)$. So we can apply any convenient architecture to this factorization to obtain the scaled marginals. But this involves first scaling of all the factors, which is in fact not necessary.

We start with the Shenoy-Shafer architecture. We may apply this architecture essentially as before, but use Theorem 4.10 to obtain the scaled marginals by

$$(\phi^{\downarrow})^{\downarrow s_i} = (\phi^{\downarrow s_i})^{\downarrow} = \phi^{\downarrow s_i} \otimes (\phi^{\downarrow \emptyset})^{-1}$$

$$= \psi_i \otimes \left(\bigotimes_{j \in ne(i)} \mu_{j \to i} \right) \otimes (\phi^{\downarrow \emptyset})^{-1}.$$

However there is a scaled version of the computations in the Shenoy-Shafer architecture which leads to more efficient computations, and which avoids the divisions by $\phi^{\downarrow \emptyset}$ in each node. In fact, we replace the valuation ψ_r at the root by

$$\psi_r \otimes (\phi^{\downarrow \emptyset})^{-1}.$$

Of course, this presupposes that all incoming messages of node r are first computed (collect phase), so that $\phi^{\downarrow r}$ and then $\phi^{\downarrow \emptyset}$ can be determined. That is why we take r as the root. Then outgoing messages from the root use this new valuation. This we call the scaled version of the Shenoy-Shafer architecture.

Theorem 4.19 *At the end of the computations in the scaled Shenoy-Shafer architecture, we obtain at node i*

$$(\phi^{\downarrow})^{\downarrow s_i} = \psi_i \otimes \left(\bigotimes_{j \in ne(i)} \mu'_{j \to i} \right) \tag{4.40}$$

Here $\mu'_{j \to i}$ are the modified messages due to the scaled computations.

Proof. The messages directed to the root, that is the messages computed in the collect phase, do not change, $\mu_{j \to i} = \mu'_{j \to i}$. We claim that for the messages directed outward from the root, we have

$$\mu'_{j \to i} = \mu_{j \to i} \otimes (\phi^{\downarrow \emptyset})^{-1}.$$

This holds surely for the root. In fact, let i be a neighbor of the root r, then (see (4.15))

$$\mu'_{r \to i} = \left(\psi_r \otimes (\phi^{\downarrow \emptyset})^{-1} \otimes \bigotimes_{k \in ne(r), k \neq i} \mu_{k \to r} \right)^{\downarrow s_r \cap s_i}$$

$$= \left(\psi_r \otimes \bigotimes_{k \in ne(r), k \neq i} \mu_{k \to r} \right)^{\downarrow s_r \cap s_i} \otimes (\phi^{\downarrow \emptyset})^{-1}$$

$$= \mu_{j \to i} \otimes (\phi^{\downarrow \emptyset})^{-1}.$$

This follows from the combination axiom (which guarantees also the existence of the marginals involved) and since the messages $\mu_{k \to r}$ in the formula above are all directed towards the root. By induction, assume that the proposition holds for all nodes $k > i$ in the numbering induced by the directed join tree. Let $h > i$ be the unique inward neighbor of i (its child). Then $k \in ne(i)$, $k \neq h$ implies k$<$ i, hence $\mu'_{k \to i} = \mu_{k \to i}$ and therefore we obtain for a neighbor $j < i$ of i,

$$
\mu'_{i \to j} = \left(\psi_i \otimes \bigotimes_{k \in ne(i), k \neq j} \mu'_{k \to i} \right)^{\downarrow s_i \cap s_j}
$$

$$
= \left(\psi_i \otimes \bigotimes_{k \in ne(i), k \neq j, k \neq h} \mu_{k \to i} \otimes \mu'_{h \to i} \right)^{\downarrow s_i \cap s_j}
$$

$$
= \left(\psi_i \otimes \bigotimes_{k \in ne(i), k \neq j, k \neq h} \mu_{k \to i} \otimes \mu_{h \to i} \otimes (\phi^{\downarrow \emptyset})^{-1} \right)^{\downarrow s_i \cap s_j}
$$

$$
= \left(\psi_i \otimes \bigotimes_{k \in ne(i), k \neq j, h} \mu_{k \to i} \otimes \mu_{h \to i} \right)^{\downarrow s_i \cap s_j} \otimes (\phi^{\downarrow \emptyset})^{-1}
$$

$$
= \mu_{i \to j} \otimes (\phi^{\downarrow \emptyset})^{-1}.
$$

But then, we conclude from (4.16), Theorem 4.10, that

$$
\psi_i \otimes \left(\bigotimes_{j \in ne(i)} \mu'_{j \to i} \right) = \psi_i \otimes \left(\bigotimes_{j \in ne(i)} \mu_{j \to i} \right) \otimes (\phi^{\downarrow \emptyset})^{-1}
$$

$$
= \phi^{\downarrow s_i} \otimes (\phi^{\downarrow \emptyset})^{-1}
$$

$$
= (\phi^{\downarrow s_i})^{\downarrow} = (\phi^{\downarrow})^{\downarrow s_i}.
$$

This is true, since exactly one incoming message in a node i different from the root is an outward directed message. And this proves the theorem. $\Box$

Assume again that the join tree G of the factorization is directed towards a root r. Whatever architecture with division we apply, we first perform an ordinary collect phase, like in the Shenoy-Shafer architecture above (only the memory organization will be different). At the end of this phase we have the marginal $\phi^{\downarrow r}$ in the root r. From this valuation we compute $\phi^{\downarrow \emptyset} = (\phi^{\downarrow r})^{\downarrow \emptyset}$. Then we start the distribute phase with

$$
(\phi^{\downarrow})^{\downarrow r} = (\phi^{\downarrow r})^{\downarrow} = \phi^{\downarrow r} \otimes (\phi^{\downarrow \emptyset})^{-1}
$$

in the root r. We call this the *scaled* computations or *scaled* versions of the different architectures. We are going to verify that these scaled versions lead in the cases of the Lauritzen-Spiegelhalter and the HUGIN architectures to the scaled marginals. The following theorem corresponds to Theorem 4.11 for the Lauritzen-Spiegelhalter architecture.

Theorem 4.20 *At the end of the scaled computations in the Lauritzen-Spiegelhalter architecture, each node i of the join tree G contains the scaled marginal $(\phi^{\downarrow s_i})^{\downarrow}$.*

Proof. At the end of the collect phase, we have, as stated above, $(\phi^{\downarrow})^{\downarrow r}$ at the root. We prove now by induction that we have $(\phi^{\downarrow})^{\downarrow v}$ at any node v of the join tree at the end. Assume that this is true up to some node w, which sends a message to its outward neighbor v during the distribute phase. The message that w sends to v is $(\phi^{\downarrow})^{\downarrow s}$, if $s = v \cap w$. On the other hand, we know that at node v we have stored a valuation η_v from the collect phase, which is the *same* as in the original Lauritzen-Spiegelhalter computation. And we know from the proof of Theorem 4.11, that

$$\eta_v \otimes \phi^{\downarrow s} = \phi^{\downarrow v}.$$

Therefore, we obtain, when we combine the new message in the scaled computation,

$$\eta_v \otimes (\phi^{\downarrow})^{\downarrow s} = \eta_v \otimes (\phi^{\downarrow s})^{\downarrow} = \eta_v \otimes \phi^{\downarrow s} \otimes (\phi^{\downarrow \emptyset})^{-1}$$
$$= \phi^{\downarrow v} \otimes (\phi^{\downarrow \emptyset})^{-1} = (\phi^{\downarrow v})^{\downarrow} = (\phi^{\downarrow})^{\downarrow v}.$$

This proves the theorem. □

The next theorem corresponds to Theorem 4.14 for the HUGIN architecture.

Theorem 4.21 *Assume*

$$\phi = \bigotimes_{v \in V} \psi_v \otimes \bigotimes_{s \in S} \psi_s^{-1}, \tag{4.41}$$

were we assume $d(\psi_v) = v$, $d(\psi_s) = s$, and moreover that

$$\gamma(\psi_s) \leq \gamma(\psi_v), \gamma(\psi_w) \tag{4.42}$$

for any separator s between two nodes v and w. Then, if we put ψ_v into node $v \in V$ and ψ_s into separator $s \in S$ at the start of the computation in the scaled HUGIN architecture, then at the end we have $(\phi^{\downarrow v})^{\downarrow}$ for $v \in V$ and $(\phi^{\downarrow s})^{\downarrow}$ for $s \in S$.

Proof. Again, at the end of the scaled collect phase, we have $(\phi^{\downarrow r})^{\downarrow}$ in the root r. We prove again by induction that at the end of the scaled computations we have $(\phi^{\downarrow v})^{\downarrow}$ in each node v and $(\phi^{\downarrow s})^{\downarrow}$ in each separator s. Assume thus that this holds for some node w with outward neighbor v and separator s between.

We recall from the Section 4.4 that, without scaling, the new valuations in s and v are

$$\eta_s^* = \eta_v^{\downarrow s} = (\phi^{\downarrow w})^{\downarrow s} = \phi^{\downarrow s},$$
$$\eta_v^* = \eta_v \otimes \phi^{\downarrow s} \otimes \eta_s^{-1} = \phi^{\downarrow v}.$$

Now, with scaled computations, according to the assumption of the induction we have

$$\eta_s^* = \eta_v^{\downarrow s} = ((\phi^\downarrow)^{\downarrow w})^{\downarrow s} = (\phi^\downarrow)^{\downarrow s}.$$

And, furthermore

$$\eta_v^* = \eta_v \otimes (\phi^\downarrow)^{\downarrow s} \otimes \eta_s^{-1} = \eta_v \otimes \phi^{\downarrow s} \otimes \eta_s^{-1} \otimes (\phi^{\downarrow \emptyset})^{-1}$$
$$= \phi^{\downarrow v} \otimes (\phi^{\downarrow \emptyset})^{-1} = (\phi^{\downarrow v})^\downarrow = (\phi^\downarrow)^{\downarrow v}$$

This proves the theorem. $\square$

These considerations show that computing scaled marginals, as is needed in many applications, does not cause a considerable additional effort. In all three main architectures, only minor changes to scaled computations are needed.

Often a new information (valuation) is added after marginals have been computed. In many cases it is in such a case not necessary to repeat the whole computation. So assume that $\phi \in \Phi$ factors into

$$\phi = \psi_1 \otimes \psi_2 \otimes \cdots \otimes \psi_m, \tag{4.43}$$

where the $d(\psi_i) = s_i$ form a join tree G. Suppose now that a new valuation ψ' with $d(\psi') \subseteq s_i$ for some domain s_i of the join tree i is to be added, so that

$$\phi' = \phi \otimes \psi'.$$

Clearly, this gives us a new join tree factorization

$$\phi = \psi_1' \otimes \psi_2' \otimes \cdots \otimes \psi_m', \tag{4.44}$$

where $\psi_i' = \psi_i \otimes \psi'$ for a i such that $d(\psi') \subseteq s_i$ (if there are several such indices, select arbitrarily one) and $\psi_j' = \psi_j$ for $j \neq i$. We want to compute $\phi'^{\downarrow s_j}$ for all nodes s_j. This is called *updating*.

We select the node i to which we assign the new valuation ψ' as the root node r of the directed join tree G. We note then that

$$\phi'^{\downarrow r} = (\phi \otimes \psi')^{\downarrow r} = \phi^{\downarrow r} \otimes \psi'. \tag{4.45}$$

Furthermore, we have by (4.23)

$$\phi' = (\phi^{\downarrow r} \otimes \psi') \otimes \bigotimes_{v \in V, v \neq r} \phi^{\downarrow v} \otimes \bigotimes_{s \in S} (\phi^{\downarrow s})^{-1}$$
$$= \phi'^{\downarrow r} \otimes \bigotimes_{v \in V, v \neq r} \tag{4.46}$$

So we may start the HUGIN architecture with $\phi^{\downarrow v}$ in the nodes $v \in \mathcal{V}, v \neq r$, $\phi'^{\downarrow r}$ in the root r and $\phi^{\downarrow s}$ in the separators $s \in \mathcal{S}$. But it is clear that in the collect phase, towards the root r, the contents of nodes v and separators s do not change. So only the *distribute phase* is necessary. Hence, in the HUGIN architecture, updating can be done simply by updating first the node where the new valuation belongs to, and then by evoking the distribute phase from this node on.

This applies in particular for idempotent valuation algebras, since the HUGIN architecture applies to this important case. But we recall from Section 4.4 that in this case the separators are not needed, such that the distribute phase simplifies, and no divisions are needed.

In the Shenoy-Shafer architecture we have a similar situation. Only the messages outward from the node where the new valuation is added have to be recomputed. They then have to be recombined to get the marginals in the nodes, just like in the ordinary computations. We remind that binary join trees should be used in this architecture in order to avoid inefficiencies.

Sometimes a factor ψ_i of a factorization (4.43) is to be replaced by another valuation ψ_i'. We assume a separative (or regular) valuation algebra. Then, if $\gamma(\psi_i') \geq \gamma(\psi_i)$, we have

$$\psi_i \otimes (\psi_i)^{-1} \otimes \psi_i' = f_{\gamma(\psi_i)} \otimes \psi_i' = \psi_i'.$$

This shows that we can treat this case as an updating with a new information $(\psi_i)^{-1} \otimes \psi_i'$, provided the new factorization has still all its marginals. For example, if we want to remove a factor ψ_i from a factorization, then this means that we want to replace it by the neutral element e_{s_i}. This works, if ψ_i is a positive element. Otherwise we have $\gamma(e_{s_i}) \leq \gamma(\psi_i)$ and the trick does not work.

5 Conditional Independence

5.1 Factorization and Graphical Models

From probability theory we know the concepts of stochastic independence and conditional independence of random variables. These notions capture the presence or absence of influence of the information on certain variables on other ones. These influence structures are both important for modeling and for computation. As may be conjectured, such questions of mutual influence of variables over other ones are not confined to the formalism of probability theory, but are more general (see for example (Cowell et. al., 1999; Studeny, 1993)). In particular the notion of conditional independence can be introduced at the level of valuation algebras, generalizing thus the notion of stochastic independence of random variables. This has already been noted in (Shenoy, 1997 a). Conditional independence regarded from the point of view of valuation algebras will be the subject of the present chapter.

Let (Φ, D) be a labeled valuation algebra. We define conditional independence of groups of variables with respect to a valuation by a factorization of this valuation.

Definition 5.1 Conditional Independence: *If $\phi \in \Phi$, and s, t, u are disjoint subsets of $d(\phi)$, we say that s is* conditionally independent *of t given u with respect to ϕ, and we write $s\perp_\phi t|u$, if there exist $\psi_1, \psi_2 \in \Phi$ such that $d(\psi_1) = s \cup u, d(\psi_2) = t \cup u$ and*

$$\phi^{\downarrow s \cup t \cup u} \;=\; \psi_1 \otimes \psi_2. \tag{5.1}$$

If $u = \emptyset$, then we write $s\perp_\phi t$ instead of $s\perp_\phi t|\emptyset$ and we say that s is *independent* of t with respect to ϕ . Note carefully, that the notion of conditional independence depends on the valuation algebra within which the factorization (5.1) is considered. That is, a conditional independence relation with respect to ϕ may or may not hold, depending on which algebra containing ϕ is considered.

Assume $s\perp_\phi t|u$. Then we obtain

$$\phi^{\downarrow s\cup u} \;=\; (\psi_1 \otimes \psi_2)^{\downarrow s\cup u} \;=\; \psi_1 \otimes \psi_2^{\downarrow u}$$

since $d(\psi_1) = s \cup u, d(\psi_2) = t \cup u$. This shows that the part of information in ϕ relating to the variables $s \cup u$, depends only on the part of information of ψ_2 relating to u.

Let's examine some examples to see what conditional independence means in different contexts. We start with probability theory, that is, with *potentials* with multiplication as combination and summing out as marginalization. So let p be a *probability distribution* for $s\cup t\cup u$, where s, t, u, are disjoint. Assume $s\perp_p t|u$ and let $\mathbf{x}, \mathbf{y}, \mathbf{z}$ be configurations for s, t, u respectively. Then there exist potentials $a(\mathbf{x}, \mathbf{z}), b(\mathbf{y}, \mathbf{z})$ such that

$$p(\mathbf{x}, \mathbf{y}, \mathbf{z}) \;=\; a(\mathbf{x}, \mathbf{z})b(\mathbf{y}, \mathbf{z}).$$

Consider any configuration $\mathbf{y}, \mathbf{z}$ for which the marginal probability

$$p^{\downarrow t\cup u}(\mathbf{y}, \mathbf{z}) \;=\; b(\mathbf{y}, \mathbf{z}) \sum_{\mathbf{x}} a(\mathbf{x}, \mathbf{z})$$

is *positive*. For this configuration the conditional probability $p(\mathbf{x}|\mathbf{y}, \mathbf{z})$ exists and we see that

$$p(\mathbf{x}|\mathbf{y}, \mathbf{z}) \;=\; \frac{p(\mathbf{x}, \mathbf{y}, \mathbf{z})}{p^{\downarrow t\cup u}(\mathbf{y}, \mathbf{z})} \;=\; \frac{a(\mathbf{x}, \mathbf{z})}{\sum_{\mathbf{x}} a(\mathbf{x}, \mathbf{z})}$$

is a function of $\mathbf{x}, \mathbf{z}$ only. So we have

$$p(\mathbf{x}|\mathbf{y}, \mathbf{z}) \;=\; p(\mathbf{x}|\mathbf{z})$$

and also, similarly,

$$p(\mathbf{y}|\mathbf{x}, \mathbf{z}) \;=\; p(\mathbf{y}|\mathbf{z}).$$

This corresponds to what is understood by *stochastic conditional independence* in probability theory: the conditional probability of any value of variables in s, given values of the variables in t and u depends only on the values of the variables in u, but not on those in t. So our definition captures the corresponding notion in probability theory. Similarly, if $s\perp_p t$ for some probability distribution, then we have $p(\mathbf{x}, \mathbf{y}) = p^{\downarrow s}(\mathbf{x})p^{\downarrow t}(\mathbf{y})$. This is stochastic independence in probability theory.

We remark that a similar result holds also for densities and Gaussian potentials. These cases will be examined in more detail in Section 5.3.

If the valuation algebra is *idempotent*, then, if $s\perp_\phi t|u$, we have

$$\phi^{\downarrow s\cup u} \;=\; \psi_1 \otimes \psi_2^{\downarrow u}, \qquad \phi^{\downarrow t\cup u} \;=\; \psi_2 \otimes \psi_1^{\downarrow u}$$

and hence

$$\phi^{\downarrow s\cup t\cup u} \;=\; \psi_1 \otimes \psi_1^{\downarrow u} \otimes \psi_2 \otimes \psi_2^{\downarrow u} \;=\; \phi^{\downarrow s\cup u} \otimes \phi^{\downarrow t\cup u}.$$

For example in possibility theory, if combination is defined by minimization (see Section 2), then $s\perp_\phi t|u$ implies that

$$p(\mathbf{x},\mathbf{y},\mathbf{z}) \;=\; \min\{a(\mathbf{x},\mathbf{z})b(\mathbf{y},\mathbf{z})\} \;=\; \min\{p^{\downarrow s\cup u}(\mathbf{x},\mathbf{z}), p^{\downarrow t\cup u}(\mathbf{y},\mathbf{z})\}$$
$$=\; \min\{\max_{\mathbf{y}} p(\mathbf{x},\mathbf{y},\mathbf{z}), \max_{\mathbf{x}} p(\mathbf{x},\mathbf{y},\mathbf{z})\}.$$

Another example of an idempotent valuation algebra is provided by subsets or relations. If C is a subset of $\Omega_{s\cup t\cup u}$, where s,t,u are disjoint sets of variables, then $s\perp_C t|u$ means that $C = C_1 \bowtie C_2$, where $C_1 \subseteq \Omega_{s\cup u}$ and $C_2 \subseteq \Omega_{t\cup u}$. That is, if C is a relation over $s \cup t \cup u$, then $s\perp_C t|u$ means that C is the join of a relation C_1 with domain $s \cup u$ and a relation C_2 with domain $t \cup u$.

Next let's turn to (quasi-) set potentials. Consider a commonality function q for $s\cup t\cup u$, for some disjoint sets s,t,u of variables. Then $s\perp_q t|u$ means that there are functions q_1 and q_2 for $s \cup u$ and $t \cup u$ respectively, such that $q(A) = q_1(A^{\downarrow s\cup u})q_2(A^{\downarrow t\cup u})$. The question here is whether for such a factorization only commonality functions q_1 and q_2 are admitted or, less stringent, arbitrary non-negative set functions q_1 and q_2 (quasi-set potentials). This leads to two different notions of conditional independence. A similar situation arises in any separative valuation algebra. We may allow only factors in Φ or also factors in Φ^*. And this is not necessarily the same (see Section 5.3 below, where we come back to this issue).

In general terms, in a valuation algebra with neutral elements, since $e_{s\cup t\cup u} = e_{s\cup u} \otimes e_{t\cup u}$ we have $s\perp_{e_{s\cup t\cup u}} t|u$ if s,t,u are disjoint sets of variables.

The following theorem states a number of properties for the ternary relation $s\perp_\phi t|u$.

Theorem 5.2 *Let $\phi \in \Phi$ and assume $s,t,u,v \subseteq d(\phi)$ to be disjoint sets of variables. Then*

(G1) Symmetry: *$s\perp_\phi t|u$ implies $t\perp_\phi s|u$.*
(G2) Decomposition: *$s\perp_\phi t \cup v|u$ implies $s\perp_\phi t|u$.*
(G3) Weak Union: *$s\perp_\phi t \cup v|u$ implies $s\perp_\phi t|u \cup v$, if the valuation algebra* (Φ, D) *has neutral elements.*

Proof. (G1) follows immediately from the definition of conditional independence.

(G2) $s\perp_\phi t \cup v|u$ says that

$$\phi^{\downarrow s\cup t\cup v\cup u} \;=\; \psi_1 \otimes \psi_2$$

where $d(\psi_1) = s \cup u, d(\psi_2) = t \cup v \cup u$. Since $s\cup u \subseteq s\cup t\cup u \subseteq s\cup t\cup v\cup u$ we have by Lemma 2.1 (3) that

$$\phi^{\downarrow s\cup t\cup u} \;=\; (\psi_1 \otimes \psi_2)^{\downarrow s\cup t\cup u} \;=\; \psi_1 \otimes \psi_2^{\downarrow t\cup u}.$$

Since $d(\psi_1) = s \cup u, d(\psi_2^{\downarrow t \cup u}) = t \cup u$ we have that $s \perp_\phi t | u$.

(G3) $s \perp_\phi t \cup v | u$ implies that

$$\phi^{\downarrow s \cup t \cup v \cup u} = \psi_1 \otimes \psi_2$$

where $d(\psi_1) = s \cup u, d(\psi_2) = t \cup v \cup u$. By Lemma 3.1 (4) we obtain that

$$\phi^{\downarrow s \cup t \cup v \cup u} = (\psi_1 \otimes e_v) \otimes \psi_2.$$

Since $d(\psi_1 \otimes e_v) = s \cup v \cup u$ and $d(\psi_2) = t \cup v \cup u$ we have proved that $s \perp_\phi t | u \cup v$.
□

In the case of probability it is well known that further properties hold. But these additional properties seem to presuppose more structure than is available in a simple valuation algebra. On the other hand, the proof (G3) depends on the existence of neutral elements. This requirement can be relaxed. We return to these questions in Sections 5.2 and 5.3.

It is also known from probability theory that conditional independence can often be represented in different ways by graphs (see for example (Cowell et. al., 1999)). Graphical models are in fact important for the specification of the structure of probability models. At the level of general valuation algebras less structure is available for graphical modeling. Nevertheless, undirected graphs can be associated to factorizations of valuations and express thus conditional independence relations. We recall that the valuation network is a graphical representation of a factorization

$$\phi = \phi_1 \otimes \phi_2 \otimes \cdots \otimes \phi_n$$

with $d(\phi_1) = s_1, d(\phi_2) = s_2, \ldots, d(\phi_n) = s_n$ (see Section 2.1). Here, another graphical representation is introduced. Variables are represented by nodes and two nodes are linked by an edge, if they are together in some domain s_i of the factorization. The following is a more formal definition of the graph associated with a factorization.

Definition 5.3 Factor Graph: *Let* $\phi = \phi_1 \otimes \phi_2 \otimes \cdots \otimes \phi_n$ *with* $d(\phi_1) = s_1, d(\phi_2) = s_2, \ldots, d(\phi_n) = s_n$, *and let* $s = s_1 \cup s_2 \cup \cdots \cup s_n$. *The undirected graph* $G = (s, E)$ *where*

$$E = \{\{X, Y\} : exists\ s_i\ such\ that\ X, Y \in s_i\} \tag{5.2}$$

is called the factor graph *of the factorization.*

Different factorizations of the same valuation ϕ may have the same factor graph G: a subset of nodes of a graph G, which are all pairwise linked by edges, is called *complete*. Each domain s_i of the a factorization of ϕ is a complete set of nodes in the factor graph G. A *maximal* complete set of nodes of G is called a *clique* of G. Each domain s_i of a factorization of ϕ is contained in a clique of its factor graph G. Consider then the cliques C of the factor graph G

of a factorization of ϕ. Associate each domain s_i to exactly one of the cliques it is contained in. Combine then all factors ϕ_i with domain s_i associated with the clique C to get a valuation ψ_C. Then

$$\phi = \bigotimes_{C \in \mathcal{C}} \psi_C$$

is a factorization of ϕ with the factor graph G.

Consider the medical example of Section 2.1 whose valuation network is given in Fig. 2.1. Its factor graph is represented in Fig. 5.1. The cliques of this graph are the following sets: $\{A,T\}$, $\{T,L,E\}$, $\{S,L\}$, $\{S,B\}$, $\{E,B,D\}$, $\{E,X\}$. And the factorization according to these cliques is

$$(\alpha \otimes o_A \otimes \tau) \otimes \lambda \otimes \beta \otimes \epsilon \otimes \xi \otimes (\delta \otimes o_D).$$

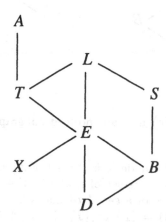

Figure 5.1: The factor graph for the medical example of Section 2.1.

A factorization represents a certain conditional independence structure and its factor graph G gives a graphical representation of this structure. In order to study graphical representations of conditional independence structures we may follow the corresponding discussions in probability theory as for example in (Lauritzen, 1996; Cowell et. al., 1999) and reproduce these developments in the framework of valuation algebras. In order to do so we start with the following definition:

Definition 5.4 Property (F): *Let G be an undirected graph with the family of cliques $\mathcal{C}$. A valuation $\phi \in \Phi$ is said to have* property (F) *with respect to G, if there are valuations $\psi_C \in \Psi$ for all $C \in \mathcal{C}$ with $d(\psi_C) = C$, such that*

$$\phi = \bigotimes_{C \in \mathcal{C}} \psi_C \qquad (5.3)$$

Note that, according to the discussion above, if ϕ has a factorization with a factor graph G, then ϕ has property (F) with respect to its factor graph.

For further reference we need a few graph theoretical notions. Let $G = (V, E)$ be an undirected graph. A graph $G_A = (A, E_A)$ is called a *subgraph* of G, if $A \subseteq V$ and $E_A \subseteq E \cap A \times A$. A subgraph may have the same node set as G but only a subset of edges. If $E_A = E \cap A \times A$, then G_A is called the subgraph *node-induced* by A. This is illustrated in Fig. 5.2.

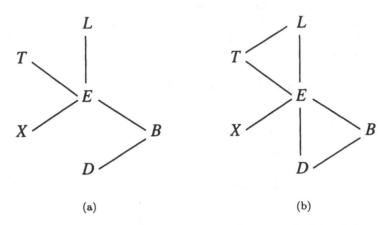

(a)   (b)

Figure 5.2: A subgraph (a) and an induced subgraph (b) of the graph of Fig. 5.1.

The *boundary* $bd(X)$ of a node X is the set of its neighbors; the boundary $bd(A)$ of a set A of nodes is the set of nodes in $V - A$ which are neighbors to nodes of A. The *closure* of a set A of nodes is $cl(A) = A \cup bd(A)$. For example, in Fig. 5.1, $bd(A) = \{T\}, bd(\{T, L\}) = \{A, E, S\}$ and $cl(\{T, L\}) = \{A, T, L, E, S\}$.

A *path* from X to Y is a sequence of distinct nodes $X_0, X_1, \ldots, X_n$ where $n > 0$, $X = X_0$ and $X_n = Y$ and $\{X_{i-1}, X_i\} \in E$. A path never crosses itself. A subset C of nodes is said to be a (X, Y) *separator* if all paths from X to Y intersect C and $X, Y \notin C$. The subset C is said to separate disjoint node sets A and B if it is a (X, Y)-separator for all pairs of nodes $X \in A$ and $Y \in B$. In Fig. 5.1, $\{L, B\}$ separates S from $\{E, D, X\}$. We write $A \perp_G B | C$ if C separates A and B in the graph G.

Associated with an undirected graph $G = (s, E)$ and a valuation ϕ with $d(\phi) = s$ there is a range of conditional independence properties. In probability theory these properties are called *Markov properties* and we will use the same terminology.

Definition 5.5 Markov Properties. *A valuation ϕ is said to obey*

- *(P) the pairwise Markov property, relative to G, if for any pair (X, Y) of non-adjacent nodes*

$$\{X\} \perp_\phi \{Y\} | s - \{X, Y\};$$

- *(L)* the local Markov property, *relative to G, if for any node* $X \in s$

$$\{X\}\bot_\phi s - cl(X)|bd(X);$$

- *(G)* the global Markov property, *relative to G, if for any triple* (t, u, v) *of disjoint subsets of s such that v separates t from u in G*

$$t\bot_\phi u|v.$$

The Markov properties are related as described in the following theorem.

Theorem 5.6 *Assume that* (Φ, D) *is a valuation algebra with neutral elements. For any undirected graph* $G = (s, E)$ *and any valuation* ϕ *with* $d(\phi) = s$ *it holds that*

$$(F) \Rightarrow (G) \Rightarrow (L) \Rightarrow (P). \tag{5.4}$$

Proof. Assume (F) and let (t, u, v) be a triple of disjoint sets of nodes such that v separates t from u. Consider the node-induced subgraph G_{s-v} and let t' be the nodes linked by a path to nodes in t in the subgraph G_{s-v}. Define $u' = s - (t' \cup v)$. t' and u' are disjoint and $t \subseteq t'$, $u \subseteq u'$, since v separates t from u. Furthermore any clique of G must either be contained in $t' \cup v$ or in $u' \cup v$. If C_t denotes the family of cliques contained in $t' \cup v$ we obtain from (F) that

$$\phi = \bigotimes_{C \in C} \psi_C = \left(\bigotimes_{C \in C_t} \psi_C\right) \otimes \left(\bigotimes_{C \in C - C_t} \psi_C\right).$$

Since $d(\bigotimes_{C \in C_t} \psi_C) = t' \cup v$ and $d(\bigotimes_{C \in C - C_t} \psi_C) = u' \cup v$ we see that $t' \bot_\phi u'|v$. If we apply (G2) of Theorem 5.2 twice, we obtain that $t\bot_\phi u|v$ and (G) holds.

(G) implies (L) because $bd(X)$ separates X from $s - cl(X)$.

Assume that (L) holds. Consider a pair of non-adjacent nodes X and Y. Then $Y \in s - cl(X)$. Hence

$$bd(X) \cup ((s - cl(X)) - \{Y\}) = s - \{X, Y\}.$$

According to (L) we have that

$$\{X\}\bot_\phi s - cl(X)|bd(X).$$

(G3) of Theorem 5.2 gives then

$$\{X\}\bot_\phi \{Y\}|s - \{X, Y\}$$

which is (P). □

It is known from probability theory that the three Markov properties are not equivalent in general (see for example (Lauritzen, 1996)).

There are cases where property (F) of a valuation with respect to a graph G can be broken down into the same property relative to subgraphs of G. This is possible with respect to a certain decomposition of graph G. This is the subject of the following definition.

Definition 5.7 Graph decomposition: *Let $G = (s, E)$ be an undirected graph, t, u, v subsets of s. The triple (t, u, v) is said to form a* decomposition *of G if*

1. $t \cup u \cup v = s$,

2. *v separates t from u in G,*

3. *v is a complete subset of nodes.*

Such a decomposition is said to be *proper* if both t and u are not empty. The next theorem shows that if ϕ has property (F) with respect to a graph G, then this carries over to the marginals relative to a decomposition of G.

Theorem 5.8 *Assume a decomposition (t, u, v) of a graph G. If $\phi \in \Phi$ has property (F) with respect to G, then $\phi^{\downarrow t \cup v}$ and $\phi^{\downarrow u \cup v}$ have property (F) with respect to the node-induced subgraphs $G_{t \cup v}$ and $G_{u \cup v}$ respectively. Furthermore*

$$\phi \otimes \phi^{\downarrow v} = \phi^{\downarrow t \cup v} \otimes \phi^{\downarrow u \cup v}. \tag{5.5}$$

Proof. Let $\mathcal{C}$ be the family of cliques of G. Since ϕ has property (F), we have

$$\phi = \bigotimes_{C \in \mathcal{C}} \psi_C$$

with $d(\psi_C) = C$. Since (t, u, v) decomposes G, all cliques of G are either contained in $t \cup v$ or in $u \cup v$ and they are cliques either of $G_{t \cup v}$ or of $G_{u \cup v}$. But no clique is contained both in $t \cup v$ and in $u \cup v$, unless v itself is a clique. Assume first that this is not the case.

Let then $\mathcal{C}_t$ and $\mathcal{C}_u$ be the family of cliques of G contained in $t \cup v$ and $u \cup v$ respectively. If we put

$$\psi_1 = \bigotimes_{C \in \mathcal{C}_t} \psi_C, \qquad \psi_2 = \bigotimes_{C \in \mathcal{C}_u} \psi_C$$

then we have either $d(\psi_1) = t \cup v$ and $u \subseteq d(\psi_2) \subseteq u \cup v$ or $t \subseteq d(\psi_1) \subseteq t \cup v$ and $d(\psi_2) = u \cup v$ and in both cases $\phi = \psi_1 \otimes \psi_2 = \psi_1 \otimes \psi_2 \otimes e_v$. Let's consider the first case. Then

$$\phi^{\downarrow t \cup v} = \psi_1 \otimes (\psi_2 \otimes e_v)^{\downarrow v}$$

and

$$\phi^{\downarrow v} = \psi_1^{\downarrow v} \otimes (\psi_2 \otimes e_v)^{\downarrow v}.$$

On the other hand,

$$\phi^{\downarrow u \cup v} = \psi_1^{\downarrow v} \otimes \psi_2.$$

So we obtain that

$$
\begin{aligned}
\phi \otimes \phi^{\downarrow v} &= \psi_1 \otimes \psi_2 \otimes \psi_1^{\downarrow v} \otimes (\psi_2 \otimes e_v)^{\downarrow v} \\
&= (\psi_1 \otimes (\psi_2 \otimes e_v)^{\downarrow v}) \otimes \psi_2 \otimes \psi_1^{\downarrow v} \\
&= \phi^{\downarrow t \cup v} \otimes \phi^{\downarrow u \cup v}.
\end{aligned}
$$

In the second of the possible cases above we obtain in a similar way the same result.

Any clique of $G_{t \cup v}$ or of $G_{u \cup v}$ is also a clique of G or it is v. And any clique of G is a clique of $G_{t \cup v}$ or of $G_{u \cup v}$. If $d(\psi_1) = t \cup v$, then v can not be a clique of $G_{t \cup v}$ (since we assume here that v is not a clique of G). It must be contained in some clique $C' \in C_t$. Since we have that

$$
\phi^{\downarrow t \cup v} = \bigotimes_{C \in C_t} \psi_C \otimes (\psi_2 \otimes e_v)^{\downarrow v}
$$

we can combine $(\psi_2 \otimes e_v)^{\downarrow v}$ with $\psi_{C'}$ and obtain thus with this new definition of ψ_C

$$
\phi^{\downarrow t \cup v} = \bigotimes_{C \in C_t} \psi_C.
$$

Thus $\phi^{\downarrow t \cup v}$ has property (F).

If v is a clique of $G_{t \cup v}$, then we have $d(\psi_1) \subset t \cup v$ and $d(\psi_2) = u \cup v$. In this case

$$
\phi^{\downarrow t \cup v} = \psi_1 \otimes \psi_2^{\downarrow v} = \bigotimes_{C \in C_t} \psi_C \otimes \psi_2^{\downarrow v}.
$$

But now $G_{t \cup v}$ has the cliques $C_t \cup \{v\}$. Thus $\phi^{\downarrow t \cup v}$ has again property (F). That $\phi^{\downarrow u \cup v}$ has property (F) with respect to $G_{u \cup v}$ is proved in the same way.

If v is itself a clique of G, then we write

$$
\phi = \psi_1 \otimes \psi_2 \otimes \psi_v,
$$

where

$$
\psi_1 = \bigotimes_{C_t - \{v\}} \psi_C, \qquad \psi_2 = \bigotimes_{C_u - \{v\}} \psi_C.
$$

Note that $t \subseteq d(\psi_1) \subseteq t \cup v$ and $u \subseteq d(\psi_2) \subseteq u \cup v$. Now we have

$$
\begin{aligned}
\phi^{\downarrow t \cup v} &= (\psi_1 \otimes \psi_v) \otimes \psi_2^{\downarrow v_2}, \quad \text{where } v_2 = d(\psi_2) \cap v, \\
\phi^{\downarrow u \cup v} &= (\psi_2 \otimes \psi_v) \otimes \psi_1^{\downarrow v_1}, \quad \text{where } v_1 = d(\psi_1) \cap v.
\end{aligned}
$$

Furthermore, we obtain

$$
\phi^{\downarrow v} = \psi_1^{\downarrow v_1} \otimes \psi_2^{\downarrow v_2} \otimes \psi_v.
$$

Therefore, in this case too

$$\phi \otimes \phi^{\downarrow v} = \psi_1 \otimes \psi_2 \otimes \psi_v \otimes \psi_1^{\downarrow v_1} \otimes \psi_2^{\downarrow v_2} \otimes \psi_v$$
$$= \phi^{\downarrow t \cup v} \otimes \phi^{\downarrow u \cup v}.$$

And, we see that $\phi^{\downarrow t \cup v}$ has property (F) with respect to $G_{t \cup v}$, when we write

$$\phi^{\downarrow t \cup v} = \bigotimes_{C \in \mathcal{C}_t - \{v\}} \psi_C \otimes (\psi_v \otimes \psi_2^{\downarrow v_2}).$$

In the same way property (F) of $\phi^{\downarrow u \cup v}$ can be verified.   □

We remark that the converse of this theorem is valid in many cases. That is, if $\phi^{\downarrow t \cup v}$ and $\phi^{\downarrow u \cup v}$ have property (F) relative to graphs $G_{t \cup v}$ and $G_{u \cup v}$ and respectively, and if (5.5) holds, then ϕ has property (F) with respect to graph G.

The concept of graph decomposition can be carried further with the following definition.

Definition 5.9 Decomposable Graph: *An undirected graph is called* decomposable, *if either (i) G is complete, or (ii) G possesses a proper decomposition (t, u, v) such that both subgraphs $G_{t \cup v}$ and $G_{u \cup v}$ are decomposable.*

Since the decomposition is proper, each subgraph $G_{t \cup v}$ an $G_{u \cup v}$ has fewer nodes than G so that this recursive definition makes sense. This is a concept which has been much studied in graph theory, we refer to (Lauritzen, 1996; Cowell et. al., 1999) for details.

Theorem 5.8 has the following corollary.

Corollary 5.10 *Let G be a decomposable graph with family of cliques $\mathcal{C}$ and and family of separators $\mathcal{S}$. If ϕ has property (F) with respect to G, then*

$$\phi \otimes \left(\bigotimes_{S \in \mathcal{S}} \phi^{\downarrow S} \right) = \bigotimes_{C \in \mathcal{C}} \phi^{\downarrow C}. \tag{5.6}$$

Proof. This result is obtained by recursive application of (5.5).   □

This concludes our discussion of graphical models and conditional independence with respect to general valuation algebras. There is much more in this domain, especially in relation to probability. We refer to (Lauritzen, 1996) for a more thorough discussion of this subject and for further references to the literature in this domain. Some of the more specific results regarding conditional independence carry over to valuation algebras, if we add some more structure. This will be the subject of the following two sections of this chapter.

5.2 Conditionals in Regular Algebras

The richest theory of conditional independence has been developed in the framework of discrete probability. The corresponding valuation algebra of potentials is a model for regular valuation algebras. And it turns out that many aspects of conditional independence in probability carry over to regular algebras. This will be examined in this section.

Let (Φ, D) be a labeled, regular valuation algebra. The valuation

$$\chi = (\phi^{\downarrow t})^{-1} \otimes \phi \qquad (5.7)$$

for $t \subseteq d(\phi)$ is of particular interest. We recall from Section 3.5 that

$$\phi = \chi \otimes \phi^{\downarrow t}.$$

A valuation χ satisfying this equation is said to *continue* ϕ from t to $d(\phi)$ (Shafer, 1991) and χ is called a *continuation* of ϕ. Of course there may be many continuations, (5.7) is only one among many others. But we have seen in Section 3.5 that χ is closely related to conditional probability distributions. Motivated by this example we introduce therefore the following definition:

Definition 5.11 Conditionals: *For $\phi \in \Phi$ and $s, t \subseteq d(\phi)$ disjoint,*

$$\phi_{s|t} = (\phi^{\downarrow t})^{-1} \otimes \phi^{\downarrow s \cup t} \qquad (5.8)$$

is called the conditional of ϕ for s given t.

$\phi_{s|t}$ is clearly a continuation of ϕ from t to $s \cup t$,

$$\phi^{\downarrow s \cup t} = \phi_{s|t} \otimes \phi^{\downarrow t}$$

We have also $d(\phi_{s|t}) = s \cup t$.

If $p(\mathbf{x}, \mathbf{y})$ is a probability potential and $\mathbf{x}$, $\mathbf{y}$ configurations in Ω_s and Ω_t respectively, s and t disjoint, then the conditional $p_{s|t}$ is

$$p_{s|t}(\mathbf{x}, \mathbf{y}) = \begin{cases} \frac{p(\mathbf{x}, \mathbf{y})}{p^{\downarrow t}(\mathbf{y})} & \text{if } p^{\downarrow t}(\mathbf{y}) > 0, \\ 0 & \text{otherwise.} \end{cases}$$

So, for all configurations $\mathbf{y}$, where $p^{\downarrow t}(\mathbf{y}) > 0$ the conditional $p_{s|t}$ defines the conditional probability

$$p_{s|t}(\mathbf{x}, \mathbf{y}) = p(\mathbf{x}|\mathbf{y}).$$

In the other cases, the potential $p_{s|t}$ is set to zero. This illustrates the close connection between conditionals and conditional probability in the case of probability potentials. And this relation is the model for the subsequent developments in the framework of regular valuation algebras. Of course, the definition of $p_{s|t}(\mathbf{x}, \mathbf{y}) = 0$ if $p^{\downarrow t}(\mathbf{y}) = 0$ is arbitrary. The freedom we have in this case reflects the fact that there are many continuations of p.

In possibility theory, if combination is taken as multiplication of possibility potentials, then the conditional $p_{s|t}$ for a potential $p(\mathbf{x}, \mathbf{y})$ is defined as

$$
p_{s|t}(\mathbf{x}, \mathbf{y}) \;=\; \begin{cases} \frac{p(\mathbf{x},\mathbf{y})}{\max_{\mathbf{x}} p(\mathbf{x},\mathbf{y})} & \text{if } \max_{\mathbf{x}} p(\mathbf{x}, \mathbf{y}) > 0, \\ 0 & \text{otherwise.} \end{cases}
$$

This is very similar to probability theory.

Idempotent valuation algebras are trivial regular algebras. Each valuation ϕ is its own inverse. So the conditional $\phi_{s|t}$ is again simply $\phi^{\downarrow s \cup t}$ itself. This is another hint to the triviality of idempotent valuation algebras as regular algebras. Hence, if in possibility theory, combination is taken as the minimum of potentials, then conditioning becomes trivial. We remark here, that, since possibility potentials with Lukasziewicz's t-norm for combination is even not separative, there is no notion of division and hence there are no conditionals in this case. Possibility theory is very disparate as far as conditioning, and therefore also as far as conditional independence is concerned.

For Spohn potentials $p(\mathbf{x}, \mathbf{y})$ conditionals are defined by

$$
p_{s|t}(\mathbf{x}, \mathbf{y}) \;=\; p(\mathbf{x}, \mathbf{y}) - \max_{\mathbf{x}} p(\mathbf{x}, \mathbf{y})
$$

since division is here subtraction. Here the notion of a conditional is unambiguous. We refer to (Studeny, 1995) for a comprehensive discussion of conditional independence with respect to Spohn potentials.

The following lemma will often be used without explicit reference.

Lemma 5.12 *For $\phi \in \Phi$ and $s, t \subseteq d(\phi)$ disjoint,*

$$
\gamma(\phi_{s|t}) \;\geq\; \gamma(\phi^{\downarrow t}).
$$

Proof. By definition we have that

$$
\gamma(\phi_{s|t}) \;=\; \gamma((\phi^{\downarrow t})^{-1} \otimes \phi^{\downarrow s \cup t}) \;=\; \gamma((\phi^{\downarrow t})^{-1}) \vee \gamma(\phi^{\downarrow s \cup t})
$$
$$
=\; \gamma(\phi^{\downarrow t}) \vee \gamma(\phi^{\downarrow s \cup t})
$$

by (3.55). But this implies that $\gamma(\phi_{s|t}) \geq \gamma(\phi^{\downarrow t})$. $\square$

Conditionals in regular algebras have many properties of conditional probabilities. The following lemma collects a few of such results.

Lemma 5.13

1. *Assume $\phi \in \Phi$ and $s, t \subseteq d(\phi)$ disjoint. Then*

$$
\phi_{s|t}^{\downarrow t} \;=\; f_{\gamma(\phi^{\downarrow t})}.
$$

2. *Assume $\phi \in \Phi$ and $s, t, u \subseteq d(\phi)$ disjoint. Then*

$$
\phi_{s \cup t | u} \;=\; \phi_{s|t \cup u} \otimes \phi_{t|u}.
$$

3. *Assume* $\phi \in \Phi$ *and* $s, t \subseteq d(\phi)$ *disjoint,* $u \subseteq s$. *Then*

$$\phi_{s|t}^{\downarrow t \cup u} = \phi_{u|t}.$$

4. *Assume* $\phi \in \Phi$ *and* $s, t, u \subseteq d(\phi)$ *disjoint. Then*

$$(\phi_{u|s \cup t} \otimes \phi_{s|t})^{\downarrow t \cup u} = \phi_{u|t}.$$

5. *Assume* $\phi, \psi \in \Phi$, $s, t \subseteq d(\phi)$ *disjoint, and* $d(\psi) = t$. *Then*

$$(\phi^{\downarrow s \cup t} \otimes \psi)_{s|t} = \phi_{s|t} \otimes f_{\gamma(\psi)}.$$

Proof. (1) From the definition of a conditional, using the combination axiom we obtain

$$\phi_{s|t}^{\downarrow t} = ((\phi^{\downarrow t})^{-1} \otimes \phi^{\downarrow s \cup t})^{\downarrow t} = (\phi^{\downarrow t})^{-1} \otimes \phi^{\downarrow t} = f_{\gamma(\phi^{\downarrow t})}.$$

(2) Again, starting with the definition of a conditional, we have

$$\phi_{s|t \cup u} \otimes \phi_{t|u} = \phi^{\downarrow s \cup t \cup u} \otimes (\phi^{\downarrow t \cup u})^{-1} \otimes \phi^{\downarrow t \cup u} \otimes (\phi^{\downarrow u})^{-1}$$
$$= \phi^{\downarrow s \cup t \cup u} \otimes (\phi^{\downarrow u})^{-1} \otimes f_{\gamma(\phi^{\downarrow t \cup u})}.$$

Since $\gamma(\phi^{\downarrow s \cup t \cup u}) \geq \gamma(\phi^{\downarrow t \cup u})$ (Lemma 3.20 (2)) we obtain

$$\phi_{s|t \cup u} \otimes \phi_{t|u} = \phi^{\downarrow s \cup t \cup u} \otimes (\phi^{\downarrow u})^{-1} = \phi_{s \cup t|u}.$$

(3) Here we see that

$$\phi_{s|t}^{\downarrow t \cup u} = (\phi^{\downarrow s \cup t} \otimes (\phi^{\downarrow t})^{-1})^{\downarrow t \cup u} = \phi^{\downarrow t \cup u} \otimes (\phi^{\downarrow t})^{-1} = \phi_{t|u}$$

using Lemma 2.1 (3).

(4) By point (2) proved above, we have that

$$\phi_{u|s \cup t} \otimes \phi_{s|t} = \phi_{s \cup u|t}.$$

Therefore, (3) just proved implies that

$$(\phi_{u|s \cup t} \otimes \phi_{s|t})^{\downarrow t \cup u} = \phi_{s \cup u|t}^{\downarrow t \cup u} = \phi_{u|t}.$$

(5) On the one hand we have

$$\phi^{\downarrow s \cup t} \otimes \psi = \phi_{s|t} \otimes \phi^{\downarrow t} \otimes \psi$$

and on the other hand

$$\phi^{\downarrow s \cup t} \otimes \psi = (\phi^{\downarrow s \cup t} \otimes \psi)_{s|t} \otimes (\phi^{\downarrow s \cup t} \otimes \psi)^{\downarrow t}$$
$$= (\phi^{\downarrow s \cup t} \otimes \psi)_{s|t} \otimes \phi^{\downarrow t} \otimes \psi.$$

This gives us the equation

$$\phi_{s|t} \otimes (\phi^{\downarrow t} \otimes \psi) \;=\; (\phi^{\downarrow s \cup t} \otimes \psi)_{s|t} \otimes (\phi^{\downarrow t} \otimes \psi)$$

from which we deduce that

$$\phi_{s|t} \otimes f_{\gamma(\phi^{\downarrow t} \otimes \psi)} \;=\; (\phi^{\downarrow s \cup t} \otimes \psi)_{s|t} \otimes f_{\gamma(\phi^{\downarrow t} \otimes \psi)}.$$

We see that

$$
\begin{aligned}
\gamma((\phi^{\downarrow s \cup t} \otimes \psi)_{s|t}) &= \gamma((\phi^{\downarrow s \cup t} \otimes \psi) \otimes ((\phi^{\downarrow s \cup t} \otimes \psi)^{\downarrow t})^{-1}) \\
&= \gamma(\phi^{\downarrow s \cup t} \otimes \psi) \vee \gamma(\phi^{\downarrow t} \otimes \psi) \\
&\geq \gamma(\phi^{\downarrow t} \otimes \psi).
\end{aligned}
$$

Therefore, from the equation above it follows that

$$(\phi^{\downarrow s \cup t} \otimes \psi)_{s|t} \;=\; \phi_{s|t} \otimes f_{\gamma(\phi^{\downarrow t})} \otimes f_{\gamma(\psi)} \;=\; \phi_{s|t} \otimes f_{\gamma(\psi)}.$$

$\square$

These are useful results for working with conditionals. It is well known from probability theory that conditional independence can be equivalently expressed in several ways in terms of conditional probabilities. Exactly the same definitions can be given in regular valuation algebras using conditionals. This is what the next theorem shows.

Theorem 5.14 *Assume $\phi \in \Phi$ and $s, t, u \subseteq d(\phi)$ disjoint. Then the following statements are all equivalent.*

1. $\phi^{\downarrow s \cup t \cup u} = \phi_{s|u} \otimes \phi_{t|u} \otimes \phi^{\downarrow u}.$

2. $\phi_{s \cup t|u} = \phi_{s|u} \otimes \phi_{t|u}.$

3. $\phi_{s \cup t|u} = \chi_1 \otimes \chi_2$ *with* $d(\chi_1) = s \cup u$ *and* $d(\chi_2) = t \cup u.$

4. $\phi^{\downarrow s \cup t \cup u} \otimes \phi^{\downarrow u} = \phi^{\downarrow s \cup u} \otimes \phi^{\downarrow t \cup u}.$

5. $\phi^{\downarrow s \cup t \cup u} = \phi_{s|u} \otimes \phi^{\downarrow t \cup u}.$

6. $\phi_{s|t \cup u} = \phi_{s|u} \otimes f_{\gamma(\phi^{\downarrow t \cup u})}$

7. $\phi_{s|t \cup u} = \chi \otimes f_{\gamma(\phi^{\downarrow t \cup u})}$ *with* $d(\chi) = s \cup u.$

8. $\phi^{\downarrow s \cup t \cup u} = \psi_1 \otimes \psi_2$ *with* $d(\psi_1) = s \cup u$ *and* $d(\psi_2) = t \cup u.$

Proof. We prove that (i) implies $(i+1)$ for $i = 1$ to 7 and then that (8) implies (1).

$(1) \Rightarrow (2)$: By (1) we have

$$\phi^{\downarrow s \cup t \cup u} = \phi_{s \cup t|u} \otimes \phi^{\downarrow u} = \phi_{s|u} \otimes \phi_{t|u} \otimes \phi^{\downarrow u}$$

From this we deduce (using Lemma 5.12)

$$\phi_{s \cup t | u} = \phi_{s|u} \otimes \phi_{t|u}.$$

(2) $\Rightarrow$ (3): Take $\chi_1 = \phi_{s|u}$ and $\chi_2 = \phi_{t|u}$.

(3) $\Rightarrow$ (4): From (3) we obtain

$$\phi^{\downarrow s \cup t \cup u} \otimes \phi^{\downarrow u} = \phi_{s \cup t | u} \otimes \phi^{\downarrow u} \otimes \phi^{\downarrow u} = (\chi_1 \otimes \phi^{\downarrow u}) \otimes (\chi_2 \otimes \phi^{\downarrow u}).$$

Furthermore, we have

$$
\begin{aligned}
\phi^{\downarrow s \cup u} &= (\phi^{\downarrow s \cup t \cup u})^{\downarrow s \cup u} &= (\phi_{s \cup t | u} \otimes \phi^{\downarrow u})^{\downarrow s \cup u} \\
&= (\chi_1 \otimes \chi_2 \otimes \phi^{\downarrow u})^{\downarrow s \cup u} &= \chi_1 \otimes (\chi_2 \otimes \phi^{\downarrow u})^{\downarrow u} \\
&= \chi_1 \otimes \chi_2^{\downarrow u} \otimes \phi^{\downarrow u}.
\end{aligned}
$$

Similarly we obtain

$$\phi^{\downarrow t \cup u} = \chi_2 \otimes \chi_1^{\downarrow u} \otimes \phi^{\downarrow u}$$

Now, we have also (Lemma 2.1 (1) and Lemma 5.13 (1))

$$\phi_{s \cup t | u}^{\downarrow u} = (\chi_1 \otimes \chi_2)^{\downarrow u} = \chi_1^{\downarrow u} \otimes \chi_2^{\downarrow u} = f_{\gamma(\phi^{\downarrow u})}.$$

So, we obtain finally

$$
\begin{aligned}
\phi^{\downarrow s \cup u} \otimes \phi^{\downarrow t \cup u} &= (\chi_1 \otimes \phi^{\downarrow u}) \otimes (\chi_2 \otimes \phi^{\downarrow u}) \otimes (\chi_1^{\downarrow u} \otimes \chi_2^{\downarrow u}) \\
&= (\chi_1 \otimes \phi^{\downarrow u}) \otimes (\chi_2 \otimes \phi^{\downarrow u}) \otimes f_{\gamma(\phi^{\downarrow u})} \\
&= (\chi_1 \otimes \phi^{\downarrow u}) \otimes (\chi_2 \otimes \phi^{\downarrow u}) \\
&= \phi^{\downarrow s \cup t \cup u} \otimes \phi^{\downarrow u}.
\end{aligned}
$$

(4) $\Rightarrow$ (5): Starting with (4) we obtain

$$\phi^{\downarrow s \cup t \cup u} \otimes \phi^{\downarrow u} = \phi^{\downarrow s \cup u} \otimes \phi^{\downarrow t \cup u} = \phi_{s|u} \otimes \phi^{\downarrow u} \otimes \phi^{\downarrow t \cup u}.$$

This gives us, if we multiply both sides with the inverse of $\phi^{\downarrow u}$

$$\phi^{\downarrow s \cup t \cup u} = \phi_{s|u} \otimes \phi^{\downarrow t \cup u}.$$

(5) $\Rightarrow$ (6): We have

$$\phi^{\downarrow s \cup t \cup u} = \phi_{s|t \cup u} \otimes \phi^{\downarrow t \cup u}$$

and also, by (5)

$$\phi^{\downarrow s \cup t \cup u} = \phi_{s|u} \otimes \phi^{\downarrow t \cup u}$$

From this we conclude that

$$\phi_{s|t \cup u} = \phi_{s|u} \otimes f_{\gamma(\phi^{\downarrow t \cup u})}.$$

(6) $\Rightarrow$ (7): Take $\chi = \phi_{s|u}$.

(7) $\Rightarrow$ (8): We have

$$\phi^{\downarrow s \cup t \cup u} \;=\; \phi_{s|t \cup u} \otimes \phi^{\downarrow t \cup u} \;=\; \chi \otimes f_{\gamma(\phi^{\downarrow t \cup u})} \otimes \phi^{\downarrow t \cup u} \;=\; \chi \otimes \phi^{\downarrow t \cup u}$$

where $d(\chi) = s \cup u$ and $d(\phi^{\downarrow t \cup u}) = t \cup u$. Thus, take $\psi_1 = \chi_1$ and $\psi_2 = \phi^{\downarrow t \cup u}$.

(8) $\Rightarrow$ (1): (8) implies that

$$\phi^{\downarrow u} \;=\; \psi_1^{\downarrow u} \otimes \psi_2^{\downarrow u}$$

(see Lemma 2.1 (1)). Furthermore, conditionals are continuations,

$$\psi_1 \;=\; \psi_{1s|u} \otimes \psi_1^{\downarrow u}, \qquad \psi_2 \;=\; \psi_{2t|u} \otimes \psi_2^{\downarrow u}.$$

Therefore, we obtain that

$$
\begin{aligned}
\phi^{\downarrow s \cup t \cup u} &\;=\; \psi_{1s|u} \otimes \psi_{2t|u} \otimes (\psi_1^{\downarrow u} \otimes \psi_2^{\downarrow u}) \\
&\;=\; \psi_{1s|u} \otimes \psi_{2t|u} \otimes \phi^{\downarrow u}.
\end{aligned}
$$

We obtain also from (8) that

$$
\begin{aligned}
\phi^{\downarrow s \cup u} &\;=\; (\psi_1 \otimes \psi_2)^{\downarrow s \cup u} \;=\; \psi_1 \otimes \psi_2^{\downarrow u} \\
&\;=\; \psi_{1s|u} \otimes (\psi_1^{\downarrow u} \otimes \psi_2^{\downarrow u}) \;=\; \psi_{1s|u} \otimes \phi^{\downarrow u}.
\end{aligned}
$$

Therefore we have the equation

$$\phi_{s|u} \otimes \phi^{\downarrow u} \;=\; \psi_{1s|u} \otimes \phi^{\downarrow u},$$

hence

$$\phi_{s|u} \;=\; \psi_{1s|u} \otimes f_{\gamma(\phi^{\downarrow u})}.$$

In the same way, we find that

$$\phi_{t|u} \;=\; \psi_{2t|u} \otimes f_{\gamma(\phi^{\downarrow u})}.$$

From this we find finally that

$$
\begin{aligned}
\phi^{\downarrow s \cup t \cup u} &\;=\; (\psi_{1s|u} \otimes f_{\gamma(\phi^{\downarrow u})}) \otimes (\psi_{2t|u} \otimes f_{\gamma(\phi^{\downarrow u})}) \otimes \phi^{\downarrow u} \\
&\;=\; \phi_{s|u} \otimes \phi_{t|u} \otimes \phi^{\downarrow u}.
\end{aligned}
$$

$\square$

If we consider idempotent valuation algebras, then we have $\phi_{s|t} = \phi^{\downarrow s \cup t}$. Point (1) of this theorem says that in this case $\phi^{\downarrow s \cup t \cup u} = \phi^{\downarrow s \cup u} \otimes \phi^{\downarrow t \cup u}$ and all other points (2) to (8) of the theorem collapse essentially to the same formula.

In the case of probability potentials, we assume that $\mathbf{x}, \mathbf{y}, \mathbf{z}$ are configurations for disjoint sets s, t and u of variables. In this case the different points of the theorem above are usually written in terms of conditional probabilities:

1. (1) $p(\mathbf{x}, \mathbf{y}, \mathbf{z}) = p(\mathbf{x}|\mathbf{z})p(\mathbf{y}|\mathbf{z})p(\mathbf{z})$.

2. (2) $p(\mathbf{x}, \mathbf{y}|\mathbf{z}) = p(\mathbf{x}|\mathbf{z})p(\mathbf{y}|\mathbf{z})$.

3. (4) $p(\mathbf{x}, \mathbf{y}, \mathbf{z}) = p(\mathbf{x}, \mathbf{z})p(\mathbf{y}, \mathbf{z})/p(\mathbf{z})$.

4. (5) $p(\mathbf{x}, \mathbf{y}, \mathbf{z}) = p(\mathbf{x}|\mathbf{z})p(\mathbf{y}, \mathbf{z})$.

5. (6) $p(\mathbf{x}|\mathbf{y}, \mathbf{z}) = p(\mathbf{x}|\mathbf{z})$.

In each case, the existence of the conditional probabilities appearing in the formula is assumed. It is easily seen that these formulas correspond to the items of Theorem 5.14.

In the case of possibility theory with multiplication as combination, point (4) of the theorem is most interesting. It says that $s \perp_\phi t | u$ if, and only if,

$$p(\mathbf{x}, \mathbf{y}, \mathbf{z}) = \frac{\max_{\mathbf{x}} p(\mathbf{x}, \mathbf{y}, \mathbf{z}) \max_{\mathbf{y}} p(\mathbf{x}, \mathbf{y}, \mathbf{z})}{\max_{\mathbf{x}, \mathbf{y}} p(\mathbf{x}, \mathbf{y}, \mathbf{z})}$$

An other interesting case arises, if potentials are added for combination (instead of multiplied) and marginalization is maximization. This is indeed a regular valuation algebra. Equation (3.47) takes the form

$$f(\mathbf{x}, \mathbf{y}) = \max_{\mathbf{x}} f(\mathbf{x}, \mathbf{y}) + g(\mathbf{y}) + f(\mathbf{x}, \mathbf{y})$$

which has the solution

$$g(\mathbf{y}) = -\max_{\mathbf{x}} f(\mathbf{x}, \mathbf{y}).$$

Assuming that $\mathbf{x}, \mathbf{y}$ are configurations for the disjoint sets of variables s and t, we see that the conditional $f_{s|t}$ is defined by

$$f_{s|t}(\mathbf{x}, \mathbf{y}) = f(\mathbf{x}, \mathbf{y}) - \max_{\mathbf{x}} f(\mathbf{x}, \mathbf{y}).$$

It is the "loss" of the value $f(\mathbf{x}, \mathbf{y})$ for a fixed configuration $(\mathbf{x}, \mathbf{y})$ with respect to the situation where $\mathbf{x}$ can be freely selected and f maximized over it. $s \perp_f t | u$ means that

$$f(\mathbf{x}, \mathbf{y}, \mathbf{u}) = f_1(\mathbf{x}, \mathbf{z}) + f_2(\mathbf{y}, \mathbf{z}).$$

In this case, we see that

$$f_{s|u}(\mathbf{x}, \mathbf{z}) = f_1(\mathbf{x}, \mathbf{z}) - \max_{\mathbf{x}} f_1(\mathbf{x}, \mathbf{z})$$
$$f_{t|u}(\mathbf{y}, \mathbf{z}) = f_2(\mathbf{y}, \mathbf{z}) - \max_{\mathbf{y}} f_2(\mathbf{y}, \mathbf{z}).$$

By (1) of Theorem 5.14 we have then also

$$f(\mathbf{x}, \mathbf{y}, \mathbf{u})$$
$$= (f_1(\mathbf{x}, \mathbf{z}) - \max_{\mathbf{x}} f_1(\mathbf{x}, \mathbf{z})) + (f_2(\mathbf{y}, \mathbf{z}) - \max_{\mathbf{y}} f_2(\mathbf{y}, \mathbf{z})) + \max_{\mathbf{x}, \mathbf{y}} f(\mathbf{x}, \mathbf{y}, \mathbf{u}).$$

Furthermore,

$$\max_{\mathbf{x},\mathbf{y}} f(\mathbf{x},\mathbf{y},\mathbf{u}) \quad = \quad \max_{\mathbf{x}} f_1(\mathbf{x},\mathbf{z}) + \max_{\mathbf{y}} f_2(\mathbf{y},\mathbf{z})$$

Hence, we have

$$\max_{\mathbf{x},\mathbf{y},\mathbf{z}} f(\mathbf{x},\mathbf{y},\mathbf{z}) \quad = \quad \max_{\mathbf{z}}(\max_{\mathbf{x}} f_1(\mathbf{x},\mathbf{z}) + \max_{\mathbf{y}} f_2(\mathbf{y},\mathbf{z})).$$

This is the basic recursive equation of *dynamic programming*. Of course, this same example can also be interpreted in terms of Spohn potentials, if max is replaced by min.

In practice the following problem may arise: we want to define a valuation $\phi \in \Phi$ by specifying its marginals ψ_1 and ψ_2 relative to $s \cup u$ and $t \cup u$ and requiring that $s \perp_\phi t | u$ holds. Of course, the marginals must be *consistent*, that is $\psi_1^{\downarrow u} = \psi_2^{\downarrow u}$ must hold. Is there such a valuation ϕ? This is sometimes called the *marginal problem*. The following theorem answers this question in the affirmative:

Theorem 5.15 *Let (Φ, D) be a regular valuation algebra, $s,t,u \in D$ disjoint sets and $\psi_1, \psi_2 \in \Phi$ such that $d(\psi_1) = s \cup u$, $d(\psi_2) = t \cup u$ and*

$$\psi_1^{\downarrow u} \quad = \quad \psi_2^{\downarrow u}.$$

Then there is a $\phi \in \Phi$ such that $s \perp_\phi t | u$ and

$$\phi^{\downarrow s \cup u} \quad = \quad \psi_1, \qquad \phi^{\downarrow t \cup u} \quad = \quad \psi_2. \tag{5.9}$$

Proof. Define $\eta = \psi_1^{\downarrow u} = \psi_2^{\downarrow u}$ and

$$\phi \quad = \quad \psi_1 \otimes \psi_2 \otimes \eta^{-1}.$$

We claim that ϕ satisfies (5.9). In fact,

$$\phi^{\downarrow s \cup u} \quad = \quad \psi_1 \otimes \psi_2^{\downarrow u} \otimes \eta^{-1} \quad = \quad \psi_1 \otimes \eta \otimes \eta^{-1} \quad = \quad \psi_1,$$

since $\gamma(\psi_1) \geq \gamma(\psi_1^{\downarrow u}) = \gamma(\eta)$. The other equality of (5.9) is proved in the same way. This implies then that

$$\phi^{\downarrow u} \quad = \quad \eta,$$

hence

$$\phi \otimes \phi^{\downarrow u} \quad = \quad \psi_1 \otimes \psi_2.$$

This shows that $s \perp_\phi t | u$ (Theorem 5.14 (4)). $\square$

In regular valuation algebras as in the case of probability potentials, the relation of conditional independence satisfies not only conditions (G1) to (G3) of Theorem 5.2, but a further important property. Furthermore, the validity of (G3) no longer depends on the existence of neutral elements.

Theorem 5.16 *Let (Φ, D) be a regular valuation algebra, $\phi \in \Phi$ and s, t, u, $v \subseteq d(\phi)$ disjoint. Then*
   (G3) Weak Union: $s \perp_\phi t \cup v | u$ implies $s \perp_\phi t | u \cup v$.
   (G4) Contraction: $s \perp_\phi t | u$ and $s \perp_\phi v | t \cup u$ imply $s \perp_\phi t \cup v | u$.

Proof. (G3) $s \perp_\phi t \cup v | u$ means (Theorem 5.14 (5)) that

$$\phi^{\downarrow s \cup t \cup u \cup v} = \phi_{s|u} \otimes \phi^{\downarrow t \cup u \cup v}.$$

Furthermore, by (G2), $s \perp_\phi t \cup v | u$ implies $s \perp_\phi v | u$. This, in turn, by Theorem 5.14 (6) gives us

$$\phi_{s|v \cup u} = \phi_{s|u} \otimes f_{\gamma(\phi^{\downarrow v \cup u})}.$$

Therefore we obtain

$$\phi^{\downarrow s \cup t \cup u \cup v} = \phi_{s|u} \otimes f_{\gamma(\phi^{\downarrow v \cup u})} \otimes \phi^{\downarrow t \cup u \cup v}$$
$$= \phi_{s|v \cup u} \otimes \phi^{\downarrow t \cup u \cup v} = \phi_{s|v \cup u} \otimes \phi_{t|v \cup u} \otimes \phi^{\downarrow v \cup u}.$$

The last expression implies $s \perp_\phi t | u \cup v$ by Theorem 5.14 (1).
   (G4) $s \perp_\phi t | u$ and $s \perp_\phi v | t \cup u$ imply that

$$\phi^{\downarrow s \cup t \cup u} = \psi_1 \otimes \psi_2, \qquad \phi^{\downarrow s \cup t \cup u \cup v} = \eta_1 \otimes \eta_2$$

where $d(\psi_1) = s \cup u$, $d(\psi_2) = t \cup u$ and $d(\eta_1) = s \cup t \cup u$, $d(\eta_2) = v \cup t \cup u$. By the combination axiom we obtain

$$\phi^{\downarrow s \cup t \cup u} = (\eta_1 \otimes \eta_2)^{\downarrow s \cup t \cup u} = \eta_1 \otimes \eta_2^{\downarrow t \cup u}.$$

So

$$\phi^{\downarrow s \cup t \cup u \cup v} = \eta_1 \otimes \eta_{2v|t \cup u} \otimes \eta_2^{\downarrow t \cup u} = \psi_1 \otimes (\psi_2 \otimes \eta_{2v|t \cup u}).$$

Since $d(\psi_1) = s \cup u$ and $d(\psi_2 \otimes \eta_{2v|t \cup u}) = t \cup u \cup v$ we have proved that $s \perp_\phi t \cup v | u$. $\square$

   The properties (G1) to (G4) may be considered as a system of axioms for an *abstract calculus of conditional independence*. A model of these axioms has been termed a *semi-graphoid* by Pearl (Pearl, 1988). Conditional independence in regular valuation algebras determines thus a semi-graphoid. A range of further examples of conditional independence is described by Dawid (Dawid, 1979; Dawid, 1998). We note also that Theorem 5.6 holds in this context, since it depends only on the semi-graphoid properties.
   For positive probability potentials, the conditional independence relation is furthermore known to satisfy the so-called *intersection property*. This is also the case for positive elements ϕ of a regular valuation algebra.

Theorem 5.17 *Let Φ_p denote the positive elements of a regular valuation algebra (Φ, D) with positive elements. We assume that that Φ_p is closed under marginalization. Let $\phi \in \Phi_p$, $s, t, u, v \subseteq d(\phi)$ disjoint. Then*
   (G5) Intersection: $s \perp_\phi t | u \cup v$ and $s \perp_\phi v | t \cup u$ imply that $s \perp_\phi t \cup v | u$.

Proof. Since Φ_p is closed under marginalization, $\phi^{\downarrow t}$ is positive for all $t \subseteq d(\phi)$ if ϕ itself is positive. That is, we have $f_{\gamma(\phi^{\downarrow t})} = e_t$. Therefore, by Theorem 5.14 (6), $s \perp_\phi t | u \cup v$ implies that

$$\phi_{s|t \cup u \cup v} \;=\; \phi_{s|u \cup v} \otimes e_{t \cup u \cup v} \;=\; \phi_{s|u \cup v} \otimes e_t$$

and $s \perp_\phi v | t \cup u$ implies

$$\phi_{s|t \cup u \cup v} \;=\; \phi_{s|t \cup u} \otimes e_{t \cup u \cup v} \;=\; \phi_{s|t \cup u} \otimes e_v.$$

So we obtain

$$\phi_{s|u \cup v} \otimes e_t \;=\; \phi_{s|t \cup u} \otimes e_v.$$

If we combine both sides with $\phi^{\downarrow t \cup u} \otimes \phi^{\downarrow u \cup v}$ it follows that

$$\phi^{\downarrow s \cup u \cup v} \otimes \phi^{\downarrow t \cup u} \;=\; \phi^{\downarrow s \cup t \cup u} \otimes \phi^{\downarrow u \cup v}.$$

Next we marginalize both sides to $s \cup t \cup u$ and obtain (using Lemma 2.1 (3))

$$\phi^{\downarrow s \cup u} \otimes \phi^{\downarrow t \cup u} \;=\; \phi^{\downarrow s \cup t \cup u} \otimes \phi^{\downarrow u}.$$

By Theorem 5.14 (4) this means that $s \perp_\phi t | u$. Therefore, by the same theorem (6) we have

$$\phi_{s|t \cup u} \;=\; \phi_{s|u} \otimes e_{t \cup u} \;=\; \phi_{s|u} \otimes e_t.$$

From this it follows that

$$\phi_{s|t \cup u \cup v} \;=\; \phi_{s|u} \otimes e_t \otimes e_v \;=\; \phi_{s|u} \otimes e_{t \cup v}.$$

This in turn implies

$$\phi^{\downarrow s \cup t \cup u \cup v} \;=\; \phi_{s|t \cup u \cup v} \otimes \phi^{\downarrow t \cup u \cup v} \;=\; \phi_{s|u} \otimes \phi^{\downarrow t \cup u \cup v}.$$

By Theorem 5.14 (5) this proves that $s \perp_\phi t \cup v | u$. $\square$

The intersection property is not satisfied in general in a regular valuation algebra. A counter example for a non-positive probability potential is known (Lauritzen, 1996). Note that in the case of additive potentials with maximization or minimization as marginalization, *all* potentials are positive. Thus Theorem 5.17 applies for all potentials in this valuation algebra. This holds for example for Spohn potentials.

5.3 Conditionals in Separative Algebras

The question is, to what degree the results relative to conditional independence in a regular valuation algebra can be extended beyond regular algebras. Conditionals are related to division. So a notion of division is needed, in order to define conditionals. This consideration leads us to consider separative valuation algebras, which can be embedded into a union of groups and provide thus a notion of division.

The model for such an algebra is provided by densities (see Section 3.6). In this example there exist conditional densities, just as in discrete probability potentials. Let $f(\mathbf{x}, \mathbf{y})$ be a density with $\mathbf{x}, \mathbf{y}$ configurations for disjoint sets s, t of variables. For $\mathbf{y}$ with $f^{\downarrow t}(\mathbf{y}) > 0$ there exists the conditional density function

$$f_{s|t}(\mathbf{x}, \mathbf{y}) \;=\; \frac{f(\mathbf{x}, \mathbf{y})}{f^{\downarrow t}(\mathbf{y})}$$

We may (arbitrarily) put $f_{s|t}(\mathbf{x}, \mathbf{y}) = 0$ when $f^{\downarrow t}(\mathbf{y}) = 0$. The difference to discrete probability potentials is that in this case the conditional $f_{s|t}$ is no more a density. That is, this conditional is no more an element of the valuation algebra Φ of densities, but of its extension Φ^*. In particular, marginalization is not possible without restriction. For example,

$$\int_{-\infty}^{+\infty} f_{s|t}(\mathbf{x}, \mathbf{y})d\mathbf{x} \;=\; \frac{\displaystyle\int_{-\infty}^{+\infty} f(\mathbf{x}, \mathbf{y})d\mathbf{x}}{f^{\downarrow t}(\mathbf{y})} \;=\; \frac{f^{\downarrow t}(\mathbf{y})}{f^{\downarrow t}(\mathbf{y})} \;=\; 1$$

whenever $f^{\downarrow t}(\mathbf{y}) > 0$. So $f_{s|t}$ is not necessarily integrable (remember that we require finiteness of the integral). Marginalization does not extend to Φ^*, at least not fully. However, $f_{s|t}$ is integrable over subsets of s. If $u \subseteq s$, and $\mathbf{x} = (\mathbf{v}, \mathbf{z})$, where $\mathbf{v}, \mathbf{z}$ are configurations for $s - u$ and u respectively, then we may define

$$f_{s|t}^{\downarrow u \cup t}(\mathbf{z}, \mathbf{y}) \;=\; \frac{\displaystyle\int_{-\infty}^{+\infty} f(\mathbf{v}, \mathbf{z}, \mathbf{y})d\mathbf{v}}{f^{\downarrow t}(\mathbf{y})} \;=\; \frac{f^{\downarrow u \cup t}(\mathbf{z}, \mathbf{y})}{f^{\downarrow t}(\mathbf{y})}$$

when $f^{\downarrow t}(\mathbf{y}) > 0$. Otherwise we have $f_{s|t}^{\downarrow u \cup t}(\mathbf{z}, \mathbf{y}) = 0$. Clearly, we have that

$$f_{s|t}^{\downarrow u \cup t}(\mathbf{z}, \mathbf{y}) \;=\; f_{u|t}(\mathbf{z}, \mathbf{y}).$$

This is the situation we have in separative valuation algebras (Φ, D), which are embedded into valuation algebras (Φ^*, D) with partial marginalization (see Section 3.6).

Let therefore in this Section (Φ, D) be a *separative valuation algebra* and Φ^* the union of a semilattice of disjoint groups into which Φ is embedded as a semigroup (Theorems 3.22 and 3.24). As before let $\gamma(\phi) = \alpha$ denote the cancellative semigroup S_α and also the group G_α to which $\phi \in \Phi$ belongs, $S_\alpha \subseteq G_\alpha$. Similarly, for $\psi \in \Phi^*$ we have $\gamma(\psi) = \alpha$, if $\psi \in G_\alpha$. We define for every $\phi \in \Phi$ conditionals exactly as in the case of regular algebras.

Definition 5.18 Conditionals: *For $\phi \in \Phi$ and $s, t \subseteq d(\phi)$ disjoint,*

$$\phi_{s|t} \;=\; (\phi^{\downarrow t})^{-1} \otimes \phi^{\downarrow s \cup t} \tag{5.10}$$

is called the conditional of ϕ for s given t.

But, in contrast to regular algebras, in separative algebras a conditional $\phi_{s|t}$ belongs in general no more to Φ, but only to Φ^*. We refer to the example of densities. However, as before, a conditional $\phi_{s|t}$ is a continuation of ϕ from t to $s \cup t$,

$$\phi^{\downarrow s \cup t} \;=\; \phi_{s|t} \otimes \phi^{\downarrow t}.$$

We remember that in a separative valuation algebra, by Lemma 3.23, $\gamma(\phi^{\downarrow t}) \leq \gamma(\phi^{\downarrow s \cup t}) \leq \gamma(\phi)$.

Since $\phi_{s|t}$ does not belong to Φ, marginalization is not fully defined for conditionals in general. However, since $d(\phi^{\downarrow t}) \subseteq d(\phi^{\downarrow s \cup t})$ and $\gamma(\phi^{\downarrow t}) \subseteq \gamma(\phi^{\downarrow s \cup t})$, marginals for $\phi_{s|t}$ are defined by (3.69) for u such that $t \subseteq u \subseteq s \cup t$. It turns out in fact, that Lemma 5.13 still holds with this extended definition of conditionals and so does Lemma 5.12. The proof of the latter lemma goes as in the case of regular valuation algebra. For the sake of completeness we restate Lemma 5.13 here.

Lemma 5.19

1. *Assume $\phi \in \Phi$ and $s, t \subseteq d(\phi)$ disjoint. Then*

$$\phi_{s|t}^{\downarrow t} \;=\; f_{\gamma(\phi^{\downarrow t})}.$$

2. *Assume $\phi \in \Phi$ and $s, t, u \subseteq d(\phi)$ disjoint. Then*

$$\phi_{s \cup t | u} \;=\; \phi_{s|t \cup u} \otimes \phi_{t|u}.$$

3. *Assume $\phi \in \Phi$ and $s, t \subseteq d(\phi)$ disjoint, $u \subseteq s$. Then*

$$\phi_{s|t}^{\downarrow t \cup u} \;=\; \phi_{u|t}.$$

4. *Assume $\phi \in \Phi$ and $s, t, u \subseteq d(\phi)$ disjoint. Then*

$$(\phi_{u|s \cup t} \otimes \phi_{s|t})^{\downarrow t \cup u} \;=\; \phi_{u|t}.$$

5. *Assume $\phi, \psi \in \Phi$, $s, t \subseteq d(\phi)$ disjoint, and $d(\psi) = t$. Then*

$$(\phi^{\downarrow s \cup t} \otimes \psi)_{s|t} \;=\; \phi_{s|t} \otimes f_{\gamma(\psi)}.$$

Proof. (1) By definition of marginalization in Φ^* and by the definition of a conditional

$$\phi_{s|t}^{\downarrow t} \;=\; (\phi^{\downarrow t})^{-1} \otimes \phi^{\downarrow t} \;=\; f_{\gamma(\phi^{\downarrow t})}$$

(2) follows from the definition of conditionals as in the proof of Lemma 5.13, using this time Lemma 3.23.

(3) We have by the definitions of a marginal and of a conditional

$$\phi_{s|t}^{\downarrow u \cup t} \;=\; (\phi^{\downarrow t})^{-1} \otimes \phi^{\downarrow u \cup t} \;=\; \phi_{u|t}.$$

(4) and (5) are proved as in Lemma 5.13. □

Conditionals correspond in the case of densities to families of conditional densities, where these are defined. Similarly, in the case of Gaussian potentials, conditionals correspond to conditional Gaussian densities. Another interesting case of a separative (but not regular) valuation algebra, is provided by set potentials. If q is a commonality function on the domains $s \cup t$, where s and t are disjoint, then the conditional $q_{s|t}$ is defined by

$$q_{s|t}(A) \;=\; \frac{q(A)}{q^{\downarrow t}(A^{\downarrow t})}$$

if $q^{\downarrow t}(A^{\downarrow t}) > 0$ and $q_{s|t}(A) = 0$ otherwise.

Given that we dispose of the notion of conditionals in the framework of separative valuation algebras, the question arises whether Theorem 5.14 still holds in this more general framework. It turns out that this does not hold in general. This is a consequence of the fact, that conditionals do not belong to Φ in general. We introduce a weaker form of conditional independence, which is based on a factorization not necessarily in Φ, but only in Φ^*.

Definition 5.20 Weak Conditional Independence: *If $\phi \in \Phi$, and s, t, u are disjoint subsets of $d(\phi)$, we say that s is* weakly conditionally independent *of t given u with respect to ϕ, and we write $s \amalg_\phi t | u$, if*

$$\phi^{\downarrow s \cup t \cup u} \;=\; \phi_{s|u} \otimes \phi_{t|u} \otimes \phi^{\downarrow u}. \tag{5.11}$$

In regular valuation algebras $s \perp_\phi t | u$ and $s \amalg_\phi t | u$ are equivalent (Theorem 5.14). In the case of separative valuation algebras, we can prove that $s \perp_\phi t | u$ is at least as strong than $s \amalg_\phi t | u$.

Theorem 5.21 *Let $\phi \in \Phi$, where (Φ, D) is a separative valuation algebra and $s, t, u \subseteq d(\phi)$. Then $s \perp_\phi t | u$ implies $s \amalg_\phi t | u$.*

Proof. $s \perp_\phi t | u$ means that

$$\phi^{\downarrow s \cup t \cup u} \;=\; \psi_1 \otimes \psi_2$$

with $\psi_1, \psi_2 \in \Phi$ and $d(\psi_1) = s \cup u$, $d(\psi_2) = t \cup u$. From this we obtain that

$$\phi^{\downarrow s \cup u} = \psi_1 \otimes \psi_2^{\downarrow u}, \qquad \phi^{\downarrow t \cup u} = \psi_2 \otimes \psi_1^{\downarrow u}, \qquad \phi^{\downarrow u} = \psi_1^{\downarrow u} \otimes \psi_2^{\downarrow u},$$

and

$$\begin{aligned}
\phi^{\downarrow s \cup t \cup u} &= \psi_{1s|u} \otimes \psi_{2t|u} \otimes \psi_1^{\downarrow u} \otimes \psi_2^{\downarrow u} \\
&= \psi_{1s|u} \otimes \psi_{2t|u} \otimes \phi^{\downarrow u}.
\end{aligned}$$

But we have also that

$$\begin{aligned}
\phi_{s|u} &= \phi^{\downarrow s \cup u} \otimes (\phi^{\downarrow u})^{-1} = \psi_1 \otimes \psi_2^{\downarrow u} \otimes (\psi_1^{\downarrow u} \otimes \psi_2^{\downarrow u})^{-1} \\
&= (\psi_1 \otimes (\psi_1^{\downarrow u})^{-1}) \otimes (\psi_2^{\downarrow u} \otimes (\psi_2^{\downarrow u})^{-1}) = \psi_{1s|u} \otimes f_{\gamma(\psi_2^{\downarrow u})}.
\end{aligned}$$

Similarly, we obtain that

$$\phi_{t|u} = \psi_{2t|u} \otimes f_{\gamma(\psi_1^{\downarrow u})}.$$

Therefore, since $\gamma(\psi_{1s|u}) \geq \gamma(\psi_1^{\downarrow u})$ and $\gamma(\psi_{2t|u}) \geq \gamma(\psi_2^{\downarrow u})$, we see that

$$\begin{aligned}
\phi^{\downarrow s \cup t \cup u} &= (\psi_{1s|u} \otimes f_{\gamma(\psi_1^{\downarrow u})}) \otimes (\psi_{2t|u} \otimes f_{\gamma(\psi_2^{\downarrow u})}) \otimes \phi^{\downarrow u} \\
&= (\psi_{1s|u} \otimes f_{\gamma(\psi_2^{\downarrow u})}) \otimes (\psi_{2t|u} \otimes f_{\gamma(\psi_1^{\downarrow u})}) \otimes \phi^{\downarrow u} \\
&= \phi_{s|u} \otimes \phi_{t|u} \otimes \phi^{\downarrow u}.
\end{aligned}$$

But this means that $s \, \mathrm{II}_\phi \, t|u$. $\square$

Weak conditional independence is not only in regular valuation algebras equivalent to conditional independence. This is the case for example for densities. Suppose further that the following two conditions are satisfied:

1. $\phi^{\downarrow u} = \chi_1 \otimes \chi_2$ with $\chi_1, \chi_2 \in \Phi$ and $d(\chi_1) = d(\chi_2) = u$,

2. $\psi_1 = \phi_{s|u} \otimes \chi_1 \in \Phi$ and $\psi_2 = \phi_{t|u} \otimes \chi_2 \in \Phi$.

Then we have $d(\psi_1) = s \cup u$ and $d(\psi_2) = t \cup u$ and

$$\begin{aligned}
\phi^{\downarrow s \cup t \cup u} &= \phi_{s|u} \otimes \phi_{t|u} \otimes \phi^{\downarrow u} = (\phi_{s|u} \otimes \chi_1) \otimes (\phi_{t|u} \otimes \chi_2) \\
&= \psi_1 \otimes \psi_2.
\end{aligned}$$

In this case we see that $s \, \mathrm{II}_\phi \, t|u$ implies $s \perp_\phi t|u$.

Note that these two conditions are only sufficient for the equivalence of $\perp_\phi$ and II_ϕ. Therefore, even if these two conditions are not satisfied, equivalence may still hold.

We show now that in any case there are equivalent conditions for weak conditional independence similar as in Theorem 5.14.

Theorem 5.22 *Assume $\phi \in \Phi$ and $s, t, u \subseteq d(\phi)$ disjoint. The following statements are all equivalent,*

1. $\phi^{\downarrow s\cup t\cup u} = \phi_{s|u} \otimes \phi_{t|u} \otimes \phi^{\downarrow u}$.

2. $\phi_{s\cup t|u} = \phi_{s|u} \otimes \phi_{t|u}$.

3. $\phi^{\downarrow s\cup t\cup u} \otimes \phi^{\downarrow u} = \phi^{\downarrow s\cup u} \otimes \phi^{\downarrow t\cup u}$.

4. $\phi^{\downarrow s\cup t\cup u} = \phi_{s|u} \otimes \phi^{\downarrow t\cup u}$.

5. $\phi_{s|t\cup u} = \phi_{s|u} \otimes f_{\gamma(\phi^{\downarrow t\cup u})}$.

Proof. We prove $(i) \Rightarrow (i+1)$ for $i = 1$ to 5 and $(5) \Rightarrow (1)$. All of the proofs are more or less similar to the corresponding proofs in Theorem 5.14.

$(1) \Rightarrow (2)$ is proved just as in Theorem 5.14.

$(2) \Rightarrow (3)$: Since $\phi_{s\cup t|u}$ is a continuation, we have

$$\phi^{\downarrow s\cup t\cup u} = \phi_{s\cup t|u} \otimes \phi^{\downarrow u}.$$

Therefore, using (2) we obtain that

$$\phi^{\downarrow s\cup t\cup u} \otimes \phi^{\downarrow u} = \phi_{s|u} \otimes \phi_{t|u} \otimes \phi^{\downarrow u} \otimes \phi^{\downarrow u}$$
$$= (\phi_{s|u} \otimes \phi^{\downarrow u}) \otimes (\phi_{t|u} \otimes \phi^{\downarrow u})$$
$$= \phi^{\downarrow s\cup u} \otimes \phi^{\downarrow t\cup u}.$$

$(3) \Rightarrow (4)$ is proved just like $(4) \Rightarrow (5)$ in Theorem 5.14.

$(4) \Rightarrow (5)$: Since $\phi_{s|t\cup u}$ is a continuation, we have

$$\phi^{\downarrow s\cup t\cup u} = \phi_{s|t\cup u} \otimes \phi^{\downarrow t\cup u}.$$

Therefore, by (4), we have the equation

$$\phi_{s|t\cup u} \otimes \phi^{\downarrow t\cup u} = \phi_{s|u} \otimes \phi^{\downarrow t\cup u}$$

If we multiply both sides of this equation by $(\phi^{\downarrow t\cup u})^{-1}$ we obtain (5).

$(5) \Rightarrow (1)$: Using (5) we obtain

$$\phi^{\downarrow s\cup t\cup u} = \phi_{s|t\cup u} \otimes \phi^{\downarrow t\cup u} = \phi_{s|u} \otimes f_{\gamma(\phi^{\downarrow t\cup u})} \otimes \phi^{\downarrow t\cup u}$$
$$= \phi_{s|u} \otimes \phi^{\downarrow t\cup u} = \phi_{s|u} \otimes \phi_{t|u} \otimes \phi^{\downarrow u}.$$

$\square$

Theorem 5.15 concerning the marginal problem for regular valuation algebras carries over to separative valuation algebras and weak conditional independence.

Theorem 5.23 *Let* (Φ, D) *be a separative valuation algebra, $s, t, u \in D$ disjoint sets and $\psi_1, \psi_2 \in \Phi$ such that $d(\psi_1) = s \cup u$, $d(\psi_2) = t \cup u$ and*

$$\psi_1^{\downarrow u} = \psi_2^{\downarrow u}.$$

Then there is a $\phi \in \Phi$ such that $s \amalg_\phi t|u$ and

$$\phi^{\downarrow s\cup u} = \psi_1, \qquad \phi^{\downarrow t\cup u} = \psi_2. \qquad (5.12)$$

This theorem is proved just as Theorem 5.15.

According to (Studeny, 1993), in the case of *set potentials*, there might be no set potential ϕ solving the marginal problem for strong conditional independence. This implies that $\perp_\phi$ and $\amalg_\phi$ are not always equivalent in the separative valuation algebra of set potentials. This is an explanation, why conditionals are of little interest in relation to set potentials (or Dempster-Shafer theory of belief functions): factorization with conditionals do not allow in general to impose (strong) conditional independence.

Even so $\amalg_\phi$ is possibly not equivalent to $\perp_\phi$, the relation still forms a semigraphoid, as the next theorem shows.

Theorem 5.24 *Assume* (Φ, D) *a separative valuation algebra. Let* $\phi \in \Phi$ *and* $s, t, u, v \subseteq d(\phi)$ *disjoint sets of variables. Then*

(G1) Symmetry: $s \amalg_\phi t|u$ *implies* $t \amalg_\phi s|u$.

(G2) Decomposition: $s \amalg_\phi t \cup v|u$ *implies* $s \amalg_\phi t|u$.

(G3) Weak Union: $s \amalg_\phi t \cup v|u$ *implies* $s \amalg_\phi t|u \cup v$.

(G4) Contraction: $s \amalg_\phi t|u$ *and* $s \amalg_\phi v|t \cup u$ *imply* $s \amalg_\phi t \cup v|u$.

Proof. (G1) follows directly from the definition of $s \amalg_\phi t|u$.

(G2) $s \amalg_\phi t \cup v|u$ means, according to Theorem 5.22 (3) that

$$\phi^{\downarrow s \cup t \cup u \cup v} \otimes \phi^{\downarrow u} \; = \; \phi^{\downarrow s \cup u} \otimes \phi^{\downarrow t \cup u \cup v}.$$

If we apply twice Lemma 2.1 (3), then we obtain

$$
\begin{aligned}
\phi^{\downarrow s \cup t \cup u} \otimes \phi^{\downarrow u} \; &= \; (\phi^{\downarrow s \cup t \cup u \cup v} \otimes \phi^{\downarrow u})^{\downarrow s \cup t \cup u} \\
&= \; (\phi^{\downarrow s \cup u} \otimes \phi^{\downarrow t \cup u \cup v})^{\downarrow s \cup t \cup u} \\
&= \; \phi^{\downarrow s \cup u} \otimes \phi^{\downarrow t \cup u}.
\end{aligned}
$$

By Theorem 5.22 (3) this implies that $s \amalg_\phi t|u$.

(G3) is proved like in Theorem 5.16.

(G4) $s \amalg_\phi v|t \cup u$ means that

$$\phi^{\downarrow s \cup t \cup u \cup v} \; = \; \phi_{s|t \cup u} \otimes \phi_{v|t \cup u} \otimes \phi^{\downarrow t \cup u}.$$

Using the fact, that conditionals are continuations, we obtain

$$\phi_{v|t \cup u} \otimes \phi^{\downarrow t \cup u} \; = \; \phi^{\downarrow t \cup u \cup v} \; = \; \phi_{t \cup v|u} \otimes \phi^{\downarrow u}.$$

Further, $s \amalg_\phi t|u$ implies by Theorem 5.22 (5) that

$$\phi_{s|t \cup u} \; = \; \phi_{s|u} \otimes f_{\gamma(\phi^{\downarrow t \cup u})}.$$

From this we find

$$
\begin{aligned}
\phi^{\downarrow s \cup t \cup u \cup v} \; &= \; \phi_{s|u} \otimes f_{\gamma(\phi^{\downarrow t \cup u})} \otimes \phi_{v|t \cup u} \otimes \phi^{\downarrow t \cup u} \\
&= \; \phi_{s|u} \otimes \phi_{v|t \cup u} \otimes \phi^{\downarrow t \cup u} \\
&= \; \phi_{s|u} \otimes \phi_{t \cup v|u} \otimes \phi^{\downarrow u}.
\end{aligned}
$$

This holds since $\gamma(\phi_{v|t\cup u}) \geq \gamma(\phi^{\downarrow t\cup u})$. But the last equation means that $s \amalg_\phi t \cup v|u$. $\qquad\qquad\square$

Assume that the valuation algebra (Φ^*, D) has neutral elements e_x for all domains $x \in D$ (see Section 3.6). Elements ϕ of Φ with support $\gamma(\phi) = \gamma(e_x)$ for some domain x are called *positive* as in regular valuation algebras. If marginals of positive elements are still positive, we say that the separative valuation algebra (Φ, D) has *positive elements*. Of course, this notion of positiveness generalizes the corresponding notion introduced for regular valuation algebras. It is clear that densities have positive elements: the minimal supports correspond to supports which equal the domains $\Omega_x = \mathbf{R}^{|x|}$. A density f is positive, if $f(\mathbf{x}) > 0$ for (almost) all $\mathbf{x}$. The neutral elements are functions $e_x(\mathbf{x}) = 1$ for all $\mathbf{x} \in \Omega_x = \mathbf{R}^{|x|}$. Note that these neutral elements are not densities, hence not in Φ. In the case of Gaussian densities, all elements are positive, $\Phi_p = \Phi$, since each sub-semigroup Φ_x is itself cancellative. The neutral elements are the same as for densities. This is no surprise since Gaussian potentials form a subalgebra of densities.

In the case of set potentials we have neutral elements e_x in the valuation algebra itself. The positive elements are commonality functions q with $q(A) > 0$ for all subsets A. In particular, we have then $q(\Omega_x) = m(\Omega_x) > 0$.

It turns out that for positive valuations the intersection property of weak conditional independence is satisfied.

Theorem 5.25 *Assume that (Φ, D) is a separative valuation algebra with positive elements. Let $\phi \in \Phi_p$ be a positive valuation and $s, t, u, v \subseteq d(\phi)$ disjoint sets of variables. Then*

(G5) Intersection: $s \amalg_\phi t|u \cup v$ and $s \amalg_\phi v|t \cup u$ imply $s \amalg_\phi t \cup u|v$.

Proof. From $s\amalg_\phi t|u\cup v$ we conclude, using Theorem 5.22 (5) and the neutrality axiom, that

$$\phi_{s|t\cup u\cup v} = \phi_{s|u\cup v} \otimes e_{t\cup u\cup v} = \phi_{s|u\cup v} \otimes e_t \otimes e_{u\cup v} = \phi_{s|u\cup v} \otimes e_t$$

$s \amalg_\phi v|t \cup u$ similarly implies that

$$\phi_{s|t\cup u\cup v} = \phi_{s|t\cup u} \otimes e_v.$$

We obtain thus the equation

$$\phi_{s|u\cup v} \otimes e_t = \phi_{s|t\cup u} \otimes e_v.$$

If we multiply both sides by $\phi^{\downarrow t\cup u} \otimes \phi^{\downarrow u\cup v}$ and note that conditionals are continuations, then we find that

$$\phi^{\downarrow s\cup u\cup v} \otimes \phi^{\downarrow t\cup u} = \phi^{\downarrow s\cup t\cup u} \otimes \phi^{\downarrow u\cup v}.$$

Now, we marginalize both sides to $s \cup t \cup u$ and apply Lemma 2.1 (3). This gives

$$\phi^{\downarrow s\cup u} \otimes \phi^{\downarrow t\cup u} = \phi^{\downarrow s\cup t\cup u} \otimes \phi^{\downarrow u}.$$

According to Theorem 5.22 (3) this means that $s \amalg_\phi t|u$. From, this we obtain by Lemma 5.22 (5) that

$$\phi_{s|t \cup u} \quad = \quad \phi_{s|u} \otimes e_{t \cup u} \quad = \quad \phi_{s|u} \otimes e_t.$$

This gives us

$$\phi_{s|t \cup u \cup v} \quad = \quad \phi_{s|u} \otimes e_t \otimes e_v \quad = \quad \phi_{s|u} \otimes e_{t \cup v}.$$

And this allows us finally to conclude that

$$\phi^{\downarrow s \cup t \cup u \cup v} \quad = \quad \phi_{s|t \cup u \cup v} \otimes \phi^{\downarrow t \cup u \cup v} \quad = \quad \phi_{s|u} \otimes \phi^{\downarrow t \cup u \cup v}. \tag{5.13}$$

Therefore, by Theorem 5.22 (4) we see that indeed $s \amalg_\phi t \cup v|u$. $\square$

This concludes our discussion of conditional independence in valuation algebras. We have shown that many properties of conditional independence of probability extends to the abstract framework of valuation algebras. This has some importance for local computation schemes based on conditional independence and related graphical representations, which extend in the same way from probability theory to valuation algebras and thus to many frameworks different from discrete probability theory.

6 Information Algebras

6.1 Idempotency

On several occasions we encountered idempotent valuation algebras. In the labeled version this means that

$$\phi \otimes \phi^{\downarrow t} = \phi \qquad (6.1)$$

whenever $t \subseteq d(\phi)$. This a very appealing property. In fact, if a valuation ϕ represents some piece of information about a domain $d(\phi) = s$, then its marginal $\phi^{\downarrow t}$ represents a part of ϕ. The idempotency says that combining a piece of information with a part of it, gives nothing new! This really is a property which is characteristic of information. We may repeat a piece of information as often as we want, we never get something new. That is why we call valuation algebras which possess the idempotency property an *information algebra*. The justification of this name will be later reinforced by other considerations (see Sections 6.2, 6.3 and 6.4 below).

Idempotency has very strong and profound consequences as we shall see in this chapter. The typical example of an idempotent valuation algebra is a relational database. It turns out that this example can be generalized and then becomes a universal representation of information algebras in the sense that any information algebra can be embedded into a generalized relational database or, in other words, into an algebra of subsets of some appropriate reference set (see Section 6.3). Thus, an "information" can always be represented by a set (file) of tuples, that is a relation in the generalized sense. In another direction, we shall see that information algebras arise from systems composed of a "language" and an entailment relation or a consequence operator (Section 6.4). This is another possible representation of information algebras, which places these algebras into the realm of logic. In this respect, information algebras become related to Scott's information systems and domain theory, an important subject of theoretical computer science which is concerned with the subject of "information". Computation in information algebras can then, based on this relation, be transformed into computation in information systems or deduction in logic. Domain theory is also concerned with approximation, and it is therefore no surprise that approximation of "infinite" information by "finite" information can also be discussed in the context of information algebras (see Section 6.6).

Information algebras and information systems provide finally also a very natural framework to model uncertainty. This is done in an abstract version of "assumption-based" modeling and reasoning. Thereby logic and probability are combined in a new way. It turns out that this leads to a general theory of Dempster-Shafer belief functions. In other words, information algebras are also the natural framework for this theory of uncertainty. This is discussed in Chapter 7.

It is convenient to place information algebras into a slightly more general context than that of valuations referring to sets of variables. In fact, by saying that a valuation ϕ refers to some subset s of variables, we mean that ϕ tells us something about the possible configuration in Ω_s. These frames are ordered by a partial order such that $\Omega_t \leq \Omega_s$ if $t \subseteq s$. With respect to this partial order, two frames Ω_s and Ω_t have both an *infimum*, that is a largest frame, smaller than both frames, written as $\Omega_t \wedge \Omega_s = \Omega_{s \cap t}$ and a *supremum*, that is a smallest frame larger than both frames, written as $\Omega_t \vee \Omega_s = \Omega_{s \cup t}$. Infimum and supremum are also called *meet* and *join*. A partially ordered family such that meet and join exist for any pair of elements is called a *lattice*.

So the domains of valuations considered so far are elements of a particular lattice. It is now an easy, but useful generalization to assume that from now on, the domains of valuations are elements of an arbitrary lattice D provided with a partial order $\leq$, and the operations of meet ($\wedge$) and join($\vee$). Let then Φ be a set of valuations and D a lattice of domains. Suppose there are, as before, three operations defined:

1. *Labeling*: $\Phi \to D; \phi \mapsto d(\phi)$,

2. *Combination*: $\Phi \times \Phi \to \Phi; (\phi, \psi) \mapsto \phi \otimes \psi$,

3. *Marginalization*: $\Phi \times D \to \Phi; (\phi, x) \mapsto \phi^{\downarrow x}$, for $x \leq d(\phi)$.

We now impose the following axioms on Φ and D:

1. *Semigroup*: Φ is associative and commutative under combination. For all $s \in D$ there is an element e_s with $d(e_s) = s$ such that for all $\phi \in \Phi$ with $d(\phi) = s, e_s \otimes \phi = \phi \otimes e_s = \phi$.

2. *Labeling*: For $\phi, \psi \in \Phi$,

$$d(\phi \otimes \psi) \quad = \quad d(\phi) \vee d(\psi). \tag{6.2}$$

3. *Marginalization*: For $\phi \in \Phi, x \in D, x \leq d(\phi)$,

$$d(\phi^{\downarrow x}) \quad = \quad x. \tag{6.3}$$

4. *Transitivity*: For $\phi \in \Phi$ and $x \subseteq y \subseteq d(\phi)$,

$$(\phi^{\downarrow y})^{\downarrow x} = \phi^{\downarrow x}. \tag{6.4}$$

5. *Combination.* For $\phi, \psi \in \Phi$ with $d(\phi) = x, d(\psi) = y$,

$$(\phi \otimes \psi)^{\downarrow x} = \phi \otimes \psi^{\downarrow x \wedge y}. \tag{6.5}$$

6. *Stability.* For $x, y \in D, x \leq y$,

$$e_y^{\downarrow x} = e_x. \tag{6.6}$$

7. *Idempotency.* For $\phi \in \Phi$ and $x \in D, x \leq d(\phi)$,

$$\phi \otimes \phi^{\downarrow x} = \phi. \tag{6.7}$$

These are the usual axioms for a valuation algebra as given in Section 2.2, slightly reformulated for a general lattice of domains D. In addition we assume *stability* instead of only neutrality (which is implied by stability) and, most important *idempotency*. We call a valuation algebra (Φ, D) satisfying these axioms a *labeled information algebra*. With D a general lattice, instead of the lattice of subsets of a set of variables, most results derived so far remain true. An important exception is Lemma 2.1 (3), which holds only, if the lattice D is *modular*, that is, if for $x \geq z$ we have always that

$$x \wedge (y \vee z) = (x \wedge y) \vee z.$$

So, for example, many of the results on conditional independence and conditionals in Chapter 5 depend on Lemma 2.1 (3) and are thus valid only for modular lattices of domains. However, these results are of little interest for idempotent valuation algebras.

For an example of a labeled information algebra we refer to relations as explained in Subsection 2.3.2. There are many other examples, but most of them are related to information systems and will be introduced later.

A labeled information algebra may have *null elements* z_x for each domain x (see Section 3.4). Note that this element is always idempotent. We replace the nullity axiom of Section 3.4 in the context of information algebras by a simpler version:

(8) *Nullity*: For all $x \in D$ there is an element $z_x \in \Phi$ such that $z_x \otimes \phi = z_x$ for all $\phi \in \Phi$ with $d(\phi) = x$. If $x, y \in D$ and $x \leq y$ then

$$z_x \otimes e_y = z_y. \tag{6.8}$$

The stability and idempotency of an information algebra allows then to prove the following lemma:

Lemma 6.1 *Let (Φ, D) be an information algebra. Then*

1. For $x, y \in D$, and $x \leq y$ we have $z_y^{\downarrow x} = z_x$.

2. If $\phi \in \Phi$ such that $d(\phi) = y \geq x$ and $\phi^{\downarrow x} = z_x$, then $\phi = z_y$.

3. For $x, y \in D$ we have $z_x \otimes z_y = z_{x \vee y}$.

Proof. (1) By the nullity axiom (6.8), and the combination and stability axioms,

$$z_y^{\downarrow x} \;=\; (z_x \otimes e_y)^{\downarrow x} \;=\; z_x \otimes e_y^{\downarrow x} \;=\; z_x \otimes e_x \;=\; z_x.$$

(2) By the idempotency and nullity axioms,

$$\phi \;=\; \phi \otimes \phi^{\downarrow x} \;=\; \phi \otimes z_x \;=\; \phi \otimes e_y \otimes z_x \;=\; \phi \otimes z_y \;=\; z_y.$$

(3) The nullity axiom gives

$$z_x \otimes z_y \;=\; (z_x \otimes e_{x \vee y}) \otimes (z_y \otimes e_{x \vee y}) \;=\; z_{x \vee y} \otimes z_{x \vee y} \;=\; z_{x \vee y}.$$

$\square$

Since labeled information algebras are stable, they have an associated domain-free version. We recall from Section 3.1 that their elements are the equivalence classes $[\phi]_\sigma$ of the congruence σ defined by $\phi \equiv \psi \pmod{\sigma}$ if $\phi^{\uparrow x \cup y} = \psi^{\uparrow x \cup y}$ if $d(\phi) = x, d(\psi) = y$. We have then, if $d(\phi) = y$,

$$[\phi]_\sigma \otimes [\phi]_\sigma^{\Rightarrow x} \;=\; [\phi \otimes \phi^{\downarrow x \cap y} \otimes e_x]_\sigma \;=\; [\phi]_\sigma$$

because of the idempotency. This identity is the translation of the idempotency property from labeled information algebras to their domain-free counter part.

We introduce in correspondence with this result the following concept of a domain-free information algebra. Let Ψ be a set of elements, called domain-free valuations, and D a lattice of domains. Suppose two operations defined:

1. *Combination*: $\Psi \times \Psi \to \Psi; (\phi, \psi) \mapsto \phi \otimes \psi$,

2. *Focusing*: $\Psi \times D \to \Psi; (\psi, x) \mapsto \psi^{\Rightarrow x}$.

We impose the following axioms on Ψ and D:

1. *Semigroup*: Ψ is associative and commutative under combination. There is a neutral element $e \in \Psi$ such that $e \otimes \psi = \psi \otimes e = \psi$ for all $\psi \in \Psi$.

2. *Transitivity*: For $\psi \in \Psi$ and $x, y \in D$,

$$(\psi^{\Rightarrow x})^{\Rightarrow y} \;=\; \psi^{\Rightarrow x \wedge y}. \tag{6.9}$$

3. *Combination*: For $\psi, \phi \in \Psi$ and $x \in D$

$$(\psi^{\Rightarrow x} \otimes \phi)^{\Rightarrow x} \;=\; \psi^{\Rightarrow x} \otimes \phi^{\Rightarrow x}. \tag{6.10}$$

4. *Support*: For $\psi \in \Psi$, there is an $x \in D$ such that

$$\psi^{\Rightarrow x} = \psi. \tag{6.11}$$

5. *Idempotency*: For $\psi \in \Psi$ and $x \in D$,

$$\psi \otimes \psi^{\Rightarrow x} = \psi. \tag{6.12}$$

These are essentially the axioms of Section 3.2 for a domain-free valuation algebra, complemented with the *idempotency axiom*. We call this valuation algebra (Ψ, D) a *domain-free information algebra*. Note that in a general lattice D of domains there exists not necessarily a *top* element, a largest element. Therefore, the axiom of support has been changed slightly in comparison to the system of axioms of Section 3.2. This is not an essential change, especially since nearly all examples have a lattice of domains with a greatest element. Then the support axiom and Lemma 3.6 (6) imply that $\top$ is a support for all $\phi \in \Phi$. Note furthermore, that the axiom of neutrality is no more needed. It follows from idempotency,

$$e^{\Rightarrow x} = e^{\Rightarrow x} \otimes e = e. \tag{6.13}$$

A domain-free information algebra may possess a *null element* z, such that $\psi \otimes z = z$ for all $\psi \in \Psi$. Then we require the domain-free version of the nullity axiom:

(8) *Nullity*: There is an element $z \in \Phi$ such that $z \otimes \phi = z$ for all $\phi \in \Phi$. Further, for all $x \in D$, $z^{\Rightarrow x} = z$.

We note that $\phi^{\Rightarrow x} = z$ implies that $\phi = z$, since $\phi = \phi \otimes \phi^{\Rightarrow x} = \phi \otimes z = z$.

Standard example of a domain-free information algebra is the system of subsets of a frame Ω_r. The lattice of domains is here the lattice of frames Ω_s for $s \subseteq r$. This lattice has the top-element Ω_r (see also Section 3.2). The null element of the algebra is the empty set. If for each variable X the frame is the set of real numbers, $\Omega_X = \mathbf{R}$, then the information algebra of subsets of $\mathbf{R}^{|r|}$ has interesting subalgebras. One is the information algebra of *linear manifolds* in $\mathbf{R}^{|r|}$, another one is the algebra of *convex subsets*, or of *convex polyhedra*. Other examples of domain-free information algebras are known under the name of *cylindric algebras* (Henkin, Monk & Tarski, 1971). These algebras were introduced for the algebraic study of first order logic.

Of course we may derive from a domain-free information algebra its labeled version exactly the same way as in the more general case of valuation algebras (see Section 3.2).

6.2 Partial Order of Information

The idempotency of information algebras permits the ordering of information. Let (Φ, D) be a domain-free information algebra. We say that an information ϕ from Φ is *more informative*, and write $\phi \geq \psi$, than another information ψ

form Φ, if the latter, combined with first one, does not change the first one. Formally, $\phi \geq \psi$ if, and only if,

$$\phi \otimes \psi \;=\; \phi. \tag{6.14}$$

The idea is, that a less informative information does not add anything to a more informative one. This defines a *reflexive partial order* on Φ:

1. *Reflexivity:* $\phi \leq \phi$. This follows from idempotency $\phi \otimes \phi = \phi$.

2. *Antisymmetry:* $\phi \leq \psi$ and $\psi \leq \phi$ implies $\phi = \psi$. This follows from $\psi = \psi \otimes \phi = \phi$.

3. *Transitivity:* $\phi \leq \psi$ and $\psi \leq \eta$ imply $\phi \leq \eta$. This follows since $\psi = \psi \otimes \phi$ and $\eta = \eta \otimes \psi$ imply that $\eta \otimes \phi = (\eta \otimes \psi) \otimes \phi = \eta \otimes (\psi \otimes \phi) = \eta \otimes \psi = \eta$.

The following lemma contains a number of simple results concerning this ordering.

Lemma 6.2

1. $e \leq \phi \leq z$ *for all* $\phi \in \Phi$.

2. $\phi, \psi \leq \phi \otimes \psi$.

3. $\phi \otimes \psi = \sup\{\phi, \psi\}$.

4. $\phi^{\Rightarrow x} \leq \phi$.

5. $\phi \leq \psi$ *implies* $\phi^{\Rightarrow x} \leq \psi^{\Rightarrow x}$.

6. $\phi_1 \leq \phi_2$ *and* $\psi_1 \leq \psi_2$ *imply* $\phi_1 \otimes \psi_1 \leq \phi_2 \otimes \psi_2$.

7. $\phi^{\Rightarrow x} \otimes \psi^{\Rightarrow x} \leq (\phi \otimes \psi)^{\Rightarrow x}$.

8. $x \leq y$ *implies* $\phi^{\Rightarrow x} \leq \phi^{\Rightarrow y}$.

Proof. (1) Follows since $e \otimes \phi = \phi$ and $\phi \otimes z = z$.

(2) $\phi \otimes (\phi \otimes \psi) = (\phi \otimes \phi) \otimes \psi = \phi \otimes \psi$ shows that $\phi \leq \phi \otimes \psi$ and in the same way with ψ at the place of ϕ we obtain $\psi \leq \phi \otimes \psi$.

(3) According to (2) $\phi \otimes \psi$ is an upper bound of ϕ and ψ. If η is another upper bound of ϕ and ψ, that is, if $\phi \otimes \eta = \eta$ and $\psi \otimes \eta = \eta$, then $(\phi \otimes \psi) \otimes \eta = \phi \otimes (\psi \otimes \eta) = \phi \otimes \eta = \eta$, thus $\phi \otimes \psi \leq \eta$ and $\phi \otimes \psi$ is in fact the least upper bound of ϕ and ψ.

(4) This follows from idempotency $\phi^{\Rightarrow x} \otimes \phi = \phi$.

(5) Let $\phi \leq \psi$. Then we have (using idempotency) $\phi^{\Rightarrow x} \otimes \psi^{\Rightarrow x} = (\phi^{\Rightarrow x} \otimes \psi)^{\Rightarrow x} = (\phi^{\Rightarrow x} \otimes \phi \otimes \psi)^{\Rightarrow x} = (\phi \otimes \psi)^{\Rightarrow x} = \psi^{\Rightarrow x}$.

(6) We have $(\phi_1 \otimes \psi_1) \otimes (\phi_2 \otimes \psi_2) = (\phi_1 \otimes \phi_2) \otimes (\psi_1 \otimes \psi_2) = \phi_2 \otimes \psi_2$.

(7) We have by (4) $\phi^{\Rightarrow x} \leq \phi$. Then, by (6), $\phi^{\Rightarrow x} \otimes \psi \leq \phi \otimes \psi$ and thus $\phi^{\Rightarrow x} \otimes \psi^{\Rightarrow x} = (\phi^{\Rightarrow x} \otimes \psi)^{\Rightarrow x} \leq (\phi \otimes \psi)^{\Rightarrow x}$ by (5).

(8) $x \leq y$ implies $x = x \wedge y$ and therefore $\phi^{\Rightarrow x} \otimes \phi^{\Rightarrow y} = \phi^{\Rightarrow x \wedge y} \otimes \phi^{\Rightarrow y} = (\phi^{\Rightarrow y})^{\Rightarrow x} \otimes \phi^{\Rightarrow y} = \phi^{\Rightarrow y}$ because of the idempotency. □

Property (3) of this lemma means that any finite set of elements from Φ has a least upper bound in Φ, that is, Φ is a *semilattice*. We write sometimes also $\phi \otimes \psi = \phi \vee \psi$ to emphasize the fact that the combination of two pieces of information is their *join*.

Order may be introduced in a similar way into a *labeled* information algebra (Φ, D). As before we define $\phi \geq \psi$ if and only if (6.14) holds. As before it can easily be verified that this defines a partial order in Φ. In this case we have the following elementary results:

Lemma 6.3

1. *If $\phi \leq \psi$, then $d(\phi) \leq d(\psi)$.*

2. *If $x \leq y$ then $e_x \leq e_y$ and $z_x \leq z_y$.*

3. *If $x \leq d(\phi)$, then $e_x \leq \phi$ and if $d(\phi) \leq x$, then $\phi \leq z_x$.*

4. *$\phi, \psi \leq \phi \otimes \psi$.*

5. *$\phi \otimes \psi = \sup\{\phi, \psi\}$.*

6. *If $x \leq d(\phi)$, then $\phi^{\downarrow x} \leq \phi$.*

7. *If $d(\phi) \leq y$, then $\phi \leq \phi^{\uparrow y}$.*

8. *If $x \leq y = d(\phi)$, then $(\phi^{\downarrow x})^{\uparrow y} \leq \phi$.*

9. *$\phi_1 \leq \phi_2$ and $\psi_1 \leq \psi_2$ imply $\phi_1 \otimes \psi_1 \leq \phi_2 \otimes \psi_2$.*

10. *If $x \leq d(\phi) \wedge d(\psi)$, then $\phi^{\downarrow x} \otimes \psi^{\downarrow x} \leq (\phi \otimes \psi)^{\downarrow x}$.*

11. *If $x \leq d(\phi)$, then $\phi \leq \psi$ implies $\phi^{\downarrow x} \leq \psi^{\downarrow x}$.*

12. *If $d(\phi) \leq y$, then $\phi \leq \psi$ implies $\phi^{\uparrow y} \leq \psi^{\uparrow y}$.*

13. *If $d(\phi) \leq x \leq d(\psi)$, then $\phi \leq \psi$ implies $\phi \leq \psi^{\downarrow x}$.*

We leave the proof, which is similar to the one of Lemma 6.2, to the reader.

Information may be asserted, that is, a piece of information may be stated to hold, to be true. Suppose that a body of information represented by an element ϕ from a domain-free information algebra Φ is asserted to hold. Then, intuitively, all pieces of information ψ which are contained in ϕ, that is for which $\psi \leq \phi$ hold, should also be true. Moreover, if two pieces of information ϕ and ψ are asserted, then their combination $\phi \otimes \psi$ should also hold. So, an assertion of an information should be a set of elements from Φ, which satisfy these conditions. This leads to the following definition:

Definition 6.4 Ideals: *A non-empty set $I \subseteq \Phi$ is called an ideal of the domain-free information algebra (Φ, D), if the following holds:*

1. $\phi \in I$ *and* $\psi \in \Phi, \psi \leq \phi$ *imply* $\psi \in I$.

2. $\phi \in I$ *and* $\psi \in I$ *imply* $\phi \otimes \psi \in I$.

Note that this concept is a notion related to a semi-lattice rather than to an information algebra. However, it follows immediately, that the neutral element e is contained in every ideal, and that, since $\phi^{\Rightarrow x} \leq \phi$, $\phi \in I$ implies also $\phi^{\Rightarrow x} \in I$ for all domains x and every ideal I. This relates the notion of ideals more specifically to information algebras. Φ itself is an ideal. Any ideal I different from Φ is called a *proper* ideal. The set of all elements $\psi \in \Phi$ which are less informative than ϕ form also an ideal. It is called the *principal* ideal $I(\phi)$ of ϕ. An ideal, as stated above, represents an assertion of an information. The principal ideal $I(\phi)$ represents the simple information asserted by the body of information ϕ. The order-relation $\psi \leq \phi$ can, in this respect, be considered as a relation saying that ψ in implied or entailed by ϕ. This consideration will be pursued in Section 6.4.

The intersection of an arbitrary number of ideals is clearly still an ideal. Thus we may define the ideal generated by a subset X of Φ by

$$I(X) = \bigcap \{I : I \text{ ideal in } \Phi, X \subseteq I\}. \tag{6.15}$$

This represents the information generated, if the bodies of information ϕ in the set X are asserted. There is also an alternative representation of $I(X)$.

Lemma 6.5 *If $X \subseteq \Phi$, then*

$$\begin{aligned} I(X) = \{\phi \in \Phi : \phi \leq \phi_1 \otimes \cdots \otimes \phi_m \text{ for some} \\ \text{finite set of elements } \phi_1, \ldots, \phi_m \in X\}. \end{aligned} \tag{6.16}$$

Proof. If $\phi_1, \ldots, \phi_m \in X$, then $\phi_1, \ldots, \phi_m \in I(X)$, hence $\phi_1 \otimes \cdots \otimes \phi_m \in I(X)$, hence $\phi \leq \phi_1 \otimes \cdots \otimes \phi_m$ implies $\phi \in I(X)$. Conversely, the set on the right-hand side in the lemma is clearly an ideal which contains X. Thus this set contains $I(X)$, hence it equals it. $\square$

As a corollary we obtain that if the set X is *finite*, then

$$I(X) = I(\vee X), \tag{6.17}$$

The ideal generated from X is the principal ideal of the element $\vee X$ of Φ.

It is well known in lattice theory, that a system of sets which is closed under intersection, such as ideals, form a lattice (see for example (Davey & Priestley, 1990)), where the supremum of two ideals I_1 and I_2 is defined as

$$I_1 \vee I_2 = \bigcap \{I : I \text{ ideal in } \Phi, I_1 \cup I_2 \subseteq I\}.$$

It is then obvious that this is the natural definition of the *aggregation* or the *combination* of two assertions of information. So we equip the family of ideals of an information algebra Φ with the operation of combination

$$I_1 \otimes I_2 \;=\; I_1 \vee I_2. \qquad (6.18)$$

There is an alternative, equivalent definition of this operation between ideals.

Lemma 6.6

$$I_1 \otimes I_2 \;=\; \{\psi \in \Phi : \psi \leq \phi_1 \otimes \phi_2 \text{ for some } \phi_1 \in I_1, \phi_2 \in I_2\}. \qquad (6.19)$$

Proof. The right-hand side of equation (6.19) is clearly an ideal and it contains both I_1 and I_2. Hence it contains $I_1 \vee I_2$. Assume then ψ an element of the right-hand side of equation (6.19), that is, $\psi \leq \phi_1 \otimes \phi_2$ and $\phi_1 \in I_1$, $\phi_2 \in I_2$. But, if I is an ideal containing I_1 and I_2, we have $\phi_1 \in I$ and $\phi_2 \in I$, hence $\phi_1 \otimes \phi_2 \in I$ and therefore $\psi \in I$. This tells us that the right-hand side of equation (6.19) is contained in $I_1 \vee I_2$. $\qquad \square$

As a corollary note that $\phi_1 \in I_1$ and $\phi_2 \in I_2$ imply that $\phi_1 \otimes \phi_2 \in I_1 \otimes I_2$.

We may also equip the family of ideals of Φ with a *focusing* operation. If $x \in D$ define

$$I^{\Rightarrow x} \;=\; \{\psi \in \Phi : \psi \leq \phi^{\Rightarrow x} \text{ for some } \phi \in I\}. \qquad (6.20)$$

It is easy to verify that this is indeed an ideal. Of course, we have $\phi^{\Rightarrow x} \in I^{\Rightarrow x}$, if $\phi \in I$.

The family I_Φ of ideals, equipped with the operations of combination and focusing as defined above is nearly a domain-free information algebra. This statement is made more precise in the following theorem.

Theorem 6.7 *The algebraic structure (I_Φ, D), where the operations $\otimes$ and $\Rightarrow$ are defined by (6.18) or (6.19) and (6.20), satisfies all axioms of a domain-free information algebra, except the support axiom. If the lattice D has a top element, then (I_Φ, D) is an information algebra.*

Proof. (1) The semi-lattice axioms (idempotent semi-group) are satisfied since I_Φ is a lattice, as remarked above. The neutral element of combination is $\{e\} = I(e)$.

(2) Here is the verification of transitivity:

$$\begin{aligned}
(I^{\Rightarrow x})^{\Rightarrow y} &= \{\psi : \psi \leq \phi'^{\Rightarrow y}, \phi' \in I^{\Rightarrow x}\} \\
&= \{\psi : \psi \leq (\phi^{\Rightarrow x})^{\Rightarrow y} = \phi^{\Rightarrow x \cap y}, \phi \in I\} \\
&= I^{\Rightarrow x \cap y}.
\end{aligned}$$

(3) To verify the combination axiom, we use equation (6.19):

$$(I_1^{\Rightarrow x} \otimes I_2)^{\Rightarrow x} \;=\; \{\psi : \psi \leq \phi^{\Rightarrow x}, \phi \in I_1^{\Rightarrow x} \otimes I_2\}$$

$$
\begin{aligned}
&= \{\psi : \psi \le (\phi_1' \otimes \phi_2)^{\Rightarrow x}, \phi_1' \in I_1^{\Rightarrow x}, \phi_2 \in I_2\} \\
&= \{\psi : \psi \le (\phi_1^{\Rightarrow x} \otimes \phi_2)^{\Rightarrow x} = \phi_1^{\Rightarrow x} \otimes \phi_2^{\Rightarrow x}, \phi_1 \in I_1, \phi_2 \in I_2\} \\
&= \{\psi : \psi \le \phi_1' \otimes \phi_2', \phi_1' \in I_1^{\Rightarrow x}, \phi_2' \in I_2^{\Rightarrow x}\} \\
&= I_1^{\Rightarrow x} \otimes I_2^{\Rightarrow x}.
\end{aligned}
$$

(4) Idempotency can be verified as follows:

$$
\begin{aligned}
I \otimes I^{\Rightarrow x} &= \{\psi : \psi \le \phi \otimes \phi', \phi \in I, \phi' \in I^{\Rightarrow x}\} \\
&= \{\psi : \psi \le \phi \otimes \phi'^{\Rightarrow x}, \phi, \phi' \in I, \}.
\end{aligned}
$$

This shows that $I \otimes I^{\Rightarrow x} \subseteq I$. The converse inclusion is evident.

(5) If D has a top element $\top$, then this is a support for all elements of Φ. Hence we have for every ideal

$$
I^{\Rightarrow \top} = \{\psi : \psi \le \phi^{\Rightarrow \top} = \phi, \phi \in I\} = I.
$$

$\square$

The structure (I_Φ, D) with the operations $\otimes$ and $\Rightarrow$ as defined above, is called the *completion* of Φ. This terminology is justified by the following theorem:

Theorem 6.8 *The domain-free information algebra* (Φ, D) *can be embedded into the structure* (I_Φ, D).

Proof. We show that the mapping $I : \Phi \to I_\Phi$ defined by $I(\phi)$ is an embedding. That is we show that

1. $I(\phi \otimes \psi) = I(\phi) \otimes I(\psi)$.

2. $I(\phi^{\Rightarrow x}) = I^{\Rightarrow x}(\phi)$.

3. $I(e) = \{e\}$.

4. $I(\phi) = I(\psi)$ implies $\phi = \psi$.

(1) is a direct consequence of equation (6.19) applied to the principal ideals $I(\phi)$ and $I(\psi)$.

(2) We have that $\psi \in I^{\Rightarrow x}(\phi)$ if, and only if, $\psi \le \phi^{\Rightarrow x}$ which holds if, and only if, $\psi \in I(\phi^{\Rightarrow x})$.

(3) holds since there are no elements different from e, for which $\phi \le e$ holds.

(4) $\psi \in I(\phi)$ means that $\psi \le \phi$ and $\phi \in I(\psi)$ means that $\phi \le \psi$, thus $\phi = \psi$. $\square$

We shall reconsider the structure (I_Φ, D) in Section 6.4 in a different context.

6.3   File Systems

A relational database is a typical example of a labeled information algebra.
This is essentially the same as a system of subsets. Here we shall show that this
example can be generalized and, what is more, that any information algebra
can be embedded in such a subset system. So this will be in some sense a
universal system. An alternative universal system will be introduced in the
next section.

In the view of generalized relational systems, information is expressed by
listing atomic propositions (tuples) in files (relations). We start by defining a
general notion of tuples and show then that sets of tuples (files, relations) form
a labeled information algebra. Combination corresponds to join, marginalization to projection. We show how bodies of information of an abstract information algebra can be understood as relations over an abstract tuple system.
Throughout this section we assume that D is a lattice and we work with labeled
information systems.

Definition 6.9 Tuple System. *A tuple system over D is a set T together
with two operations $d : T \to D$ and $\cdot[\cdot] : T \times D \to T$, defined for $x \le d(f)$,
which satisfies the following axioms for $f, g \in T$ and $x, y \in D$:*

1. *If $x \le d(f)$, then $d(f[x]) = x$.*

2. *If $x \le y \le d(f)$, then $f[y][x] = f[x]$.*

3. *If $d(f) = x$, then $f[x] = f$.*

4. *If $d(f) = x$, $d(g) = y$ and $f[x \wedge y] = g[x \wedge y]$, then there exists a $h \in T$
   such that $d(h) = x \vee y$ and $h[x] = f$, $h[y] = g$.*

5. *If $d(f) = x$ and $x \le y$, then there exists a $g \in T$ such that $d(g) = y$ and
   $g[x] = f$.*

The elements of T are called *tuples* and $f[x]$ is called the *projection* of f
to x and $d(f)$ is the *domain* of f. Obviously, the mathematical tuples from
the examples of relations in Section 2.3.2 form a tuple system. There are,
however, other examples where the elements of T are not tuples in the usual
sense.

Let (T, D) be a tuple system with the operations d and $\cdot[\cdot]$. Generalizing
relational databases we define a *relation* over x to be a subset R of T such
that $d(f) = x$ for all $f \in R$. The *domain* of f is supposed to be attached to
R. It is denoted by $d(R)$. For a relation R and $x \le d(R)$, the *projection* of R
onto x is defined as follows:

$$\pi_x(R) \;=\; \{f[x] : f \in R\}. \tag{6.21}$$

The *join* of a relation R over x and a relation S over y is defined as follows:

$$R \bowtie S \;=\; \{f \in T : d(f) = x \vee y, f[x] \in R, f[y] \in S\}. \tag{6.22}$$

It is easily possible, that the set on the right-hand side above is empty. We attach then this empty set with the domain $x \vee y$, in correspondence with the philosophy of labeled relations, and call it $Z_{x \vee y}$, the *empty* relation on $x \vee y$. We assign it the domain $d(Z_{x \vee y}) = x \vee y$.

Finally, for $x \in D$, the full relation over x is

$$E_x \;=\; \{f \in T : d(f) = x\}. \tag{6.23}$$

This is the neutral element for the join operation, $E_x \bowtie R = R$, if $d(R) = x$. Let $\mathcal{R}_T$ be the set of all relations. Then these relations form a labeled information algebra.

Theorem 6.10 $(\mathcal{R}_T, D)$ *is a labeled information algebra with the join $\bowtie$ as combination and the projection π as marginalization.*

We leave the easy proof to the reader. Note also that the empty relations Z_x satisfy the nullity axiom. So, the information algebra of relations over a generalized tuple systems is an algebra with *null elements*.

Elements of a labeled information algebra can be understood as tuples as the following lemma shows.

Lemma 6.11 *Let (Φ, D) be a labeled information algebra. Then (Φ, D) is a tuple system with domain d and projection $\downarrow$.*

Proof. We have to show that the five axioms of a tuple system are satisfied.
  (1) if $x \leq d(\phi)$, then $d(\phi^{\downarrow x}) = x$ (marginalization axiom).
  (2) If $x \leq y \leq d(\phi)$, then $(\phi^{\downarrow y})^{\downarrow x} = \phi^{\downarrow x}$ (transitivity axiom).
  (3) If $d(\phi) = x$, then $\phi^{\downarrow x} = \phi$. This is a consequence of stability.
  (4) Assume that $d(\phi) = x$, $d(\psi) = y$ and $\phi^{\downarrow x \wedge y} = \psi^{\downarrow x \wedge y}$. Then $d(\phi \otimes \psi) = x \vee y$ and

$$(\phi \otimes \psi)^{\downarrow x} \;=\; \phi \otimes \psi^{\downarrow x \wedge y} \;=\; \phi \otimes \phi^{\downarrow x \wedge y} \;=\; \phi,$$
$$(\phi \otimes \psi)^{\downarrow y} \;=\; \phi^{\downarrow x \wedge y} \otimes \psi \;=\; \psi^{\downarrow x \wedge y} \otimes \psi \;=\; \psi.$$

  (5) Assume that $d(\phi) = x$ and $x \leq y$. Then $d(\phi^{\uparrow y}) = y$ and $(\phi^{\uparrow y})^{\downarrow x} = (\phi \otimes e_y)^{\downarrow x} = \phi \otimes e_y^{\downarrow x} = \phi \otimes e_x = \phi$ by the stability axiom. $\qquad \square$

Now, an element ϕ of a labeled information algebra Φ can be identified with the set of all "tuples" $\psi \in \Phi$ such that $\phi \leq \psi$ and $d(\psi) = d(\phi)$. In this way we obtain an isomorphism of the information algebra (Φ, D) onto a subalgebra of the relational information algebra $\mathcal{R}_\Phi$ generated by the tuple system (Φ, D).

Theorem 6.12 *Let (Φ, D) be a labeled information algebra. Then (Φ, D) can be embedded into the relational information algebra $\mathcal{R}_\Phi$ generated by the tuple system (Φ, D).*

Proof. We define $F(\phi) = \{\psi \in \Phi : d(\psi) = d(\phi), \phi \leq \psi\}$. Then we show that the mapping F has the following properties:

1. $d(F(\phi)) = d(\phi)$.

2. $F(\phi \otimes \psi) = F(\phi) \bowtie F(\psi)$.

3. If $x \leq d(\phi)$, then $F(\phi^{\downarrow x}) = \pi_x(F(\phi))$.

4. $F(e_x) = E_x$.

5. $F(\phi) = F(\psi)$ implies $\phi = \psi$.

The first four properties show that F is a *homomorphism* and the fifth property says that this homomorphism is an *embedding*.

We prove now the properties above.

(1) Let $d(\phi) = x$. Then $F(\phi)$ is a relation over x. Thus, $d(F(\phi)) = x$.

(2) Let $d(\phi) = x$ and $d(\psi) = y$. Assume that $\eta \in F(\phi \otimes \psi)$. Then $\phi \otimes \psi \leq \eta$ and $d(\eta) = x \vee y$. Since $\phi \leq \phi \otimes \psi$ and $\psi \leq \phi \otimes \psi$, we obtain that $\phi \leq \eta$ and $\psi \leq \eta$. Hence, $\phi = \phi^{\downarrow x} \leq \eta^{\downarrow x}$ and $\psi = \psi^{\downarrow y} \leq \eta^{\downarrow y}$ (Lemma 6.3 (11)). Therefore, $\eta^{\downarrow x} \in F(\phi)$ and $\eta^{\downarrow y} \in F(\psi)$. By the definition of the join this means that $\eta \in F(\phi) \bowtie F(\psi)$. So we have shown that $F(\phi \otimes \psi) \subseteq F(\phi) \bowtie F(\psi)$. For the converse inclusion assume that $\eta \in F(\phi) \bowtie F(\psi)$. This means that $d(\eta) = x \vee y$, $\eta^{\downarrow x} \in F(\phi)$ and $\eta^{\downarrow y} \in F(\psi)$. We obtain that $\phi \leq \eta^{\downarrow x} \leq \eta$ and $\psi \leq \eta^{\downarrow y} \leq \eta$ (Lemma 6.3 (6)). Hence $\phi \otimes \psi \leq \eta$ and $\eta \in F(\phi \otimes \psi)$. Therefore, $F(\phi \otimes \psi) \supseteq F(\phi) \bowtie F(\psi)$.

(3) Let $x \leq d(\phi) = y$. Assume that $\eta \in F(\phi^{\downarrow x})$. By definition, this means that $d(\eta) = x$ and $\phi^{\downarrow x} \leq \eta$. We obtain that $\phi = \phi^{\downarrow x} \otimes \phi \leq \eta \otimes \phi$ (Lemma 6.3 (9)). Since $d(\eta \otimes \phi) = x \vee y = y$, the tuple $\eta \otimes \phi$ is in $F(\phi)$. By the definition of the projection, $(\eta \otimes \phi)^{\downarrow x}$ is in $\pi_x(F(\phi))$. Since $(\eta \otimes \phi)^{\downarrow x} = \eta \otimes \phi^{\downarrow x} = \eta$, we obtain that $\eta \in \pi_x(F(\phi))$. So we have shown that $F(\phi^{\downarrow x})$ is contained in $\pi_x(F(\phi))$. For the converse inclusion assume that $\eta \in \pi_x(F(\phi))$. This means that there exists a ν in $F(\phi)$ such that $\eta = \nu^{\downarrow x}$. Since $\nu \in F(\phi)$, we obtain that $\phi \leq \nu$ and $d(\nu) = y$. Thus, $\phi^{\downarrow x} \leq \nu^{\downarrow x}$ (Lemma 6.3 (11)) and we obtain that $\eta \in F(\phi^{\downarrow x})$.

(4) We have $F(e_x) = \{\psi \in \Phi : d(\psi) = x\} = E_x$.

(5) Assume that $F(\phi) = F(\psi)$. Since $\phi \in F(\phi)$, we obtain that $\phi \in F(\psi)$ and thus $\psi \leq \phi$. Since $\psi \in F(\psi)$, we obtain that $\psi \in F(\phi)$, and $\phi \leq \psi$. Hence $\phi = \psi$. $\qquad\square$

In an information $\mathcal{R}_T$ algebra over a generalized tuple system T, the individual tuples $f \in T$, or more strictly speaking, the one-tuple relations $\{f\}$, represent the most informative information on the domain $d(f)$. In fact, any relation R which contains a tuple f satisfies $R \leq \{f\}$, since $R \bowtie \{f\} = \{f\}$. No other information can be contained in $\{f\}$. Such information is rightly called *atomic* for the domain. This notion can be defined relative to any labeled information algebra with null elements:

Definition 6.13 Atoms: *An element $\alpha \in \Phi$ with $d(\alpha) = x$ is called an* atom *on x, if*

1. $\alpha \neq z_x$.

2. *For all $\phi \in \Phi$, $d(\phi) = x$ and $\alpha \leq \phi$ implies either $\alpha = \phi$ or $\phi = z_x$.*

This says that no information in a domain, except the null information, can be more informative than an atom; no information, except the null information, can be contained in an atom. Clearly, the one-tuple relations are exactly the atoms in the relational algebra $\mathcal{R}_T$. By abuse of language, we shall simply say that tuples are atoms in the relational algebras. On the other hand, an information algebra may well have no atoms.

Here are some elementary results about atoms:

Lemma 6.14   1. *If α is an atom on x, $d(\phi) \leq x$, then $\alpha \otimes \phi = \alpha$ or z_x.*

2. *If α is an atom on x, $y \leq x$, then $\alpha^{\downarrow y}$ is an atom on y.*

3. *If α is an atom on x and $d(\phi) = x$, then either $\phi \leq \alpha$ or $\alpha \otimes \phi = z_x$.*

4. *If α, β are atoms on x, then either $\alpha = \beta$ or $\alpha \otimes \beta = z_x$.*

Proof. (1) Let $\alpha \otimes \phi = \alpha \otimes \phi^{\uparrow x} = \eta$ such that $\alpha \otimes (\phi^{\uparrow x} \otimes \eta) = \phi^{\uparrow x} \otimes \eta$. This means that $\alpha \leq \phi^{\uparrow x} \otimes \eta$. Since α is an atom we have either

$$\alpha \;=\; \phi^{\uparrow x} \otimes \eta \quad \text{or} \quad \phi^{\uparrow x} \otimes \eta \;=\; z_x.$$

In the first case we obtain that

$$\alpha \otimes \phi \;=\; \alpha \otimes \phi^{\uparrow x} \otimes \eta \;=\; \phi^{\uparrow x} \otimes \eta \;=\; \alpha.$$

And in the second case we obtain that

$$\alpha \otimes \phi \;=\; \alpha \otimes \phi^{\uparrow x} \otimes \eta \;=\; \phi^{\uparrow x} \otimes \eta \;=\; z_x.$$

(2) Since $\alpha \neq z_x$ we have also that $\alpha^{\downarrow y} \neq z_y$ (Lemma 6.1 (2)). Assume that $\alpha^{\downarrow y} \leq \phi$ and $d(\phi) = y$. We have

$$(\alpha \otimes \phi)^{\downarrow y} \;=\; \alpha^{\downarrow y} \otimes \phi \;=\; \phi.$$

From (1) of this lemma it follows that either

$$\alpha \otimes \phi \;=\; \alpha \quad \text{or} \quad \alpha \otimes \phi \;=\; z_x.$$

In the first case we obtain

$$\alpha^{\downarrow y} \;=\; (\alpha \otimes \phi)^{\downarrow y} \;=\; \phi$$

and in the second case

$$\phi \;=\; (\alpha \otimes \phi)^{\downarrow y} \;=\; z_x^{\downarrow y} \;=\; z_y.$$

(see Lemma 6.1 (1)). This proves that $\alpha^{\downarrow y}$ is an atom on y.

(3) is an immediate consequence of (1) of this lemma.

(4) Since α is an atom it follows from (1) of this lemma that either $\alpha \otimes \beta = \alpha$ or $\alpha \otimes \beta = z_x$. In the first case we have that $\beta \leq \alpha$, hence $\alpha = \beta$, since β is also an atom. □

Denote by $At_x(\Phi)$ the set of all atoms on the domain x in the information algebra Φ and denote by $At(\Phi)$ the set of all atoms in Φ. Furthermore define

$$At(\phi) \; = \; \{\alpha \in At(\Phi) : d(\alpha) = d(\phi), \phi \leq \alpha\}. \tag{6.24}$$

If $\alpha \in At(\phi)$ we say also that α is an atom in ϕ or contained in ϕ. Clearly, in the relational algebra over a tuple system, a relation R over domain x contains all the tuples $f \in R$. Strictly speaking this means that $At(R) = \{\{f\} : f \in R\}$.

In the example of the relational algebra $\mathcal{R}_T$, any element contains at least one atom, except the empty relations Z_x. More than that, any relation R which is not empty, is a lower bound of all the tuples it contains. More precisely, $R \leq \{f\}$ for all $f \in R$. But if a relation S on the same domain is another lower bound of $At(R)$, then this means that $S \leq \{f\}$, that is, $f \in S$, for all $f \in R$, hence $R \subseteq S$. So we have for any other lower bound $S \leq R$. Hence, R is the largest lower bound, the *infimum* of $At(R)$, $R = \wedge At(R)$. Moreover, any set of tuples R has an infimum in $\mathcal{R}_T$, namely

$$\wedge \{\{f\} : f \in R\} \; = \; R.$$

Any element of the algebra $\mathcal{R}_T$ is determined by the atoms it contains, and any set of atoms defines an element of $\mathcal{R}_T$. These are strong properties which show that the relational algebra $\mathcal{R}_T$ really is essentially an algebra of sets.

This motivates the following definitions:

Definition 6.15 Atomic Algebras.

1. *A labeled information algebra Φ with null elements is called* atomic, *if for all $\phi \in \Phi$, if ϕ is not a null element, the set of atoms $At(\phi)$ is not empty.*

2. *It is called* atomic composed, *if it is atomic and if for all $\phi \in \Phi$*

$$\phi \; = \; \wedge At(\phi). \tag{6.25}$$

3. *It is called* atomic closed, *if it is atomic composed and if for every subset $A \subseteq At_x(\Phi)$ the infimum exists and belongs to Φ.*

Thus, any relational information algebra on a tuple system is atomic closed and the tuples (or the one-tuple relations) are its atoms. Not surprisingly, there are close relations between relational information algebras and atomic information algebra and between tuples and atoms. We begin by showing that if an information algebra is atomic, then its atoms form a tuple system.

Lemma 6.16 *If the labeled information algebra (Φ, D) is atomic, then $(At(\Phi), D)$ is a tuple system with domain d and projection $\downarrow$.*

Proof. We have to prove the five axioms of a tuple system. The first two axioms correspond simply to the marginalization and transitivity axioms of the information algebra and the third one is a direct consequence of stability.

In order to prove axiom (4) assume $d(\alpha) = x$, $d(\beta) = y$ and $\alpha^{\downarrow x \wedge y} = \beta^{\downarrow x \wedge y}$ for two atoms α and β. Suppose first that $\alpha \otimes \beta = z_{x \vee y}$. Then,

$$\alpha = (\alpha \otimes \beta)^{\downarrow x} = z_{x \vee y}^{\downarrow x} = z_x. \qquad (6.26)$$

But $\alpha \neq z_x$ and therefore $\alpha \otimes \beta \neq z_{x \vee y}$.

Since Φ is atomic, the set $At(\alpha \otimes \beta)$ is not empty. Let $\gamma \in At(\alpha \otimes \beta)$. Then $d(\gamma) = x \vee y$ and $\alpha \otimes \beta \leq \gamma$. It follows, using Lemma 6.3 (11), that

$$\alpha = (\alpha \otimes \beta)^{\downarrow x} \leq \gamma^{\downarrow x}.$$

But α is an atom, hence $\gamma^{\downarrow x} = \alpha$ or $\gamma^{\downarrow x} = z_x$. The latter case is excluded, since $\gamma \neq z_{x \vee y}$ (Lemma 6.1 (2)). So we have $\gamma^{\downarrow x} = \alpha$. $\gamma^{\downarrow y} = \beta$ is obtained in the same way. So axiom (4) of a tuple system is satisfied.

Let α be an atom with $d(\alpha) = x \leq y$. Suppose that $\alpha \otimes e_y = \alpha^{\uparrow y} = z_y$. But then we obtain that $\alpha = (\alpha \otimes e_y)^{\downarrow x} = z_y^{\downarrow x} = z_x$. But this is not possible since α is an atom. Therefore $\alpha^{\uparrow y} \neq z_y$.

Because Φ is atomic, the set $At(\alpha^{\uparrow y})$ is not empty. Let $\beta \in At(\alpha^{\uparrow y})$. Then $d(\beta) = y$ and $\alpha^{\uparrow y} \leq \beta$. It follows that

$$\alpha = (\alpha^{\uparrow y})^{\downarrow x} \leq \beta^{\downarrow x}.$$

But α is an atom. Therefore we have either $\beta^{\downarrow x} = \alpha$ or $\beta^{\downarrow x} = z_x$. The latter case is excluded since $\beta \neq z_y$. This proves axiom (5) of tuple systems. □

The tuple system $(At(\Phi), D)$ is a subsystem of the tuple system $(\Phi, D, d, \downarrow)$. In other words, if Φ is atomic, the atoms in Φ form already a tuple system. We call the relational algebra associated with this tuple system $\mathcal{R}_{At(\Phi)}$. $At(\phi)$ belongs to $\mathcal{R}_{At(\Phi)}$ for every element ϕ in Φ. So, At can be seen as a mapping from Φ into $\mathcal{R}_{At(\Phi)}$.

Theorem 6.17 *If Φ is an atomic information algebra, then $At : \Phi \to \mathcal{R}_{At(\Phi)}$ is a* homomorphism.

Proof. We have to show the following:

1. $At(\phi \otimes \psi) = At(\phi) \bowtie At(\psi)$.

2. $At(\phi^{\downarrow x}) = \pi_x(At(\phi))$.

3. $At(e_x) = E_x$.

(1) Let $d(\phi) = x$ and $d(\psi) = y$. Assume that $\alpha \in At(\phi \otimes \psi)$. This means that $\phi \otimes \psi \leq \alpha$, hence $\phi \leq \alpha$ and therefore by the monotonicity of marginalization $\phi = \phi^{\downarrow x} \leq \alpha^{\downarrow x}$ But $\alpha^{\downarrow x}$ is an atom on x (Lemma 6.14 (2))

and thus $\alpha^{\downarrow x} \in At(\phi)$. In the same way we find that $\alpha^{\downarrow y} \in At(\psi)$. Since $d(\alpha) = x \vee y$ this shows that $\alpha \in At(\phi) \bowtie At(\psi)$.

Conversely, assume that $\alpha \in At(\phi) \bowtie At(\psi)$. Then $d(\alpha) = x \vee y$ and $\phi \leq \alpha^{\downarrow x} \leq \alpha$ and also $\psi \leq \alpha^{\downarrow y} \leq \alpha$. This implies $\phi \otimes \psi \leq \alpha$ and hence $\alpha \in At(\phi \otimes \psi)$. This proves (1).

(2) Assume first that $\alpha \in At(\phi^{\downarrow x})$. This means that $d(\alpha) = x$ and $\phi^{\downarrow x} \leq \alpha$. Let $d(\phi) = y$. Then we have $\phi \leq \phi \otimes \alpha$. Suppose that $\phi \otimes \alpha = z_y$. Then we obtain

$$\alpha = \phi^{\downarrow x} \otimes \alpha = (\phi \otimes \alpha)^{\downarrow x} = z_y^{\downarrow x} = z_x.$$

But this is not possible, since α is an atom. Thus $\phi \otimes \alpha \neq z_y$. Hence, because Φ is atomic, $At(\phi \otimes \alpha) \neq \emptyset$. Let $\beta \in At(\phi \otimes \alpha)$ such that $d(\beta) = y$ and $\phi \leq \phi \otimes \alpha \leq \beta$. This shows that $\beta \in At(\phi)$. But we have also that

$$\alpha = (\phi \otimes \alpha)^{\downarrow x} \leq \beta^{\downarrow x}.$$

Since α and $\beta^{\downarrow x}$ are both atoms, we must have $\alpha = \beta^{\downarrow x}$. This shows that $\alpha \in \pi_x(At(\phi))$.

Assume now that $\beta \in At(\phi)$, hence $\beta^{\downarrow x} \in \pi_x(At(\phi))$. We have $\phi \leq \beta$, hence $\phi^{\downarrow x} \leq \beta^{\downarrow x}$ and thus $\beta^{\downarrow x} \in At(\phi^{\downarrow x})$. This proves (2).

(3) follows directly from the definition of At. $\qquad\square$

This theorem has two important results as corollaries:

Corollary 6.18

1. If Φ is atomic composed, *then* At *is an* embedding.

2. If Φ *is atomic closed and if for* $X, Y \subseteq At_x(\Phi)$, $X \neq Y$ *implies* $\wedge X \neq \wedge Y$, *then* At *is an* isomorphism.

Proof. (1) Assume that $At(\phi) = At(\psi)$. Then $\phi = \wedge At(\phi) = \wedge At(\psi) = \psi$.

(2) Let $X \subseteq At_x(\Phi)$ for some domain x. That is, X is a relation in $\mathcal{R}_{At(\Phi)}$. Since Φ is atomic closed $\wedge X$ exists in Φ and $\wedge X = \wedge At(\wedge X)$. But this implies that $X = At(\wedge X)$, hence the mapping At is onto $\mathcal{R}_{At(\Phi)}$. $\qquad\square$

The second part of the corollary says that an atomic closed information algebra is under some additional condition essentially identical to a relational algebra over a tuple system. It is a system of subsets. As such it has much more structure than an information algebra. The first part of the corollary says that an atomic composed information algebra is embedded in such a subset system. This line of inquiry is best continued in the framework of domain free algebra. This necessitates first a translation of the theory of atoms into domain-free information algebras. We leave this as an exercise for the reader.

6.4   Information Systems

Here we introduce a different way to represent information and information algebras. Data and facts, that is information, are described in formal systems like computers in a formal language. We do not want here to go into the syntactical structure of a language. It is sufficient for our purposes to consider a language as a set $\mathcal{L}$ of well-formed sentences without taking care of the syntactic structure of sentences. Data, facts and information are then expressed by subsets X of $\mathcal{L}$.

Certain sentences can be deduced from others. Moreover, the same information can be expressed by different sets of sentences. In order to express the information contained in a set of sentences we need therefore an *entailment relation* defined between sets of sentences $X \subseteq \mathcal{L}$ and sentences $A \in \mathcal{L}$. The notation $X \vdash A$ says that the sentence A can be derived from the sentences X, that X entails A. This relation must satisfy the following conditions:

(E1) $X \vdash A$ for each $A \in X$.

(E2) If $X \vdash B$ for each $B \in Y$ and $Y \vdash A$, then $X \vdash A$.

As an immediate consequence of (E1) and (E2) we obtain that the entailment relation is monotone.

Lemma 6.19 *If* $Y \subseteq X$ *and* $Y \vdash A$, *then* $X \vdash A$.

Proof. By (E1) $X \vdash B$ for all $B \in Y$ and from (E2) follows that $X \vdash A$.   □

As an example we cite linear equations. The language $\mathcal{L}$ is formed here over a set of variables $X_1, \ldots, X_n$ and consists of sentences of the form

$$\sum_{j \in J} a_j X_j = a_0,$$

where $a_0, a_1, \ldots, a_n$ are real (or rational or integer) values and J is a subset of the index set $\{1, 2, \ldots, n\}$. The entailment relation corresponds to linear dependence. Thus, a linear equation E as above is entailed by a set E_i of linear equations

$$\sum_{j=1}^{n} a_{i,j} X_j = a_{i,0} \qquad \text{for } i = 1, \ldots, m$$

if E is a *linear combination* of the equations E_i, that is, if

$$a_j = \sum_{i=1}^{m} \lambda_i a_{i,j}, \qquad \text{for all } j = 0, 1 \ldots, m.$$

It is easy to verify that this relation satisfies the requirements (E1) and (E2) for an entailment relation. A similar example is provided by systems of linear inequalities, where entailment corresponds to *non-negative linear combination*. Of course, logic systems like classical propositional logic, predicate logic, etc.

have also entailment relations and provide thus further examples (see for example Section 6.5.1).

Instead of the entailment relation we shall use an associated *consequence operator* C which is a mapping form the power set $\mathcal{P}(\mathcal{L})$ of the language $\mathcal{L}$ into itself. For a subset $X \subseteq \mathcal{L}$ we define $C(X)$ to be the set of all sentences derivable from X, That is,

$$C(X) \;=\; \{A \in \mathcal{L} : X \vdash A\}. \tag{6.27}$$

This operator satisfies three fundamental properties which are stated in the following lemma:

Lemma 6.20 *(C1)* $X \subseteq C(X)$ *for every subset* $X \subseteq \mathcal{L}$.
(C2) $C(C(X)) = C(X)$ *for every subset* $X \subseteq \mathcal{L}$.
(C3) If $X \subseteq Y$, *then* $C(X) \subseteq C(Y)$.

Proof. (C1) follows from (E1).

(C2) We need only to show that $C(C(X)) \subseteq C(X)$ since the converse inclusion holds by (C1). So, if $A \in C(C(X))$, then $C(X) \vdash A$. Since $X \vdash B$ for all $B \in C(X)$, (E2) implies $X \vdash A$, hence $A \in C(X)$.

(C3) $A \in C(X)$ means that $X \vdash A$. By Lemma 6.19, $X \subseteq Y$ implies thus $Y \vdash A$, hence $A \in C(Y)$. $\qquad\square$

An operator C satisfying conditions (C1), (C2) and (C3) is called a *consequence operator* (or also a *closure operator*). We refer to (Wojcicki, 1988) for a systematic discussion of consequence operators and their relation to logic. Two sets of sentences X and Y are called *equivalent*, if each one is among the consequences of the other one. This means that $X \subseteq C(Y)$ and $Y \subseteq C(X)$. Apply in both inclusions the operator C. Then it follows from (C3) and (C2) that $C(X) = C(Y)$. Thus, X and Y are equivalent if and only if $C(X) = C(Y)$. The sentences in $C(\emptyset)$ are called *tautologies*. A set X of sentences is called *closed* if $X = C(X)$.

The following property is used often:

Lemma 6.21 $C(X \cup Y) = C(C(X) \cup Y)$.

Proof. We have only to prove that $C(C(X) \cup Y) \subseteq C(X \cup Y)$ since the converse inclusion is obvious from (C1) and (C3). From $X, Y \subseteq X \cup Y$ we deduce that $C(X) \subseteq C(X \cup Y)$, $Y \subseteq C(X \cup Y)$ and thus $C(X) \cup Y \subseteq C(X \cup Y)$. From this it follows, using (C3) and (C2) that $C(C(X) \cup Y) \subseteq C(X \cup Y)$. $\qquad\square$

It is often the case that we are interested only in certain aspects of a set of sentences $X \subseteq \mathcal{L}$. For example, if X is a system of linear equations, we may be interested only in the consequences of this system for a certain selected group of variables, possibly even only for the value of a single variable. So we are interested in the set $C(X) \cap M$, where M is a certain subset of $\mathcal{L}$ and not so much in the set $C(X)$ of all consequences of X. Therefore we introduce for arbitrary subsets $X, M \subseteq \mathcal{L}$ the following notation:

$$C_M(X) \;=\; C(X) \cap M. \tag{6.28}$$

We can consider C_M as an operator from $\mathcal{P}(\mathcal{L})$ into $\mathcal{P}(\mathcal{L})$. It has properties similar to the original operator C.

Lemma 6.22 *Let M be a subset of $\mathcal{L}$. Then we have:*

1. *If $X \subseteq M$, then $X \subseteq C_M(X)$.*

2. *$C_M(C_M(X)) = C_M(X)$.*

3. *If $X \subseteq Y$, then $C_M(X) \subseteq C_M(Y)$.*

4. *If $X \subseteq M$, then $C(C_M(X)) = C(X)$.*

Proof. (1) Condition (C1) says that $X \subseteq C(X)$. If $X \subseteq M$, then $X \subseteq C(X) \cap M = C_M(X)$.

(2) Since $C_M(X) \subseteq M$, it follows from (1) that $C_M(X) \subseteq C_M(C_M(X))$. For the other inclusion we use the fact that $C_M(X) \subseteq C(X)$. By the monotonicity property (C3) and (C2) we obtain that $C(C_M(X)) \subseteq C(X)$. Now we intersect both sides with M and obtain $C_M(C_M(X)) = C_M(X)$.

(3) If $X \subseteq Y$, then $C(X) \subseteq C(Y)$. This is what (C3) says. Therefore $C(X) \cap M \subseteq C(Y) \cap M$.

(4) Assume that $X \subseteq M$. From (1) we obtain that $X \subseteq C_M(X)$ and, by monotonicity, it follows that $C(X) \subseteq C(C_M(X))$. Since $C_M(X) \subseteq C(X)$, we obtain $C(C_M(X)) \subseteq C(C(X)) = C(X)$. Hence, we have $C(C_M(X)) = C(X)$. $\square$

So far the set M was an arbitrary subset of $\mathcal{L}$. We are now going to restrict the sublanguages M we are interested in, to a given class $\mathcal{S}$. We assume that $\mathcal{S}$ is closed under intersection; with any two sublanguages L_1 and L_2 in $\mathcal{S}$, the intersection $L_1 \cap L_2$ must be in $\mathcal{S}$. Furthermore, we require that $\mathcal{S}$ is equipped with a lattice structure with respect to set inclusion, and that $\mathcal{L}$ belongs itself to $\mathcal{S}$. In other words, we impose the following conditions on $\mathcal{S}$:

1. $\mathcal{L} \in \mathcal{S}$.

2. If $L_1 \in \mathcal{S}$ and $L_2 \in \mathcal{S}$, then $L_1 \cap L_2 \in \mathcal{S}$.

3. If $L_1 \in \mathcal{S}$ and $L_2 \in \mathcal{S}$, then there exists the supremum $L_1 \vee L_2$ of L_1 and L_2 in $\mathcal{S}$ (with respect to set inclusion).

Note that $L_1 \cap L_2$ is the infimum with respect to set inclusion, whereas $L_1 \cup L_2 \subseteq L_1 \vee L_2$. The lattice $\mathcal{S}$ has the top-element $\mathcal{L}$. As an example consider the language $\mathcal{L}$ whose sentences are linear equations over a set of variables $X_1, \ldots, X_n$. If M is a subset of the index set $\{1, 2, \ldots, n\}$ we define the sublanguage $\mathcal{L}_M$ of linear equations over variables $\{X_j : j \in M\}$. Clearly the family $\mathcal{S}$ of these sublanguages $\mathcal{L}_M$ for all subsets M of $\{1, 2, \ldots, n\}$ satisfies the conditions above and the lattice $\mathcal{S}$ is isomorph to the lattice of subsets of $\{1, 2, \ldots, n\}$. This is a typical case of a language $\mathcal{L}$ defined over a set of variables and of the class of sublanguages $\mathcal{S}$ defined over subsets of variables. Many examples are of this type.

The triple $(\mathcal{L}, C, \mathcal{S})$ is called an *information system*. In domain theory a related notion has been introduced by Scott (see for example (Davey & Priestley, 1990)). The terminology used here is motivated by Scott's concept of an information system. There are however some differences between the two concepts: Scott considers only "consistent" sets of sentences, which in our case means that $C(X) \neq \mathcal{L}$. We do not exclude such sets of sentences. Furthermore, no sublanguages are introduced in Scott's information system. This is explained by the fact that information systems in domain theory are introduced for other purposes than here. Finally, Scott's information systems are "compact". This is an important property which we add to our notion of information systems in Section 6.6. All these differences are not really essential, so the use of the same name seems justified.

Given the language $\mathcal{L}$, the consequence operator $C : \mathcal{P}(\mathcal{L}) \to \mathcal{P}(\mathcal{L})$ and the class of sublanguages $\mathcal{S} \subseteq \mathcal{P}(\mathcal{L})$ we now define an associated domain-free information algebra. We already mentioned that bodies of information are encoded as sets of sentences of the language $\mathcal{L}$. There are, however, several equivalent ways to do this. Different, but equivalent sets of sentences X, X', X'', etc. describe the *same* information. Therefore bodies of information have to be identified with closed sets $C(X)$. So let Φ_C be the sets of closed subsets of $\mathcal{L}$

$$\Phi_C = \{E \subseteq \mathcal{L} : C(E) = E\} = \{C(X) : X \subseteq \mathcal{L}\}. \qquad (6.29)$$

The operation of combination and focusing are defined in Φ_C in the following way:

1. Combination: $E_1 \otimes E_2 = C(E_1 \cup E_2)$ for $E_1, E_2 \in \Phi_C$.

2. Focusing: $E^{\Rightarrow L} = C(E \cap L)$ for $E \in \Phi_C$ and $L \in \mathcal{S}$.

Combination is defined in such a way that the combination of two closed sets E_1 and E_2 is the least closed set of sentences that contains both E_1 and E_2. This corresponds to the semi-lattice property of information algebras. The information E focused to a sublanguage $L \in \mathcal{S}$ is the least closed set of sentences that contains the intersection of E and L. For arbitrary subsets X and Y which are not necessarily closed, we have (using Lemma 6.21)

1. Combination: $C(X) \otimes C(Y) = C(X \cup Y)$.

2. Focusing: $C(X)^{\Rightarrow L} = C(C_L(X))$.

This gives another justification for the definitions above: The combined information of the two sets of sentences X and Y is simply the information of the union $X \cup Y$ of the two sets. The focused information of a set of sentences X to a sublanguage L is the information obtained, when the information of X is intersected with L.

Given this definition of combination and focusing we have to examine whether the axioms of a domain-free information algebra are satisfied. We start with the semi-group axiom.

Lemma 6.23 *The combination on Φ_C is associative, commutative and $C(\emptyset)$ is the neutral element.*

Proof. That the combination operation is associative can be seen as follows (using Lemma 6.21):

$$\begin{aligned}
(E_1 \otimes E_2) \otimes E_3 &= C(C(E_1 \cup E_2) \cup E_3) = C(E_1 \cup E_2 \cup E_3) \\
&= C(E_1 \cup C(E_2 \cup E_3)) = E_1 \otimes (E_2 \otimes E_3).
\end{aligned}$$

That the combination is commutative follows directly from the definition. And again using Lemma 6.21 we obtain

$$C(\emptyset) \otimes E = C(C(\emptyset) \cup E) = C(\emptyset \cup E) = C(E) = E.$$

$\square$

Next we prove idempotency and support:

Lemma 6.24 *Combination and focusing in Φ_C is idempotent and $\mathcal{L}$ is a support for every closed set.*

Proof. Idempotency can be verified as follows (using Lemma 6.21 and Lemma 6.22 (4)):

$$\begin{aligned}
C(X)^{\Rightarrow L} \otimes C(X) &= C(C_L(X)) \otimes C(X) = C(C_L(X) \cup X) \\
&= C(C(C_L(X)) \cup C(X)) = C(C(X)) = C(X).
\end{aligned}$$

That $\mathcal{L}$ is a support for every closed set follows from $C(X)^{\Rightarrow \mathcal{L}} = C(C(X) \cap \mathcal{L}) = C(C(X)) = C(X)$.

$\square$

The transitivity and combination axioms, however, are not necessarily valid for arbitrary consequence operators and arbitrary classes of sublanguages. We have to impose further conditions on the consequence operator which will be elaborated in the following.

Transitivity: For an information $C(X)$ of Φ_C and sublanguages $M, L \in \mathcal{S}$ the transitivity axiom imposes

$$(C(X)^{\Rightarrow L})^{\Rightarrow M} = C(X)^{\Rightarrow L \cap M}.$$

On the left-hand side we have

$$(C(X)^{\Rightarrow M})^{\Rightarrow L} = C(C_M(X))^{\Rightarrow L} = C(C_L(C_M(X))).$$

On the right-hand side we have

$$C(X)^{\Rightarrow L \cap M} = C(C_{L \cap M}(X)).$$

So we see that the structure $(\Phi_C, \mathcal{S}, \otimes, \Rightarrow)$ satisfies the transitivity axiom if, and only if, the operator C satisfies the following condition:

(C4) $C(C_L(C_M(X))) = C(C_{L \cap M}(X))$

for $X \subseteq \mathcal{L}$ and $L, M \in \mathcal{S}$.   (6.30)

Combination: For $C(X)$ and $C(Y)$ in Φ_C and the sublanguages $L \in \mathcal{S}$ the combination axiom requires that

$$(C(X)^{\Rightarrow L} \otimes C(Y))^{\Rightarrow L} = C(X)^{\Rightarrow L} \otimes C(Y)^{\Rightarrow L}.$$

On the left-hand side we have

$$(C(X)^{\Rightarrow L} \otimes C(Y))^{\Rightarrow L} = C(C_L(X) \cup Y)^{\Rightarrow L}$$
$$= C(C_L(C_L(X) \cup Y)).$$

On the right hand side we have

$$C(X)^{\Rightarrow L} \otimes C(Y)^{\Rightarrow L} = C(C_L(X)) \otimes C(C_L(Y))$$
$$= C(C_L(X) \cup C_L(Y)).$$

Therefore, the combination axioms means that

$$C(C_L(C_L(X) \cup Y)) = C(C_L(X) \cup C_L(Y)).   (6.31)$$

If we intersect both sides with L and apply Lemma 6.22 (2), then we obtain

$$C_L(C_L(X) \cup Y) = C_L(C_L(X) \cup C_L(Y)).   (6.32)$$

Conversely, if we take this equation and apply operator C on both sides, then using Lemma 6.22 (4), we obtain equation (6.31). Hence, equations (6.31) and (6.32) are equivalent. The structure $(\Phi_C, \mathcal{S})$ satisfies the combination axiom if, and only if, the consequence operator satisfies the following condition:

C(5) $C_L(C_L(X) \cup Y) = C_L(C_L(X) \cup C_L(Y))$

for $X, Y \subseteq \mathcal{L}$ and $L \in \mathcal{S}$.   (6.33)

So we see that $(\Phi_C, \mathcal{S})$ is a domain-free information algebra if and only if the consequence operator C satisfies conditions (C4) and (C5). Note that these conditions are not just properties of a consequence operator C but properties of a consequence operator C with respect to a class of sublanguages $\mathcal{S}$.

Property (C4) has a natural characterization in terms of interpolation. We say that a consequence operator C has the *interpolation property* with respect to a class of sublanguages $\mathcal{S}$, if, for all sublanguages $L, M \in \mathcal{S}$ from

$$A \in L, \quad X \subseteq M \quad \text{and} \quad A \in C(X)$$

it follows that there exists a set of sentences $Y \subseteq L \cap M$ such that $Y \subseteq C(X)$ and $A \in C(Y)$. The set of sentences Y is called the *interpoland* between X and A.

Theorem 6.25 *A consequence operator C has the interpolation property with respect to a class of sublanguages $S \subseteq \mathcal{P}(\mathcal{L})$ if, and only if, C satisfies (C4).*

Proof. Assume that C has the interpolation property with respect to S. Note that $C_{L \cap M}(X) \subseteq C_L(C_M(X))$. It is therefore sufficient to show that $C_L(C_M(X)) \subseteq C(C_{L \cap M}(X))$ for $L, M \in S$. Assume therefore that $A \in C_L(C_M(X))$. Since $A \in L$, $C_M(X) \subseteq M$ and $A \in C(C_M(X))$, there exists an interpoland $Y \subseteq L \cap M$ such that $Y \subseteq C(C_M(X))$ and $A \in C(Y)$. Since $C(C_M(X)) \subseteq C(X)$, we obtain that $Y \subseteq C_{L \cap M}(X)$ and $A \in C(C_{L \cap M}(X))$.

For the converse direction assume that $X \subseteq M$, $A \in L$ and $A \in C(X)$. Since $X \subseteq C_M(X)$ we obtain that $A \in C_L(C_M(X))$ and $A \in C(C_L(C_M(X)))$. Condition (C4) now implies that $A \in C(C_{L \cap M}(X))$. We set $Y = C_{L \cap M}(X)$ and have $Y \subseteq L \cap M$, $Y \subseteq C(X)$ and $A \in C(Y)$. Thus Y is an interpoland between X and A.   □

The next question is whether we can find a similar property that is equivalent to property (C5). This problem is harder. Only partial results are known and this question will not be pursued here. We refer however to (Mengin & Wilson, 1999) for a discussion of information algebras related to various logics.

We have seen that an information system induces under some additional conditions on the consequence operator C an information algebra. Now we show that an information algebra generates inversely in a natural way an information system. So let (Φ, D) be a domain-free information algebra. As a first step in the direction of a construction of an associated information system we take the set Φ as the language of the information system. Next, we consider subalgebras

$$\Phi^{\Rightarrow x} \;=\; \{\dot{\phi} \in \Phi : \phi = \phi^{\Rightarrow x}\}.$$

We introduce an additional, natural condition on the information algebra, namely that

$$\Phi^{\Rightarrow x} \;=\; \Phi^{\Rightarrow y} \quad \text{implies} \quad x \;=\; y. \tag{6.34}$$

This is not necessarily the case. But $\Phi^{\Rightarrow x} = \Phi^{\Rightarrow y}$ means that all bodies of information which are supported by x are also supported by y and vice versa. So all information relative to one domain is also an information relative to the other one. Hence, in the light of the bodies of information considered in the algebra Φ there is no distinction to be made between the two domains and they can as well be identified. So in the following this condition will be assumed.

We take now the class $S = \{\Phi^{\Rightarrow x} : x \in D\}$ of sublanguages of Φ. This is a lattice, isomorph to D, under inclusion. That is what the following lemma says.

Lemma 6.26 $S = \{\Phi^{\Rightarrow x} : x \in D\}$ *is a lattice under inclusion, isomorph to D. That is,*

1. $\Phi^{\Rightarrow x \wedge y} = \Phi^{\Rightarrow x} \cap \Phi^{\Rightarrow y}$.

2. $x \leq y$ if, and only if, $\Phi^{\Rightarrow x} \subseteq \Phi^{\Rightarrow y}$.

3. $\Phi^{\Rightarrow x \vee y}$ is the supremum of $\Phi^{\Rightarrow x}$ and $\Phi^{\Rightarrow y}$ in S, that is $\Phi^{\Rightarrow x \vee y} = \Phi^{\Rightarrow x} \vee \Phi^{\Rightarrow y}$.

Proof. (1) Assume $\phi \in \Phi^{\Rightarrow x \wedge y}$. This means that $x \wedge y$ is a support of ϕ. By Lemma 3.6 (6) this implies that both x and y are supports of ϕ. Hence $\phi \in \Phi^{\Rightarrow x} \cap \Phi^{\Rightarrow y}$. Conversely, assume $\phi \in \Phi^{\Rightarrow x} \cap \Phi^{\Rightarrow y}$ such that x and y are supports of ϕ. Then, by Lemma 3.6 (4), $x \wedge y$ is a support of ϕ, hence $\phi \in \Phi^{\Rightarrow x \wedge y}$.

(2) Let $x \leq y$. Then, since by Lemma 3.6 (6) if x is a support of ϕ, then y is a support of ϕ, we have that $\Phi^{\Rightarrow x} \subseteq \Phi^{\Rightarrow y}$. Assume then that $\Phi^{\Rightarrow x} \subseteq \Phi^{\Rightarrow y}$. Using (1) of this lemma we obtain that $\Phi^{\Rightarrow x} = \Phi^{\Rightarrow x} \cap \Phi^{\Rightarrow y} = \Phi^{\Rightarrow x \wedge y}$. But then (by the extra assumption introduced above), $x = x \wedge y$, hence $x \leq y$.

(3) $\Phi^{\Rightarrow x \vee y}$ contains both $\Phi^{\Rightarrow x}$ and $\Phi^{\Rightarrow y}$. This follows from (2) of this lemma. Assume that $\Phi^{\Rightarrow z}$ is another upper bound of $\Phi^{\Rightarrow x}$ and $\Phi^{\Rightarrow y}$. Then, again by (2) of this lemma, $x, y \leq z$ or $x \vee y \leq z$. But this implies $\Phi^{\Rightarrow x \vee y} \subseteq \Phi^{\Rightarrow z}$ and $\Phi^{\Rightarrow x \vee y}$ is the least upper bound of $\Phi^{\Rightarrow x}$ and $\Phi^{\Rightarrow y}$. $\square$

Now, if D has a top element $\top$, then $\Phi = \Phi^{\Rightarrow \top}$ and $\Phi \in S$. So in this case, the class of sublanguages S satisfies the conditions imposed on the sublanguages of an information system.

In the next step we define a consequence operator on the language Φ. If $X \subseteq \Phi$ is a set of "sentences" of the language, then we define

$$C_\Phi(X) = \left\{ \begin{array}{l} \psi \in \Phi : \psi \leq \phi_1 \vee \cdots \vee \phi_m \text{ for some} \\ \text{finite set of elements } \phi_1, \ldots, \phi_m \in X. \end{array} \right\} \quad (6.35)$$

If we compare this with equation (6.16), then we see that $C_\Phi(X) = I(X)$, the ideal generated by X. This is indeed a consequence operator as the following theorem shows.

Theorem 6.27 C_Φ is a consequence operator on Φ.

Proof. (1) If $\phi \in X$, then $\phi \in C_\Phi(X)$, hence $X \subseteq C_\Phi(X)$. So (C1) holds.
(2) We have

$$C_\Phi(C_\Phi(X)) = \{\psi : \psi \leq \vee Y, Y \subseteq C_\Phi(X), Y \text{ finite set}\}$$

But, for every element $\psi_i \in Y$ we have that $\psi_i \leq \vee Y_i$ for a finite set $Y_i \subseteq X$. This implies

$$\psi \leq \vee(\cup\{Y_i : \psi_i \in Y\})$$

Since $\cup\{Y_i : \psi_i \in Y\}$ is still finite and contained in the set X, it follows that $\psi \in C_\Phi(X)$. This shows that $C_\Phi(C_\Phi(X)) \subseteq C_\Phi(X)$. The converse inclusion holds by (C1) already proved. Hence we obtain that $C_\Phi(C_\Phi(X)) = C_\Phi(X)$ and (C2) holds.

(3) Assume $X \subseteq Y$. If $\psi \in C_\Phi(X)$, then $\psi \leq \vee Z$ for some finite set $Z \subseteq X$. But we have also $Z \subseteq Y$, hence $\psi \in C_\Phi(Y)$. This proves that (C3) holds also.

$\square$

So, for example $C_\Phi(\{\phi\}) = I(\phi)$ is the set of all "consequences" of ϕ. This justifies the idea that all $\psi \leq \phi$ can be asserted, if ϕ is asserted to be true. We have shown that $(\Phi, C_\Phi, \mathcal{S})$ is an *information system*. As such it has associated with it the information algebra of its closed sets $C_\Phi(X) = I(X)$, provided that the consequence operator C_Φ satisfies also conditions (C4) and (C5). We have seen in Section 6.2 that the ideals of Φ indeed form an information algebra. It remains to verify that this information algebra is identical to the information algebra associated with the information system $(\Phi, C_\Phi, \mathcal{S})$. If this is true, then it follows automatically that C_Φ satisfies (C4) and (C5) and also that the information algebra Φ is embedded in the algebra generated by the information system $(\Phi, C_\Phi, \mathcal{S})$ it induces (see Theorem 6.8).

The two algebras are indeed identical as the following lemma shows.

Lemma 6.28

1. $C_\Phi(X_1 \cup X_2) = I(X_1) \otimes I(X_2)$.

2. $C_\Phi^{\Rightarrow \Phi^{\Rightarrow x}}(X) = I^{\Rightarrow x}(X)$.

Proof. (1) We have by definition of C_Φ

$$C_\Phi(X_1 \cup X_2) = \{\psi : \psi \leq \vee Y, Y \subseteq X_1 \cup X_2, Y \text{ finite}\}.$$

Let $Y_1 = X_1 \cap Y$ and $Y_2 = X_2 \cap Y$, such that $Y_1 \subseteq X_1$ and $Y_2 \subseteq X_2$. Define $\phi_1 = \vee Y_1$ and $\phi_2 = \vee Y_2$ (put $\vee \emptyset = e$). Then we have $\phi_1 \in I(X_1) = C_\Phi(X_1)$ and $\phi_2 \in I(X_2) = C_\Phi(X_2)$. Furthermore $\psi \leq \phi_1 \otimes \phi_2$ which shows that $\psi \in I(X_1) \otimes I(X_2)$.

Conversely, assume $\psi \in I(X_1) \otimes I(X_2)$. This means that $\psi \leq \phi_1 \otimes \phi_2$ where $\phi_1 \leq \vee Y_1, Y_1 \subseteq X_1$ is a finite set and where $\phi_2 \leq \vee Y_2, Y_2 \subseteq X_2$ is a finite set. We obtain from that $\psi \leq \vee Y$ where $Y = Y_1 \cup Y_2 \subseteq X_1 \cup X_2$ and $Y_1 \cup Y_2$ is still finite. But this says that $\psi \in C_\Phi(X_1 \cup X_2)$.

(2) We have

$$\begin{aligned} C_\Phi^{\Rightarrow \Phi^{\Rightarrow x}}(X) &= C_\Phi(C_\Phi(X) \cap \Phi^{\Rightarrow x}) \\ &= \{\psi : \psi \leq Y, Y \subseteq C_\Phi(X) \cap \Phi^{\Rightarrow x}, Y \text{ finite}\}. \quad (6.36) \end{aligned}$$

Therefore, $\psi \in C_\Phi^{\Rightarrow \Phi^{\Rightarrow x}}(X)$ means that there exist a finite number of elements $\phi_1, \ldots, \phi_m \in C_\Phi(X)$ and $\phi_i = \phi_i^{\Rightarrow x}$ for $i = 1, \ldots, m$ such that

$$\psi \leq \phi_1 \otimes \cdots \otimes \phi_m = \phi.$$

We have then $\phi = \phi^{\Rightarrow x}$ and $\phi_i \leq \vee Y_i$ where $Y_i \in X$ are finite sets. This implies $\phi \leq \vee Y$, where $Y = \cup Y_i \subseteq X$ and hence $\phi \in I(X)$. So we obtain $\psi \leq \phi^{\Rightarrow x}$ for a $\phi \in I(X)$. This shows that $\psi \in I^{\Rightarrow x}(X)$.

Conversely, assume $\psi \in I^{\Rightarrow x}(X)$. Then $\psi \leq \phi^{\Rightarrow x}$ for some $\phi \in I(X)$. We have that $\phi \leq \vee Y$ where $Y \subseteq X$ is a finite set. Now, $\phi^{\Rightarrow x} \leq \phi$ implies that $\phi^{\Rightarrow x} \in C_{\Phi}(X) \cap \Phi^{\Rightarrow x}$ and $\psi \leq \phi^{\Rightarrow x}$ means that $\psi \in C_{\Phi}(C_{\Phi}(X) \cap \Phi^{\Rightarrow x}) = C_{\Phi}^{\Rightarrow \Phi^{\Rightarrow x}}(X)$. $\qquad\square$

This proves finally the following theorem:

Theorem 6.29 *If (Φ, D) is a domain-free information algebra and D has a top element, then (Φ, C_{Φ}, S) is an information system. The information algebra (Φ, D) is embedded into the information algebra induced by the information system (Φ, C_{Φ}, S) and the latter algebra is the* completion *of the algebra (Φ, D).*

So we have now two complementary representations of information algebras, file systems and information systems. In the former case, elements ϕ of an information algebra are represented by all elements ψ, which imply ϕ, that is $\psi \geq \phi$ and in the latter case by all elements ψ which are implied by ϕ, that is $\psi \leq \phi$.

6.5 Examples

6.5.1 Propositional Logic

A first class of examples of information systems is provided by propositional logic. The language of this logic is constructed over sets of propositional symbols or variables $P = \{p_1, p_2, \ldots, p_n\}$, which we assume to be finite. The language $\mathcal{L}$ consists of *wff* (or *well formed formula*) defined as follows:

1. Each element of P, $\top$ and $\bot$ are wffs (a so-called *atomic formula*).

2. If f and g are wffs, then so are $\neg f$, $f \wedge g$, $f \vee g$, and $f \rightarrow g$ and $f \leftrightarrow g$ are wffs.

3. All wffs are generated from atomic formulas by finitely often applying rule (2).

We call the set of wffs over a set P of propositional variables also $\mathcal{L}_P$. If Q is a subset of P, then $\mathcal{L}_Q$ is a sublanguage of $\mathcal{L}_P = \mathcal{L}$. The family $\mathcal{S}$ of sublanguages $\mathcal{L}_Q$ for all subsets $Q \subseteq P$ forms a lattice with respect to set inclusion, such that

1. $\mathcal{L} \in \mathcal{S}$.

2. $\mathcal{L}_O \cap \mathcal{L}_Q = \mathcal{L}_{O \cap Q} \in \mathcal{L}_S$.

3. $\mathcal{L}_O \vee \mathcal{L}_Q = \mathcal{L}_{O \cup Q} \in \mathcal{L}_S$.

Note that $\mathcal{L}_\emptyset = \{\bot, \top\}$.

There are several equivalent ways to introduce an entailment relation into these propositional languages. We use the so-called semantic approach. We start with a map i from P in to $\{\bot, \top\}$ and extend this map inductively to any wff in $\mathcal{L}$ by the following rules:

1. $i(\neg f) = \top$ if, and only if, $i(f) = \bot$.

2. $i(f \wedge g) = \top$, if, and only if, $i(f) = \top$ and $i(g) = \top$.

3. $i(f \vee g) = \top$, if, and only if, $i(f) = \top$ or $i(g) = \top$.

4. $i(f \rightarrow g) = \top$, unless $i(f) = \top$ and $i(g) = \bot$.

5. $i(f \leftrightarrow g) = \top$ if, and only if, $i(f) = i(g) = \top$ or $i(f) = i(g) = \bot$.

Such a map i from $\mathcal{L}$ to $\{\bot, \top\}$ is called an *interpretation*. An assignment of truth-values i to the propositional variables P is a *tuple* $i : P \rightarrow \{\bot, \top\}$ with domain P and the truth-values $\{\bot, \top\}$ as attributes. If f is a wff from $\mathcal{L}$, then we denote by $I(f)$ all tuples i, which induce interpretations such that $i(f) = \top$. So, $I(f)$ is the set of tuples (the relation) under which f is interpreted as "true" ($\top$). If X is a set of wffs from $\mathcal{L}_Q$, then $I(X)$ is the relation of all tuples i such that all formulas in X are interpreted as true.

We define an entailment relation, denoted here by $\models$, in $\mathcal{L}$ as follows: If X is a set of wffs from $\mathcal{L}$ and $f \in \mathcal{L}$, then

$$X \models f \qquad \text{if, and only if,} \qquad I(X) \subseteq I(f). \tag{6.37}$$

It is easy to verify that this relation satisfies the conditions (E1) and (E2) of an entailment relation (see Subsection 6.4). This relation induces a consequence operator C of propositional logic,

$$C(X) \;=\; \{f \in \mathcal{L} : I(X) \subseteq I(f)\}. \tag{6.38}$$

Hence, $(\mathcal{L}, C, \mathcal{S})$ form an information system.

Two sets X and Y of wffs are *equivalent*, if $C(X) = C(Y)$, which is equivalent to $I(X) = I(Y)$. Classes of equivalent sets of wffs are thus determined by sets of tuples. As we have seen, classes of equivalent sets form the elements of the information algebra, associated with an information system. So we may identify the elements of this information algebra with subsets of tuples with domain P, that is with relations on P with domains $\{\bot, \top\}$. If two sets of wffs X and Y represent two pieces of information, then the union $X \cup Y$ represents the combined information. Clearly,

$$I(X \cup Y) \;=\; I(X) \cap I(Y).$$

So, combination translates into intersection of sets of tuples. If X is a set of wffs, and Q a subset of P defining a sublanguage $\mathcal{L}_Q$, then

$$C(X)^{\Rightarrow \mathcal{L}_Q} \;=\; C(C(X) \cap \mathcal{L}_Q) \;=\; \{f \in \mathcal{L} : I(X)^{\Rightarrow Q} \subseteq I(f)\}.$$

So, focusing of a piece of information onto a sublanguage over propositional variables Q translates into projection in the set of tuples.

This proves that the information algebra associated to propositional logic over a set of propositional variables P corresponds to the relational algebra of tuples over domains $Q \subseteq P$ and with attributes $\{\bot, \top\}$ ("true" and "false"). Clearly this algebra is atomic closed with the tuples corresponding to the atoms. Of course, this information algebra is closely related to the Lindenbaum algebra associated with propositional logic, see also the following Subsection 6.5.2. We refer to (Kohlas, Haenni & Moral, 1999) for a discussion of propositional information systems. We mention also that (Mengin & Wilson, 1999) discuss in detail the relation between logic and local computation.

6.5.2 Boolean Information Algebras

In a domain-free information algebra Φ is a semi-lattice. But sometimes it is more. For example subset systems or relational algebras allow for further operations like union or complementation. Such an example we encountered just in the previous subsection on propositional logic (Subsection 6.5.1). Information algebras which such additional operations are *Boolean information algebras*.

Definition 6.30 Boolean Information Algebras: *A domain-free informa-tion algebra* (Φ, D) *with null element is called* Boolean, *if*

1. Φ *is a* distributive lattice. *That is, besides the supremum (combination) of two elements ϕ and ψ, the infimum $\phi \wedge \psi$ exists too, and the distributive laws hold:*

$$\phi \wedge (\psi \vee \eta) = (\phi \vee \psi) \wedge (\phi \vee \eta),$$
$$\phi \vee (\psi \wedge \eta) = (\phi \wedge \psi) \vee (\phi \wedge \eta).$$

2. *For all $\phi \in \Phi$ exists an element ϕ^c, called the* complement of ϕ *such that*

$$\phi \wedge \phi^c = e, \qquad \phi \vee \phi^c = z.$$

This means essentially, that Φ is a *Boolean lattice*, see (Davey & Priestley, 1990).

The following results are well known for Boolean lattices (Davey & Priestley, 1990):

1. $\phi \leq \psi$ if, and only if, $\phi \wedge \psi = \phi$.

2. Every ϕ has a *unique* complement.

3. $(\phi^c)^c = \phi$.

4. $e^c = z$ and $z^c = e$.

5. $\phi \leq \psi$ if, and only if, $\phi^c \geq \psi^c$.

6. $\phi \vee \psi^c = z$ if, and only if, $\phi \geq \psi$.

7. De Morgan laws:

$$(\phi \wedge \psi)^c = \phi^c \vee \psi^c,$$
$$(\phi \vee \psi)^c = \phi^c \wedge \psi^c.$$

Here follow a few elementary results on focusing in Boolean information algebras:

Lemma 6.31

1. *$\phi \leq \psi$ if, and only if, $\psi^c \leq \phi^c$.*

2. *$\phi \vee \psi^c = z$ if, and only if, $\phi \geq \psi$.*

3. *x is a support of ϕ, if, and only if x is a support of ϕ^c.*

4. *x is a support of ϕ and ψ implies that x is a support of $\phi \wedge \psi$.*

5. *$(\phi \wedge \psi)^{\Rightarrow x} = \phi^{\Rightarrow x} \wedge \psi^{\Rightarrow x}$.*

6. *Let I be an arbitrary set of indices such that $\bigwedge_{i \in I} \phi_i$ exists. Then*

$$\left(\bigwedge_{i \in I} \phi_i \right)^{\Rightarrow x} = \bigwedge_{i \in I} \phi_i^{\Rightarrow x}.$$

7. *$(\phi \otimes (\psi^{\Rightarrow x})^c)^{\Rightarrow x} = \phi^{\Rightarrow x} \otimes (\psi^{\Rightarrow x})^c$.*

We leave the proof as an exercise for the reader.

In Boolean lattices there is a duality theory, which can be carried over to Boolean information algebras. If (Φ, D) is a Boolean information algebra, then define the following *dual operations* of combination $\otimes_d$ and focusing $\Rightarrow_d$:

$$\phi \otimes_d \psi = (\phi^c \otimes \psi^c)^c,$$
$$\phi^{\Rightarrow_d x} = ((\phi^c)^{\Rightarrow x})^c.$$

We note that, by de Morgan's law,

$$\phi \otimes_d \psi = \phi \wedge \psi.$$

This defines a new Boolean information algebra, which is called the *dual* of the original one.

Theorem 6.32 (Φ, D) *with the operations $\otimes_d$ and $\Rightarrow_d$ is a Boolean informa-tion algebra. The mapping $\phi \mapsto \phi^c$ is an isomorphism between dual Boolean information algebras.*

Proof. We note that the mapping $\phi \mapsto \phi^c$ is one-to-one, since every ϕ has a unique complement ϕ^c. We have the following relations:

1. $\phi^c \mapsto (\phi^c)^c = \phi$,

2. $\phi \vee \psi \mapsto (\phi \vee \psi)^c = \phi^c \wedge \psi^c$,

3. $\phi \wedge \psi \mapsto (\phi \wedge \psi)^c = \phi^c \vee \psi^c$,

4. $\phi^{\Rightarrow x} \mapsto (\phi^{\Rightarrow x})^c = (\phi^c)^{\Rightarrow_d x}$.

This shows that the mapping is isomorph relative to the operations of complementation, supremum and infimum and focusing. This proves that (Φ, D) is a Boolean information algebra with respect to operations $\otimes_d$ and $\Rightarrow_d$. $\square$

In the case of subset systems, dual combination corresponds to union. Focusing consists of projecting the complement and taking the complement of the result. This illustrates that dual operations are generally not very attractive. More on Boolean information algebras, and more specifically on the related cylindric algebras, can be found in (Henkin, Monk & Tarski, 1971).

6.5.3 Linear Equations

We consider in this section systems of linear equations over subsets s of a set r of variables. We denote in this section the variables by $X_1, X_2, \ldots, X_{|r|}$. Correspondingly, we identify the set r with the index set $\{1, 2, \ldots, |r|\}$ and any subset s of r with the corresponding subset of the index set $\{1, 2, \ldots, |r|\}$. A linear equation over s is then an expression of the form

$$a_0 + \sum_{j \in s} a_j X_j = 0. \tag{6.39}$$

The coefficients a_i are assumed to be real numbers in $\mathbf{R}$, but they could be elements of any field. These are the formulas of the languages $\mathcal{L}_s$ we consider. Any equation in $\mathcal{L}_s$ can also be looked at as an equation in $\mathcal{L}_t$ for some superset $t \supseteq s$. We have simply to put the coefficients a_i in $i \in t - s$ to zero. We include for s also the empty set. For $s = \emptyset$, the formulas are $a_0 = 0$. When we consider a set Γ of linear equations, then we number often the equation with $i = 1, 2, \ldots, |\Gamma| = m$. The order of the numbering plays no role. We then write for the system of linear equations

$$a_{i,0} + \sum_{j \in s} a_{i,j} X_j = 0, \qquad i = 1, 2, \ldots, m. \tag{6.40}$$

The reference language is $\mathcal{L} = \mathcal{L}_r$. Clearly, the sublanguages $\mathcal{L}_s$ form a lattice $\mathcal{S}$ under set inclusion and we have that

1. $\mathcal{L} \in \mathcal{S}$.

2. $\mathcal{L}_s \cap \mathcal{L}_t = \mathcal{L}_{s \cap t} \in \mathcal{S}$.

3. $\mathcal{L}_s \vee \mathcal{L}_t = \mathcal{L}_{s \cup t} \in \mathcal{S}$.

If X is a set of linear equations over r, and A a single linear equation over r, then we say $X \vdash A$ if there is a finite subset $Y = \{Y_1, \ldots, Y_m\}$ of X such that A is a linear combination of the equations in Y,

$$A = \lambda_1 Y_1 + \cdots + \lambda_m Y_m.$$

It can easily be verified that this is an entailment relation satisfying (E1) and (E2). Therefore, the system of linear equations defines an information system.

A real-valued vector $\mathbf{x} = (x_1, \ldots, x_{|r|})$ is called a solution of an equation

$$a_0 + \sum_{j=1}^{|r|} a_j X_j = 0, \tag{6.41}$$

if

$$a_0 + \sum_{j=1}^{|r|} a_j x_j = 0 \tag{6.42}$$

holds. $\mathbf{x}$ is a solution of a set of equations $X \subseteq \mathcal{L}$, if it is a solution of every equation of the set. We call $S(X)$ the set of solutions of a set of equations X. Of course, this set may be empty, if the system X of equations has no solution. Clearly, $X \vdash A$ if, and only if, $S(X) \subseteq S(\{A\})$. If X is a set of linear equations, then $C(X)$ is the set of all linear equations, which are a linear combination of a finite subset of equations of X. Two sets of equations X and Y are equivalent, if $S(X) = S(Y)$. In particular, we have $S(C(X)) = S(X)$. Therefore, we may identify a class of equations, equivalent with a set X by $S(X)$.

$S(X)$ defines a *linear manifold* in the $|r|$-dimensional space $\mathbf{R}^{|r|}$. And, inversely, any such manifold defines a class of equivalent sets of equations. Combination of sets X and Y of equations means taking the union $X \cup Y$. Clearly we have $S(X \cup Y) = S(X) \cap S(Y)$. If X is a set of equations, and $s \subseteq r$ a subset of variables, focusing corresponds to the elimination of the variables of $r - s$ from the equations in X. The solution manifold of this reduced system of equations corresponds to the set $S(X)^{\Rightarrow s}$, which is the projection of the manifold $S(X)$ to the subspace of the variables s. Therefore, we can identify the information algebra, associated with the information system of linear equations, with the information algebra of linear manifolds in the corresponding linear space, where combination corresponds to intersection and focusing to projection. This algebra is clearly atomic composed, since the points of the $|s|$-dimensional space of the variables in set s are atoms. Any manifold is the meet of its points. But it is not atomic closed since sets of points do not necessarily form a linear manifold.

6.5.4 Linear Inequalities

The case of linear inequalities is similar to the one of linear equations. As before we consider a set r of variables denoted by $X_1, X_2, \ldots, X_{|r|}$. If s is a subset of r, then inequalities of the form

$$a_0 + \sum_{j \in s} a_j X_j \geq 0 \tag{6.43}$$

are the formula of the sublanguage $\mathcal{L}_s$. The coefficients are assumed to be real numbers, but could more generally belong to any field. This can also be considered as an element of $\mathcal{L}_t$ for a $t \supseteq s$, when the coefficients a_j for $X_j \in t - s$ are put to zero. So all inequalities above belong to the reference language $\mathcal{L} = \mathcal{L}_r$. The sublanguages $\mathcal{L}_s$ form, as before, a lattice under set inclusion.

If X is a set of linear inequalities over r, and A a single linear inequality over r, then we say $X \vdash A$ if there is a finite subset $Y = \{Y_1, \ldots, Y_m\}$ of X such that A is a non-negative linear combination of the equations in Y,

$$A = \lambda_1 Y_1 + \cdots + \lambda_m Y_m, \qquad \text{with } \lambda_1, \ldots, \lambda_m \geq 0.$$

It can easily be verified that this is an entailment relation satisfying (E1) and (E2). Therefore, the system of linear equations defines an information system.

A real-valued vector $\mathbf{x} = (x_1, \ldots, x_{|r|})$ is called a solution of an inequality

$$a_0 + \sum_{j=1}^{|r|} a_j X_j \geq 0, \tag{6.44}$$

if

$$a_0 + \sum_{j=1}^{|r|} a_j x_j \geq 0 \tag{6.45}$$

holds. $\mathbf{x}$ is a solution of a set of inequalities $X \subseteq \mathcal{L}$, if it is a solution of every equation of the set. We call $S(X)$ the set of solutions of a set of inequalities X. Of course, this set may be empty, if the system X of inequalities has no solution. Clearly, $X \vdash A$ if, and only if, $S(X) \subseteq S(\{A\})$. If X is a set of linear inequalities, then $C(X)$ is the set of all linear inequalities, which are a non-negative linear combination of a finite subset of inequalities of X. Two sets of equations X and Y are equivalent, if $S(X) = S(Y)$. In particular, we have $S(C(X)) = S(X)$. Therefore, we may identify a class of equations, equivalent with a set X by $S(X)$.

$S(X)$ defines a *convex polyhedron* in the $|r|$-dimensional space $\mathbf{R}^{|r|}$. And, inversely, any convex polyhedron defines a class of equivalent sets of linear inequalities. Combination of sets X and Y of linear inequalities means taking the union $X \cup Y$. Clearly we have $S(X \cup Y) = S(X) \cap S(Y)$, which is again a convex polyhedron (or possibly the empty set). If X is a set of linear inequalities, and $s \subseteq r$ a subset of variables, focusing corresponds to the elimination of the variables of $r - s$ from the inequalities in X. The solution polyhedron of this reduced system of linear inequalities corresponds to the set $S(X)^{\Rightarrow s}$, which is the projection of the polyhedron $S(X)$ to the subspace of the variables s. Therefore, we can identify the information algebra, associated with the information system of linear inequalities, with the information algebra of convex polyhedra in the corresponding linear space, where combination corresponds to intersection and focusing to projection. This algebra is again clearly atomic composed, since the points of the $|s|$-dimensional space of the variables in set s are again atoms. Any convex polyhedron is the meet of its points. But it is not atomic closed since sets of points do not necessarily form a convex polyhedron.

Note that a single linear inequality has a semi-space as solution space. The information represented by a finite set X of linear inequalities can be thought of as the combination of the pieces of information represented by each individual inequality of X,

$$C(X) \;=\; C\left(\bigcup_{A \in X} \{A\}\right) \;=\; \bigotimes_{A \in X} C(\{A\}).$$

This corresponds to the well known fact, that a convex polyhedron can be determined as the intersection of the semi-planes of its faces.

The algebra of linear manifolds, considered above, is clearly a subalgebra of the information algebra of convex polyhedra. This follows since a linear manifold is a special case of a convex polyhedra. In fact, a linear equation can be represented by two inequalities of opposite sign.

6.6   Compact Systems

In this section we model another aspect of information, which has not been considered so far in this book. Computers can treat only "finite" information. "Infinite" information can however often be approximated by "finite" elements. We shall try to model this aspect of finiteness in this section. It must be stressed that we will not capture every aspect of finiteness. For example, we shall not treat questions of computability and related issues, which are also linked to the notion of finiteness. The structures considered here bring the subject of information algebras another step nearer to domain theory. Many aspects we treat in this section, are treated in domain theory, and much more in detail than here. Therefore we shall keep the presentation short. However, as in information system, the one crucial feature not addressed in domain theory, is focusing. Also domain theory places more emphasis on order, approximation and convergence of information and less on combination. So, although the subject is similar to domain theory, it is treated here with a different emphasis and goal.

We consider a domain-free information algebra (Φ, D). In the set Φ of pieces of information we single out a subset Φ_f of elements which are considered to be finite. We shall see in a moment in what respect these elements can be considered to be finite. In any way, combination of finite information should again yield finite information. So Φ_f is assumed to be closed under combination. The neutral element e is considered to be finite and belongs thus to Φ_f. May be one would expect that focusing of finite information yields also finite information. Although this is reasonable enough in many cases, we do not assume it for the purposes of this section.

Any information ϕ in Φ should be approximated with an arbitrary degree of precision by a system of finite pieces of information. Each approximating finite information should be coarser than ϕ. The latter must then be the *supremum* of the approximating elements. Inversely, the approximating system of finite elements must, for any element of it, or more generally for any finite subset of it, contain a finer finite information. This means that ϕ is approximated by more and more informative finite elements and captures the idea of *convergence* to the supremum. This concept corresponds to a directed set:

Definition 6.33 Directed Set. *A subset X of Φ is called a* directed set, *if it contains with any finite subset $\{\phi_1, \phi_2, \ldots, \phi_m\} \subseteq X$, an element $\phi \in X$ finer than all the elements of the subset,*

$$\phi_1 \otimes \phi_2 \otimes \cdots \otimes \phi_m \leq \phi.$$

So, X is directed, if it is non-empty and for any two pieces of information ϕ and ψ in X there exists a $\eta \in X$ such that $\phi \leq \eta$ and $\psi \leq \eta$. A directed set X of Φ is said to converge in Φ, if the supremum of X exists in Φ, that is $\vee X \in \Phi$.

In order to model approximation of an information ϕ in Φ by finite elements, we require first that all directed sets of finite pieces of information *converge* in Φ. This means that the respective suprema are elements of Φ which are approximated by finite elements. But we want more: any element of Φ must be approximated in this way. The finite elements must be *dense* in Φ, that is the supremum of a directed set of finite elements. It seems natural to take the directed set of all finite elements smaller than ϕ as the converging set, and require that

$$\phi = \vee\{\psi \in \Phi_f : \psi \leq \phi\}. \tag{6.46}$$

Whereas this requirement clearly must be satisfied, we need actually a stronger requirement, namely, that an information ϕ supported by some domain x can be approximated by finite pieces of information *supported by the same domain*.

These two requirements do not yet say much about finiteness. One thing they say, is that if ϕ is finite, then, according to equation (6.46) it belongs itself to the converging set of it. This is surely an important property of finiteness. But again we want more. We may possibly approximate an element by a directed set X of finite elements which is smaller than the set of all finite elements smaller than $\vee X$. But then, if ϕ is a finite element, $\phi \leq \vee X$, if ϕ belongs not to X, there must at least exist an element ψ in X finer than ϕ. This is a so-called *compactness* property.

So we impose the following conditions on Φ and Φ_f.

1. *Convergence:* If $X \subseteq \Phi_f$ is a directed set, then the supremum $\vee X$ over X exists and belongs to Φ.

2. *Density:* For $\phi \in \Phi$ and $x \in D$

$$\phi^{\Rightarrow x} = \vee\{\psi \in \Phi_f : \psi \leq \phi, \psi = \psi^{\Rightarrow x}\}. \tag{6.47}$$

3. *Compactness:* If $X \subseteq \Phi_f$ is a directed set, and $\phi \in \Phi_f$ such that $\phi \leq \vee X$, then there exists a $\psi \in X$ such that $\phi \leq \psi$.

Definition 6.34 Compact Information Algebra. *A system (Φ, Φ_f, D), where (Φ, D) is a domain-free information algebra, the lattice D has a top element, Φ_f is closed under combination and contains the empty information e, satisfying the three conditions of convergence, density and compactness formulated above, is called a* compact information algebra.

As an example of such a structure we indicate informally, without going into a detailed verification, that the family of convex sets in n-dimensional real space is a compact information algebra. Its compact or finite elements are the convex polyhedra in this space. The approximation of a convex set by convex polyhedra is from the outside. It is left to the reader to formally verify that this system satisfies the conditions of convergence, density and compactness. Furthermore, any domain-free information algebra with a finite number of elements is trivially compact, with $\Phi_f = \Phi$.

We remark that the density, equation (6.47) implies equation (6.46). So in a compact information algebra, any element is approximated by all finite element it dominates. We leave it as an exercise to prove this. However, the converse, namely that equation (6.46) implies density, equation (6.47) is not true. As a counter-example take for Φ the set $\{0, 1, 2, \ldots\} \cup \{\omega\}$. Define combination as follows:

$$\phi \otimes \psi = \begin{cases} \max(\phi, \psi), & \text{if } \phi \neq \omega \text{ and } \psi \neq \omega, \\ \omega, & \text{otherwise.} \end{cases}$$

For the lattice take $D = \{0, 1\}$, where $0 < 1$. The focusing operation is defined as follows:

$$\phi^{\Rightarrow 1} = \phi, \qquad \phi^{\Rightarrow 0} = \begin{cases} 0, & \text{if } \phi \neq \omega, \\ \omega, & \text{otherwise.} \end{cases}$$

As finite elements take the set $\{0, 1, 2, \ldots\}$. Then equation (6.46) is satisfied, but (6.47) is not.

The first main result is stated in the next theorem. It is less a theorem of information algebra than one of lattice theory. We refer therefore to (Davey & Priestley, 1990) for similar results and more details.

Theorem 6.35 *Let* (Φ, Φ_f, D) *be a compact information algebra. Then:*

1. Φ *is a complete lattice.*

2. *An information* $\phi \in \Phi$ *belongs to* Φ_f *if, and only if, for all directed subsets* $X \subseteq \Phi$, *if* $\phi \leq \vee X$, *then there exists a* $\psi \in X$ *such that* $\phi \leq \psi$.

3. *An information* $\phi \in \Phi$ *belongs to* Φ_f *if, and only if, for all subsets* $X \subseteq \Phi$, *if* $\phi \leq \vee X$, *then there exists a finite subset* $Y \subseteq X$ *such that* $\phi \leq \vee Y$.

Proof. (1) Let X be a subset of Φ. We show first that X has a greatest lower bound, an infimum. To obtain the greatest lower bound of X we define Y to be the set of all the finite lower bounds of X, i.e.

$$Y \;=\; \{\psi \in \Phi_f : \psi \leq \phi \text{ for all } \phi \in X\}.$$

It is easy to see that Y is a directed subset of Φ_f and therefore, by the convergence axiom, the supremum of Y exists in Φ. We claim that $\vee Y$ is the infimum of X. Obviously, $\vee Y$ is a lower bound of X, since every element of X is an upper bound of Y. Let ψ be another lower bound of X. We define

$$A_\psi \;=\; \{\eta \in \Phi_f : \eta \leq \psi\}.$$

such that by the density axiom $\psi = \vee A_\psi$ (see equation (6.46)). But A_ψ is contained in Y and thus $\psi = \vee A_\psi \leq \vee Y$. So $\vee Y$ is the greatest lower bound of X.

To prove that each subset X of Φ has a supremum, we consider the set of its upper bounds. The infimum of this set exists, by what we have just proved. With standard methods of lattice theory one proves now that the infimum of the upper bounds of a set X is its supremum (see (Davey & Priestley, 1990)). So each set X has a supremum. Thus Φ is a complete lattice.

(2) Assume that $\phi \in \Phi_f$, $X \subseteq \Phi$ is a directed set and $\phi \leq \vee X$. We define

$$Y \;=\; \{\psi \in \Phi_f : \text{there exists a } \chi \in X \text{ such that } \psi \leq \chi\}.$$

Since X is directed, Y is directed too. We claim that $\vee Y$ is an upper bound of X. Let χ be an element of X. Then A_χ is contained in Y and therefore $\chi = \vee A_\chi \leq \vee Y$. So we obtain $\phi \leq \vee X \leq \vee Y$ and, by the compactness axiom, it follows that there exists a $\psi \in Y$ such that $\phi \leq \psi$. By the definition of the set Y there exists a $\chi \in X$ such that $\phi \leq \psi \leq \chi$.

For the converse direction assume that for all directed sets $X \subseteq \Phi$, if $\phi \leq \vee X$, then there exists a $\psi \in X$ such that $\phi \leq \psi$. We take the directed set A_ϕ and, since $\phi = \vee A_\phi$, there exists a $\psi \in A_\phi$ such that $\phi \leq \psi$. Because, by the definition of A_ϕ, $\psi \leq \phi$ and $\psi \in \Phi_f$, we obtain that $\phi = \psi$ and thus ϕ belongs to Φ_f.

(3) The third assertion of the theorem follows from (2) by the following observation. Let X be an arbitrary, not necessarily directed, subset of Φ. Define

$$Z \;=\; \{\vee Y : Y \subseteq X, Y \text{ finite}\}.$$

Then we claim that Z is directed and $\vee X = \vee Z$. That Z is directed can be seen as follows. First, e belongs to Z, since $e = \vee \emptyset$. If Y_1 and Y_2 are finite subsets of X, then $Y_1 \cup Y_2$ is finite too, and $\vee(Y_1 \cup Y_2)$ is an upper bound of $\vee Y_1$ and $\vee Y_2$ in Z. It is obvious that $\vee Z \leq \vee X$, since for each subset Y of $X, \vee Y \leq \vee X$. But, X is contained in Z since $\phi = \vee \{\phi\} \in Z$ for all $\phi \in X$. Hence we obtain that $\vee X \leq \vee Z$, thus $\vee Z = \vee X$.

Assume then that $\phi \in \Phi_f$ and $\phi \leq \vee X = \vee Z$. By (2) there is a $\psi \in Z$ such that $\phi \leq \psi$. But we have $\psi = \vee Y$ for some finite subset Y of X. Conversely, assume $\phi \leq \vee X = \vee Z$ and that Y is a finite subset of X such that $\phi \leq \vee Y$. Since $\vee Y \in Z$ and Z is directed, it follows from (2) that $\phi \in \Phi_f$.   □

This theorem says that the finite elements of a compact information algebra are characterized by the partial order in Φ alone. In lattice theory, elements of a complete lattice, which satisfy condition (2) of the theorem are called *finite* and those which satisfy condition (3) are called *compact*. Thus, the elements of Φ_f are both "finite" and "compact" in this sense. Furthermore, lattices in which each element is the supremum of the compact elements it dominates are called *algebraic*. Hence, a compact information algebra Φ is an algebraic lattice. We refer to (Davey & Priestley, 1990) for a discussion of these subjects.

Compact information algebras can be obtained from any information algebra (Φ, D) by *completion* (see Section 6.2). The elements of Φ are then the finite elements of the compact algebra.

Theorem 6.36 *Let (Φ, D) be a domain-free information algebra and I_Φ the ideals of Φ with the operation of combination and focusing defined by (6.18) and (6.20). If D has a top element, and $I : \Phi \to I_\Phi$ is the embedding of Φ in I_Φ (see Theorem 6.8), then $(I_\Phi, I(\Phi), D)$ is a compact information algebra with finite elements $I(\Phi)$.*

Proof. I_Φ is an information algebra according to Theorem 6.7. It remains to show that $I(\Phi)$ are its finite elements, satisfying convergence, density and compactness axioms.

To simplify notation we identify Φ with $I(\Phi)$ and write simply ϕ instead of $I(\phi)$. Note then, that $\phi \leq I$ if, and only if, $\phi \in I$. More generally, we have $I_1 \leq I_2$ if, and only if, $I_1 \subseteq I_2$.

Convergence: Suppose $X \subseteq \Phi$ to be a directed subset. We claim that $I(X)$ (the ideal generated from X see Section 6.2) equals $\vee X$ in I_Φ. That is,

$$I(X) \;=\; \left\{ \begin{array}{l} \phi \in \Phi : \phi \leq \phi_1 \otimes \cdots \otimes \phi_m \\ \text{for some finite set of elements } \phi_1, \ldots, \phi_m \in X. \end{array} \right\}$$

So, if $\phi \in X$, then $\phi \in I(X)$, hence $\phi \leq I(X)$. Thus $I(X)$ is an upper bound of X.

Assume that I is another upper bound of X. Then $X \subseteq I$, hence $I(X) \subseteq I$ or $I(X) \leq I$. Thus we proved that $I(X) = \vee X$.

Density: We have

$$I^{\Rightarrow x} \;=\; \{\psi \in \Phi : \psi \leq \phi^{\Rightarrow x} \text{ for some } \phi \in I\}.$$

We have to prove $I^{\Rightarrow x} = \vee X = I(X)$ for the set $X = \{\phi \in \Phi : \phi = \phi^{\Rightarrow x} \leq I\}$. Suppose $\psi \in I(X)$ such that

$$
\begin{aligned}
\psi \;\leq\;& \phi_1 \otimes \cdots \otimes \phi_m \;=\; \phi_1^{\Rightarrow x} \otimes \cdots \otimes \phi_m^{\Rightarrow x} \\
=\;& \phi^{\Rightarrow x} \;=\; \phi \;\leq\; I.
\end{aligned}
$$

So we have $\psi \leq \phi^{\Rightarrow x}$ for some $\phi \in I$. Hence $\psi \in I^{\Rightarrow x}$ and we have shown that $I(X) \subseteq I^{\Rightarrow x}$.

Conversely, assume $\psi \in I^{\Rightarrow x}$, that is $\psi \leq \phi^{\Rightarrow x}$, $\phi \in I$. Then $\phi^{\Rightarrow x} \in I$, since $\phi^{\Rightarrow x} \leq \phi$. We conclude that $\psi \in I(X)$, since $\phi^{\Rightarrow x} = (\phi^{\Rightarrow x})^{\Rightarrow x} \leq I$, hence $\phi^{\Rightarrow x} \in X$. So we proved that $I^{\Rightarrow x} = I(X)$.

Compactness: Let $X \subseteq \Phi$ be a directed subset, and $\phi \leq \vee X$. Define $I = \vee X = I(X)$. Then $\phi \in I(X)$. But this means that

$$
\phi \;\leq\; \phi_1 \otimes \cdots \otimes \phi_m, \qquad \text{for } \phi_1, \ldots, \phi_m \in X.
$$

Since X is directed, there is some $\psi \in X$ such that

$$
\phi_1 \otimes \cdots \otimes \phi_m \;\leq\; \psi.
$$

Hence we have $\phi \leq \psi \in X$. $\square$

Another way to obtain compact information algebras is from *compact information systems*. An information system $(\mathcal{L}, C, \mathcal{S})$ is called *compact*, if the consequence operator C satisfies condition

(C6) $C(X) = \bigcup\{C(Y) : Y \subseteq X, Y \text{ finite}\}$.

This is equivalent to the following condition on the underlying entailment relation:

(E3) If $X \vdash A$, then there exists a finite subset $Y \subseteq X$ such that $Y \vdash A$.

We assume therefore now that the consequence operator C satisfies not only conditions (C1) to (C5) but also the compactness condition (C6). Then the information algebra induced by the information system is compact. The finite elements are those closed sets, which are closures of finite sets,

$$
\Phi_{C,f} \;=\; \{C(X) : X \subseteq \mathcal{L}, X \text{ finite}\}.
$$

Such closed sets are called *finitely axiomatizable*. Before we state the theorem that the $\Phi_{C,f}$ are the finite elements of the information algebra Φ_C associated with the consequence operator C, we mention that for closed sets $E_1, E_2 \in \Phi_C$ we have $E_1 \leq E_2$ if and only if $E_1 \subseteq E_2$. Thus E_1 is less informative than E_2 if, and only if, E_2 contains more conclusions than E_1. In fact, assume that $E_1 \otimes E_2 = E_2$. Then we have

$$
E_1 \;\subseteq\; C(E_1 \cup E_2) \;=\; E_1 \otimes E_2 \;=\; E_2.
$$

Conversely, if $E_1 \subseteq E_2$, then $E_1 \otimes E_2 = C(E_1 \cup E_2) = C(E_2) = E_2$.

Theorem 6.37 *Let C be a compact consequence operator, which satisfies also conditions (C4) and (C5). Then $(\Phi_C, \Phi_{C,f}, \mathcal{S})$ is a compact information algebra. We have:*

1. $C(\emptyset) \in \Phi_{C,f}$.

2. If $E_1 \in \Phi_{C,f}$ and $E_2 \in \Phi_{C,f}$, then $E_1 \otimes E_2 \in \Phi_{C,f}$.

3. *Convergence: If $\mathcal{D} \subseteq \Phi_{C,f}$ is a directed set, then the supremum $\vee\mathcal{D}$ exists and $\vee\mathcal{D} = \cup\mathcal{D} \in \Phi_C$.*

4. *Density: For $E \in \Phi_C$ and $L \in \mathcal{S}$ we have*

$$E^{\Rightarrow L} = \bigcup\{F \in \Phi_{C,f} : F \subseteq E, F^{\Rightarrow L} = F\}.$$

5. *Compactness: If $\mathcal{D} \subseteq \Phi_{C,f}$ is a directed set, $E \in \Phi_{C,f}$ and $E \subseteq \cup\mathcal{D}$, then there exists a $F \in \mathcal{D}$ such that $E \subseteq F$.*

Proof. (1) $C(\emptyset)$ belongs to $\Phi_{C,f}$ since the empty set is finite.

(2) Assume that $E_1 = C(X_1)$ and $E_2 = C(X_2)$, where X_1 and X_2 are finite. Then $E_1 \otimes E_2 = C(C(X_1) \cup C(X_2)) = C(X_1 \cup X_2) \in \Phi_{C,f}$ since $X_1 \cup X_2$ is finite too.

(3) Assume $\mathcal{D} \subseteq \Phi_{C,f}$ is directed. We have to show that $\cup\mathcal{D} \in \Phi_C$, i.e. we must show that $C(\cup\mathcal{D}) \subseteq \cup\mathcal{D}$. Therefore assume that $A \in C(\cup\mathcal{D})$. Since C is compact, there exists a finite set $X \subseteq \cup\mathcal{D}$ such that $A \in C(X)$. Note now that each element of X is in at least one of the closed sets of $\mathcal{D}$. Therefore, since $\mathcal{D}$ is directed, there exists a closed set $E \in \mathcal{D}$ such that $X \subseteq E$. So we obtain $A \in C(X) \subseteq C(E) = E \subseteq \cup\mathcal{D}$.

(4) By Lemma 6.22 (4), it follows that if $X \subseteq L$, then $C(X)^{\Rightarrow L} = C(C_L(X)) = C(X)$. Assume that $E \in \Phi_C$ and $L \in \mathcal{S}$. Then we have, by the compactness of C,

$$E^{\Rightarrow L} = C(E \cap L) = \bigcup\{C(X) : X \subseteq E \cap L, X \text{ finite}\}$$

$$\subseteq \bigcup\{F \in \Phi_{C,f} : F \subseteq E, F^{\Rightarrow L} = F\} \subseteq E^{\Rightarrow L}.$$

We obtain thus that $E^{\Rightarrow L} = \bigcup\{F \in \Phi_{C,f} : F \subseteq E, F^{\Rightarrow L} = F\}$.

(5) Assume that $\mathcal{D} \subseteq \Phi_{C,f}$ is directed, $E \in \Phi_{C,f}$ and $E \subseteq \cup\mathcal{D}$. There exists a finite set $X \subseteq \mathcal{L}$ such that $E = C(X)$. Since $X \subseteq C(X) \subseteq \cup\mathcal{D}$ and $\mathcal{D}$ is directed, there exists an element $F \in \mathcal{D}$ such that $X \subseteq F$. Hence we obtain that $E = C(X) \subseteq C(F) = F$. $\qquad\qquad\square$

Since a compact information algebra (Φ, Φ_f, D) is also an ordinary domain-free information algebra, it induces an associated information system. It generates in fact a compact information system in a way described as follows: Let (Φ, Φ_f, D) be a compact information algebra. As basic language for the information system we take now the set of finite pieces of information Φ_f. As sublanguages we take

$$\Phi_f^{\Rightarrow x} = \Phi^{\Rightarrow x} \cap \Phi_f = \{\psi \in \Phi_f : \psi = \psi^{\Rightarrow x}\}$$

for all $x \in D$. We assume in a similar way as before in the case of general information algebras that

$$\Phi_f^{\Rightarrow x} = \Phi_f^{\Rightarrow y} \qquad \text{implies} \qquad x = y. \qquad (6.48)$$

Then the following corollary of Lemma 6.26 holds:

Lemma 6.38 $S = \{\Phi_f^{\Rightarrow x} : x \in D\}$ *is a lattice under inclusion, isomorph to* D. *That is,*

1. $\Phi_f^{\Rightarrow x \wedge y} = \Phi_f^{\Rightarrow x} \cap \Phi_f^{\Rightarrow y}$.

2. $x \leq y$ *if, and only if,* $\Phi_f^{\Rightarrow x} \subseteq \Phi_f^{\Rightarrow y}$.

3. $\Phi_f^{\Rightarrow x \vee y}$ *is the supremum of* $\Phi_f^{\Rightarrow x}$ *and* $\Phi_f^{\Rightarrow y}$ *in* S, *that is* $\Phi_f^{\Rightarrow x \vee y} = \Phi_f^{\Rightarrow x} \vee \Phi_f^{\Rightarrow y}$.

This is proved just as Lemma 6.26.

Remember, that Φ is a complete lattice. In particular, the suprema over all subsets of Φ exist in Φ. So the following definition of an operator on Φ_f makes sense:

$$C_\Phi(X) = A_{\vee X} = \{\psi \in \Phi_f : \psi \leq \vee X\}. \qquad (6.49)$$

This defines a mapping $C_\Phi : \mathcal{P}(\Phi_f) \to \mathcal{P}(\Phi_f)$. We show that C_Φ is a compact consequence operator.

Theorem 6.39 *The operator* C_Φ *is a compact consequence operator on* Φ_f.

Proof. (1) Note that $X \subseteq A_{\vee X}$ for subsets of Φ_f. Hence $X \subseteq C_\Phi(X)$ and (C1) holds.

(2) Let $\eta = \vee X$. By equation (6.46) we obtain that $\vee A_\eta = \eta$. So we have

$$C_\Phi(C_\Phi(X)) = C_\Phi(A_\eta) = A_{\vee A_\eta} = A_\eta = C_\Phi(X). \qquad (6.50)$$

This is (C2).

(3) If $X \subseteq Y$, then $\vee X \leq \vee Y$ and hence

$$C_\Phi(X) = A_{\vee X} \subseteq A_{\vee Y} = C_\Phi(Y).$$

This is (C3).

(4) It remains to prove the compactness of C_Φ. Assume that $\phi \in C_\Phi(X)$. This means $\phi \leq \vee X$. By Theorem 6.35 (3) there exists a finite set $Y \subseteq X$ such that $\phi \leq \vee Y$. So $\phi \in C_\Phi(Y)$ and we obtain that

$$C_\Phi(X) \subseteq \bigcup \{C_\Phi(Y) : Y \subseteq X, Y \text{ finite}\}.$$

But the converse inclusion follows from (C3), hence we have equality above. So C_Φ satisfies (C6) and is compact. $\qquad \square$

So the compact information algebra (Φ, Φ_f, D) induces a compact information system (Φ_f, C_Φ, S). Assuming that C_Φ satisfies conditions (C4) and

(C5), which is not yet known, but will be shown to hold soon, we consider the domain-free information algebra associated to the compact information system (Φ_f, C_Φ, S). The operations of combination and focusing are defined as follows

1. Combination: $A_\phi \otimes A_\psi = A_{\vee(A_\phi \cup A_\psi)}$.

2. Focusing: $A_\phi^{\Rightarrow \Phi_f^{\Rightarrow x}} = A_{\vee(A_\phi \cap \Phi_f^{\Rightarrow x})}$.

We claim that the structure of the sets $A_\phi, \phi \in \Phi$ equipped with these operations is isomorphic to the information algebra (Φ, Φ_f, D) under the mapping

$$A \; : \; \Phi \to \{A_\phi \subseteq \Phi_f : \phi \in \Phi\}, \; \phi \mapsto A_\phi.$$

We have to check several conditions:

1. The mapping is one-one: This follows from (6.46). If $A_\phi = A_\psi$, then $\phi = \vee A_\phi = \vee A_\psi = \psi$.

2. Isomorphism of combination: Note that $\vee(A_\phi \cup A_\psi) = \phi \otimes \psi$ since $\eta \in (A_\phi \cup A_\psi)$ means either $\eta \leq \phi$ or $\eta \leq \psi$, hence $\eta \leq \phi \otimes \psi$. On the other hand let η be an upper bound of $A_\phi \cup A_\psi$, then $\eta \geq \vee A_\phi = \phi$ and $\eta \geq \vee A_\psi = \psi$, hence $\eta \geq \phi \otimes \psi$. Therefore, $A_\phi \otimes A_\psi = A_{\vee(A_\phi \cup A_\psi)} = A_{\phi \otimes \psi}$.

3. Isomorphism of Focusing: By the density axiom we have that $\vee(A_\phi \cap \Phi_f^{\Rightarrow x}) = \phi^{\Rightarrow x}$. Therefore, $A_\phi^{\Rightarrow \Phi_f^{\Rightarrow x}} = A_{\vee(A_\phi \cap \Phi_f^{\Rightarrow x})} = A_{\phi^{\Rightarrow x}}$.

4. Isomorphism of empty information: $A_e = A_{\vee \emptyset}$ since $e = \vee \emptyset$.

This proves implicitly that the consequence operator C_Φ as defined by (6.49) satisfies conditions (C4) and (C5). A finite element $\psi \in \Phi_f$ is mapped to $A_\psi = C_\Phi(\{\psi\})$, that is to a finitely axiomatizable closed set. Inversely, if $X \subseteq \Phi_f$ is a finite set, then $\vee X \in \Phi_f$ and therefore, $C_\Phi(X) = A_{\vee X}$ is the image of a finite element in Φ_f. So finite elements map to finite elements in both structures. We have therefore proved the following theorem:

Theorem 6.40 *If (Φ, Φ_f, D) is a compact information algebra and D has a top element, then (Φ_f, C_Φ, S) where C_Φ is defined by (6.49) is a compact information system. The compact information algebra (Φ, Φ_f, D) is isomorph to the compact information algebra induced by the compact information system (Φ_f, C_Φ, S).*

Note that the compact information algebra induced by the information system (Φ_f, C_Φ, S) depends *only on the finite elements* in Φ_f. Since this algebra is isomorph to the one of Φ, this indicates that the latter is also fully determined by its finite elements. Furthermore, an information algebra (Φ, D)

induces an information system (Φ, C_Φ, S). This information system induces the completed information algebra (I, Φ). But we have seen that this is in fact a compact algebra $(I, I(\Phi), D)$. Thus the associated information system (Φ, C_Φ, S) is compact too. For finite subsets $X \subseteq \Phi$, we have $C_\Phi = I(\vee X)$. So any finite set of sentences X can be replaced by its combination $\vee X$, that is a single element of Φ.

6.7 Mappings

Let now (Φ, D_Φ) and (Ψ, D_Ψ) be two information algebras. To simplify notation we denote combination and focusing in the two algebras Φ and Ψ by the same symbols, and the same holds for the order, meet and joins in the two lattices D_Φ and D_Ψ. It will always be clear from the context, were the operations take place. We study *order-preserving* or *monotone* mappings between these two algebras. A mapping $o : \Phi \to \Psi$ is called order-preserving, if

$$\phi \le \psi \qquad \text{implies} \qquad o(\phi) \le o(\psi). \qquad (6.51)$$

Such a mapping can be thought of as a processing which takes an element of Φ as input and produces an element of Ψ as output. (6.51) says then that less informative inputs produce less informative outputs.

As an immediate consequence we see that, since $\phi, \psi \le \phi \otimes \psi$, we have also $o(\phi) \otimes o(\psi) \le o(\phi \otimes \psi)$. If equality holds here, we have a *semi-group homomorphism*. Inversely, a semi-group homomorphism is an order-preserving mapping.

A few simple examples of monotone mappings from (Φ, D_Φ) into itself are as follows:

1. Identity mapping: $o(\phi) = \phi$.

2. Constant mapping: $o(\phi) = \psi$.

3. Focusing: $o(\phi) = \phi^{\Rightarrow x}$

4. Conditioning: $o(\phi) = \psi \otimes \phi$ for a fixed $\psi \in \Phi$.

In the first and third example we have that $o(e) = e$ and if $\phi \ne z$ then $o(\phi) \ne z$. These properties do not hold in the second and fourth example, but these examples are semi-group homomorphisms.

Furthermore, compositions of monotone mappings are monotone. That is, if o_1 is an order-preserving mapping from (Φ_1, D_{Φ_1}) into (Φ_2, D_{Φ_2}) and o_2 is an order-preserving mappings from (Φ_2, D_{Φ_2}) into (Φ_3, D_{Φ_3}), then $o_1 \circ o_2$ is a monotone mapping from (Φ_1, D_{Φ_1}) into (Φ_3, D_{Φ_3}).

We denote the set of all monotone mappings from (Φ, D_Φ) into (Ψ, D_Ψ) by $[\Phi \to \Psi]$. We are going to define in this set an operation of combination. Thus, if $o_1, o_2 \in [\Phi \to \Psi]$, then $o_1 \otimes o_2$ is defined by

$$o_1 \otimes o_2(\phi) = o_1(\phi) \otimes o_2(\phi). \qquad (6.52)$$

If we look at o_1 and o_2 as two processing procedures, which apply to the same input, then they can be combined into one procedure, simply by combining the outputs. If $\phi \leq \psi$, then $o_1 \otimes o_2(\phi) = o_1(\phi) \otimes o_2(\phi) \leq o_1(\psi) \otimes o_2(\psi) = o_1 \otimes o_2(\psi)$, hence $o_1 \otimes o_2 \in [\Phi \rightarrow \Psi]$. Clearly, this operation is *associative, commutative* and the constant mapping defined by $o(\phi) = e$ for all $\phi \in \Phi$ is the neutral element of this operation.

We are next going to define focusing of monotone mappings. Note that the cartesian product of two lattices $D_\Phi \times D_\Psi$ becomes also a lattice, if we define $(x, y) \leq (x', y')$ if, and only if, $x \leq x'$ and $y \leq y'$. We then have $(x, y) \wedge (x', y') = (x \wedge x', y \wedge y')$. So we define, with respect to this product lattice, $o^{\Rightarrow(x,y)}$ by

$$o^{\Rightarrow(x,y)}(\phi) = (o(\phi^{\Rightarrow x}))^{\Rightarrow y}. \tag{6.53}$$

Note that we write also $o^{\Rightarrow y}(\phi^{\Rightarrow x})$ for $(o(\phi^{\Rightarrow x}))^{\Rightarrow y}$.

It turns out that $[\Phi \rightarrow \Psi]$ together with $D_\Phi \times D_\Psi$ becomes an information algebra with combination and focusing defined as above.

Theorem 6.41 *Suppose that D_Φ and D_Ψ have both a top element. Then $([\Phi \rightarrow \Psi], D_\Phi \times D_\Psi)$, where the operations $\otimes$ and $\Rightarrow$ are defined by (6.52) and (6.53), is a domain-free information algebra.*

Proof. (1) The semigroup properties hold as noted above.

(2) In order to verify *transitivity* assume $x, x' \in D_\Phi$ and $y, y' \in D_\Psi$. Then, using the transitivity in the two algebras (Φ, D_Φ) and (Ψ, D_Ψ), we obtain

$$
\begin{aligned}
(o^{\Rightarrow(x,y)})^{\Rightarrow(x',y')}(\phi) &= (o^{\Rightarrow(x,y)}(\phi^{\Rightarrow x'}))^{\Rightarrow y'} = ((o((\phi^{\Rightarrow x'})^{\Rightarrow x}))^{\Rightarrow y})^{\Rightarrow y'} \\
&= (o(\phi^{\Rightarrow x \wedge x'}))^{\Rightarrow y \wedge y'} = o^{\Rightarrow(x \wedge x', y \wedge y')}(\phi) \\
&= o^{\Rightarrow(x,y) \wedge (x',y')}(\phi).
\end{aligned}
$$

(3) The *combination axiom* is verified as follows:

$$
\begin{aligned}
(o_1^{\Rightarrow(x,y)} \otimes o_2)^{\Rightarrow(x,y)}(\phi) &= ((o_1^{\Rightarrow(x,y)} \otimes o_2)(\phi^{\Rightarrow x}))^{\Rightarrow y} \\
&= (o_1^{\Rightarrow(x,y)}(\phi^{\Rightarrow x}) \otimes o_2(\phi^{\Rightarrow x}))^{\Rightarrow y} \\
&= ((o_1(\phi^{\Rightarrow x}))^{\Rightarrow y} \otimes o_2(\phi^{\Rightarrow x}))^{\Rightarrow y} \\
&= (o_1(\phi^{\Rightarrow x}))^{\Rightarrow y} \otimes (o_2(\phi^{\Rightarrow x}))^{\Rightarrow y} \\
&= o_1^{\Rightarrow(x,y)}(\phi) \otimes o_2^{\Rightarrow(x,y)}(\phi) \\
&= o_1^{\Rightarrow(x,y)} \otimes o_2^{\Rightarrow(x,y)}(\phi).
\end{aligned}
$$

Here we have used the combination axiom in the algebra (Ψ, D_Ψ).

(4) $(\top, \top)$ is the top-element of the lattice $D_\Phi \times D_\Psi$. This element is a support of every mapping $o \in [\Phi \rightarrow \Psi]$ since $\top$ is a support for every element in Φ and Ψ. Thus the *support axiom* holds.

(5) We have

$$o(\phi) \leq o(\phi) \otimes o^{\Rightarrow(x,y)}(\phi) = o(\phi) \otimes (o(\phi^{\Rightarrow x}))^{\Rightarrow y}.$$

But $\phi^{\Rightarrow x} \leq \phi$ implies that $o(\phi^{\Rightarrow x}) \leq o(\phi)$. Thus we have by the idempotency axiom in Ψ

$$o(\phi) \leq o(\phi) \otimes (o(\phi))^{\Rightarrow y} = o(\phi).$$

Hence we obtain that

$$o(\phi) \leq o(\phi) \otimes o^{\Rightarrow (x,y)}(\phi) = o(\phi).$$

This proves the *idempotency axiom*.   □

Not surprisingly, mappings which describe some form of information processing (procedures) represent a form of information and indeed they form also an information algebra.

We turn now to compact information algebras (Φ, Φ_f, D_Φ) and (Ψ, Ψ_f, D_Ψ). In this context we consider *continuous* mappings, which maintain limits.

Definition 6.42 Continuous Mappings. *A mapping* $f : \Phi \to \Psi$ *from a compact algebra* (Φ, Φ_f, D_Φ) *into another compact algebra* (Ψ, Ψ_f, D_Ψ) *is called* continuous, *if for all* $\phi \in \Phi$

$$f(\phi) = \vee\{f(\phi') : \phi' \in \Phi_f, \phi' \leq \phi\}. \tag{6.54}$$

We note that a continuous mapping f is order preserving. In fact, assume $\phi_1 \leq \phi_2$. Then

$$\begin{aligned} f(\phi_1) &= \vee\{f(\phi') : \phi' \in \Phi_f, \phi' \leq \phi_1\} \\ &\leq \vee\{f(\phi') : \phi' \in \Phi_f, \phi' \leq \phi_2\} = f(\phi_2). \end{aligned}$$

The following equivalent definitions of continuity are well known in lattice theory (Davey & Priestley, 1990).

Lemma 6.43 *Let* $f : \Phi \to \Psi$ *be an order preserving mapping from a compact algebra* (Φ, Φ_f, D_Φ) *into another compact algebra* (Ψ, Ψ_f, D_Ψ). *Then, the following conditions are equivalent:*

1. f *is continuous.*

2. $f(\vee X) = \vee f(X)$ *for every directed subset* $X \subseteq \Phi$.

3. $A_{f(\phi)} \subseteq \{\psi \in \Psi : \psi \leq f(\phi') \text{ for some } \phi' \in \Phi\}$.

For the proof we refer to (Davey & Priestley, 1990).

The following theorem shows that the mapping $\phi \mapsto \phi^{\Rightarrow x}$ is continuous.

Theorem 6.44 *If X is a directed subset of Φ and $x \in D$, then*

$$(\vee X)^{\Rightarrow x} = \bigvee_{\phi \in X} \phi^{\Rightarrow x}. \tag{6.55}$$

Proof. $\phi \in X$ implies $\phi \leq \vee X$, hence $\phi^{\Rightarrow x} \leq (\vee X)^{\Rightarrow x}$ and

$$\bigvee_{\phi \in X} \phi^{\Rightarrow x} \leq (\vee X)^{\Rightarrow x}.$$

Conversely, by the density axiom,

$$(\vee X)^{\Rightarrow x} = \vee\{\psi \in \Phi_f : \psi = \psi^{\Rightarrow x} \leq \vee X\}.$$

But, if $\psi = \psi^{\Rightarrow x} \in \Phi_f$ such that $\psi \leq \vee X$, then, since X is directed, there is a $\phi \in X$ such that $\psi \leq \phi$ (see Theorem 6.35 (2)). For this element ϕ we have $\psi = \psi^{\Rightarrow x} \leq \phi^{\Rightarrow x}$. Hence

$$(\vee X)^{\Rightarrow x} \leq \bigvee_{\phi \in X} \phi^{\Rightarrow x}.$$

$\square$

Let $[\Phi \to \Psi]_c$ denote the set of continuous mappings from (Φ, Φ_f, D_Φ) into (Ψ, Ψ_f, D_Ψ). We define the operations of combination and focusing between continuous mappings in $[\Phi \to \Psi]_c$ as above for monotone mappings:

1. *Combination:* for $f, g \in [\Phi \to \Psi]_c$ define $f \otimes g$ by

$$(f \otimes g)(\phi) = f(\phi) \otimes g(\phi).$$

2. *Focusing:* for $f \in [\Phi \to \Psi]_c$ and $x \in D_\Phi$, $y \in D_\Psi$ define $f^{\Rightarrow(x,y)}$ by

$$f^{\Rightarrow(x,y)}(\phi) = (f(\phi^{\Rightarrow x}))^{\Rightarrow y}.$$

We are going to show, that with these operations, $[\Phi \to \Psi]_c$ becomes a compact information algebra. In order to do that we must first verify that combination and focusing of continuous mappings yield again continuous mappings.

Theorem 6.45 *If $f, g \in [\Phi \to \Psi]_c$, $x \in D_\Phi$, $y \in D_\Psi$ then $f \otimes g \in [\Phi \to \Psi]_c$, and $f^{\Rightarrow(x,y)} \in [\Phi \to \Psi]_c$.*

Proof. (1) For every $\phi \in \Phi$ we have

$$
\begin{aligned}
(f \otimes g)(\phi) &= f(\phi) \otimes g(\phi) \\
&= (\vee\{f(\phi') : \phi' \in \Phi_f, \phi' \leq \phi\}) \vee (\vee\{g(\phi') : \phi' \in \Phi_f, \phi' \leq \phi\}) \\
&= \vee\{f(\phi'_1) \vee g(\phi'_2) : \phi'_1, \phi'_2 \in \Phi_f, \phi'_1, \phi'_2 \leq \phi\} \\
&= \vee\{f(\phi') \vee g(\phi') : \phi' \in \Phi_f, \phi' \leq \phi\}.
\end{aligned}
$$

The last equality follows since $\phi'_1, \phi'_2 \leq \phi$ implies $\phi' = \phi'_1 \otimes \phi'_2 \leq \phi$. Hence we obtain that

$$(f \otimes g)(\phi) = \vee\{(f \otimes g)(\phi') : \phi' \in \Phi_f, \phi' \leq \phi\}.$$

Thus $f \otimes g$ is continuous.

(2) We note first that $f^{\Rightarrow(x,y)}$ is order-preserving. In fact, assume $\phi_1 \leq \phi_2$, such that $\phi_1^{\Rightarrow x} \leq \phi_2^{\Rightarrow x}$. Then

$$f^{\Rightarrow(x,y)}(\phi_1) = (f(\phi_1^{\Rightarrow x}))^{\Rightarrow y} \leq (f(\phi_2^{\Rightarrow x}))^{\Rightarrow y} = f^{\Rightarrow(x,y)}(\phi_2),$$

since $\phi_1^{\Rightarrow x} \leq \phi_2^{\Rightarrow x}$ implies $f(\phi_1^{\Rightarrow x}) \leq f(\phi_2^{\Rightarrow x})$, because f is order-preserving. This in turn implies $(f(\phi_1^{\Rightarrow x}))^{\Rightarrow y} \leq (f(\phi_2^{\Rightarrow x}))^{\Rightarrow y}$.

Suppose now X to be a directed set. Note that in this case, $\{\phi^{\Rightarrow x} : \phi \in X\}$ is also directed, and so is $\{f(\phi^{\Rightarrow x}) : \phi \in X\}$ Then, by Theorem 6.44 and Lemma 6.43 (2), we obtain

$$f^{\Rightarrow(x,y)}(\vee X) = (f((\vee X)^{\Rightarrow x}))^{\Rightarrow y} = \left(f\left(\bigvee_{\phi \in X} \phi^{\Rightarrow x} \right) \right)^{\Rightarrow y}$$

$$= \left(\bigvee_{\phi \in X} (f(\phi^{\Rightarrow x})) \right)^{\Rightarrow y} = \bigvee_{\phi \in X} f(\phi^{\Rightarrow x}))^{\Rightarrow y}$$

$$= \bigvee_{\phi \in X} f^{\Rightarrow(x,y)}(\phi).$$

According to Lemma 6.43 this proves that $f^{\Rightarrow(x,y)}$ is continuous. □

Since, by this last theorem, $[\Phi \rightarrow \Psi]_c$ is closed under the operations of combination and focusing, and the neutral mapping is continuous, $[\Phi \rightarrow \Psi]_c$ must be an information algebra, namely a subalgebra of the information algebra $[\Phi \rightarrow \Psi]$ of order-preserving mappings. We are going now to show that it is a compact algebra.

What are the finite elements of this algebra? Let Y be a *finite* subset of finite elements from Φ_f. A mapping $f : Y \rightarrow \Psi_f$ is called a *simple function*. Let S be the set of all simple functions. For $s \in S$ and $\phi \in \Phi$ we define $Y(s)$ to be the domain of s and

$$\hat{s}(\phi) = \vee\{s(\phi') : \phi' \in Y(s), \phi' \leq \phi\}.$$

We have $\hat{s}(\phi) = e$, if the set on the right hand side of this definition above is empty.

Lemma 6.46 *Let $s \in S$. Then*

1. $\hat{s} : \Phi \rightarrow \Psi_f$.

2. $\hat{s}$ *is continuous.*

Proof. (1) Let $X = \{s(\psi) : \psi \in Y(s), \psi \leq \phi\}$. Then $X \subseteq \Psi_f$ is a finite set. Hence $\hat{s} = \vee X \in \Psi_f$.

(2) Assume $\phi_1 \leq \phi_2$. Then we have that

$$\{s(\psi) : \psi \in Y(s), \psi \leq \phi_1\} \subseteq \{s(\psi) : \psi \in Y(s), \psi \leq \phi_2\}.$$

Hence $\hat{s}(\phi_1) \leq \hat{s}(\phi_2)$ which shows that $\hat{s}$ is an order-preserving mapping.

Let now $X \subseteq \Phi$ be a directed set. Then $\hat{s}(\phi) \leq \hat{s}(\vee X)$ if $\phi \in X$, hence $\vee_{\phi \in X} \hat{s}(\phi) \leq \hat{s}(\vee X)$. We claim that the inverse inequality $\hat{s}(\vee X) \leq \vee_{\phi \in X} \hat{s}(\phi)$ holds also.

Let $\psi \in Y(s)$, $\psi \leq \vee X$. Note that $\psi \in \Phi_f$. Since X is directed, there is a $\phi \in X$ such that $\psi \leq \phi$. Thus we have that $s(\psi) \leq \hat{s}(\psi) \leq \hat{s}(\phi)$, so

$$\hat{s}(\vee X) \quad = \quad \vee\{s(\psi) : \psi \in Y(s), \psi \leq \vee X\} \quad \leq \quad \bigvee_{\phi \in X} \hat{s}(\phi).$$

This shows that $\vee_{\phi \in X} \hat{s}(\phi) = \hat{s}(\vee X)$, which, by Lemma 6.43 proves that $\hat{s}$ is continuous. $\square$

Let $\hat{S} = \{\hat{s} : s \in S\}$. We have shown that $\hat{S} \subseteq [\Phi \to \Psi]_c$. We claim that $\hat{S}$ are the finite elements of $[\Phi \to \Psi]_c$.

We show first that $\hat{S}$ is closed under combination. Let s_1 and s_2 be two simple mappings and $Y_1 = Y(s_1)$, $Y_2 = Y(s_2)$. Define $s : Y_1 \cup Y_2 \to \Psi_f$ by $\psi \mapsto \hat{s}_1(\psi) \vee \hat{s}_2(\psi)$. Clearly, s is a simple mapping. We claim that $\hat{s} = \hat{s}_1 \otimes \hat{s}_2$. In fact,

$$
\begin{aligned}
\hat{s}(\phi) \quad &= \quad \vee\{\hat{s}_1(\psi) \vee \hat{s}_2(\psi) : \psi \in Y_1 \cup Y_2, \psi \leq \phi\} \\
&= \quad (\vee\{s_1(\psi_1) : \psi_1 \in Y_1, \psi_1 \leq \psi \in Y_1 \cup Y_2, \psi \leq \phi\}) \\
&\quad \vee (\vee\{s_2(\psi_2) : \psi_2 \in Y_2, \psi_2 \leq \psi \in Y_1 \cup Y_2, \psi \leq \phi\}) \\
&= \quad (\vee\{s_1(\psi_1) : \psi_1 \in Y_1, \psi_1 \leq \phi\}) \vee (\vee\{s_2(\psi_2) : \psi_2 \in Y_2, \psi_2 \leq \phi\}) \\
&= \quad \hat{s}_1(\phi) \vee \hat{s}_2(\phi). \quad\quad\quad\quad\quad\quad\quad\quad\quad\quad\quad\quad\quad\quad (6.56)
\end{aligned}
$$

The vacuous mapping $e(\phi) = e$ belongs to $\hat{S}$ since $e = \hat{f}$ where $Y(f) = \{e\}$, $f(e) = e$.

Next we verify the axioms of convergence, density and compactness for $\hat{S}$ relative to $[\Phi \to \Psi]_c$.

Convergence follows from the fact that for all $X \subseteq [\Phi \to \Psi]_c$ we have that $\vee X \in [\Phi \to \Psi]_c$. In fact, let $f = \vee X$, that is, $f(\phi) = \vee_{g \in X} g(\phi)$. If $Y \subseteq \Phi$ is directed, then, by the continuity of g and the associative law for the supremum in the lattice Ψ,

$$
\begin{aligned}
f(\vee Y) \quad &= \quad \bigvee_{g \in X} g(\vee Y) \quad = \quad \bigvee_{g \in X}\left(\bigvee_{\phi \in Y} g(\phi)\right) \\
&= \quad \bigvee_{\phi \in Y}\left(\bigvee_{g \in X} g(\phi)\right) \quad = \quad \bigvee_{\phi \in Y} f(\phi).
\end{aligned}
$$

Density: We have to show that

$$f^{\Rightarrow(x,y)} \quad = \quad \vee\{\hat{s} : s \in S, \hat{s} = \hat{s}^{\Rightarrow(x,y)} \leq f\}.$$

By definition and the continuity of f,

$$f^{\Rightarrow(x,y)}(\phi) \;=\; (f(\phi^{\Rightarrow x}))^{\Rightarrow y} \;=\; (\vee\{f(\alpha) : \alpha \in \Phi_f, \alpha = \alpha^{\Rightarrow x} \leq \phi\})^{\Rightarrow y}.$$

This set is directed, hence, by Theorem 6.44,

$$f^{\Rightarrow(x,y)}(\phi) \;=\; \vee\{(f(\alpha))^{\Rightarrow y} : \alpha \in \Phi_f, \alpha = \alpha^{\Rightarrow x} \leq \phi\}.$$

We are going to show that for $\alpha \in \Phi_f$, $\alpha = \alpha^{\Rightarrow x} \leq \phi$ we have

$$(f(\alpha))^{\Rightarrow y} \;=\; \vee\{\hat{s}(\alpha) : s \in S, \hat{s} = \hat{s}^{\Rightarrow(x,y)} \leq f\}. \tag{6.57}$$

This implies then

$$f^{\Rightarrow(x,y)} \;\leq\; \vee\{\hat{s} : s \in S, \hat{s} = \hat{s}^{\Rightarrow(x,y)} \leq f\}.$$

Since the inverse inequality clearly holds, we are done.

To prove (6.57) note that by the density in Ψ we have

$$(f(\alpha))^{\Rightarrow y} \;=\; \vee\{\beta \in \Psi_f : \beta = \beta^{\Rightarrow y} \leq f(\alpha)\}.$$

Consider $\beta \in \Psi_f$ such that $\beta = \beta^{\Rightarrow y} \leq f(\alpha)$. Let $s : \{\alpha\} \to \Psi_f$ defined by $s(\alpha) = \beta$. Then s is a simple mapping in S. And,

$$\hat{s}(\phi) \;=\; \begin{cases} \beta, & \text{if } \alpha \leq \phi, \\ e, & \text{otherwise.} \end{cases}$$

So we have $\hat{s}(\phi) \leq f(\phi)$ for all ϕ, that is, $\hat{s} \leq f$.

It remains to show that $\hat{s}^{\Rightarrow(x,y)} = \hat{s}$. Consider $\gamma \in \Phi$. We have to show that $(\hat{s}(\gamma^{\Rightarrow x}))^{\Rightarrow y} = \hat{s}(\gamma)$. If $\alpha = \alpha^{\Rightarrow x}$, then $\alpha = \alpha^{\Rightarrow x} \leq \gamma$ holds if, and only if, $\alpha \leq \gamma^{\Rightarrow x}$. We have two cases:

case 1: $\alpha \leq \gamma$. Then

$$(\hat{s}(\gamma^{\Rightarrow x}))^{\Rightarrow y} \;=\; (s(\alpha))^{\Rightarrow y} \;=\; \beta^{\Rightarrow y} \;=\; \beta \;=\; \hat{s}(\gamma),$$

since $\alpha \leq \gamma^{\Rightarrow x}$ implies $\hat{s}(\gamma^{\Rightarrow x}) = s(\alpha)$.

case 2: $\alpha \not\leq \gamma$. In this case

$$(\hat{s}(\gamma^{\Rightarrow x}))^{\Rightarrow y} \;=\; e \;=\; \hat{s}(\gamma).$$

So we have indeed $\hat{s}^{\Rightarrow(x,y)} = \hat{s}$. This proves that for $\alpha = \alpha^{\Rightarrow x} \leq \phi$,

$$(f(\alpha))^{\Rightarrow y} \;\leq\; \vee\{\hat{s}(\alpha) : s \in S, \hat{s} = \hat{s}^{\Rightarrow(x,y)} \leq f\}.$$

The inverse inequality is obvious. This proves (6.57).

Compactness: Let $X \subseteq [\Phi \to \Psi]_c$ be a directed set and $\hat{s} \leq \vee X$. If $\psi \in Y(s)$, then $s(\psi) \in \Psi_f$ and $s(\psi) = \hat{s}(\psi) \leq (\vee X)(\psi) = \vee_{f \in X} f(\psi)$. The set of $f(\psi)$ for $f \in X$ is directed in Φ. Therefore, there is a $f \in X$ such that $s(\psi) \leq f(\psi)$. Since $Y(s)$ is finite and X directed, there is a $g \in X$ such that $s(\psi) \leq g(\psi)$ for all $\psi \in Y$. Hence, $\hat{s} \leq g \in X$.

Thus, we have proved the following theorem:

Theorem 6.47 $[\Phi \to \Psi]_c$ *is a compact information algebra with* $\hat{S}$ *as finite elements.*

7 Uncertain Information

7.1 Algebra of Random Variables

7.1.1 Random Variables

Information can be uncertain. Uncertainty may arise because events depend on random elements or uncertainty may be due to incomplete knowledge. In any case, information can often be asserted only with a certain probability. In this chapter we show that information algebras provide a natural framework to describe uncertainty. We start by modeling uncertainty in information algebras by (generalized) random variables, taking values in information algebras. This extends the fundamental work of A. Dempster (Dempster, 1967) and G. Shafer (Shafer, 1976) to the framework of information algebras. In the following Section 7.2 we show, how generalized random variables arise in a natural way from assumption-based reasoning in information systems. This is an extension of the idea of "probability of provability" introduced by Pearl (Pearl, 1988) and also discussed by (Laskey & Lehner, 1989). This idea has evolved into so-called *probabilistic argumentation systems* (Haenni, Kohlas & Lehmann, 2000). Closely related to such systems are structures called *hints* . The notion of a hint has been introduced in (Kohlas & Monney, 1995).

The modeling of uncertain information, discussed in the two first sections of this chapter, is based on an explicit representation of the mechanism introducing uncertainty. An alternative way to describe uncertainty in information algebra, more in the spirit of subjective belief, starts by defining allocations of probability on information algebras . The notion of allocations of probability has been introduced in (Shafer, 1979) as a fundamental concept for the so-called Dempster-Shafer theory of evidence. Again it turns out that information algebras provide the natural framework for this concept. This is discussed in Section 7.3.

Often it is assumed that uncertain information comes from "independent" sources. This case is discussed in Section 7.4. It is shown how random variables and related support functions, together with a notion of independence of sources of information, lead to valuation algebras. This then closes the circle in the sense that it shows how valuation algebras arise from information algebras, if uncertainty is present.

We introduce first the notion of a *probability algebra*. Let $\mathcal{B}$ be a Boolean σ-algebra. A Boolean σ-algebra $\mathcal{B}$ satisfies the so-called *countable chain condition*, if every family $\{b_i, i \in I\}$ of *disjoint* elements is *countable*. It is well-known that a Boolean σ-algebra $\mathcal{B}$ which satisfies the countable chain condition is *complete* (see for example (Halmos, 1963)). This means that every subset E of $\mathcal{B}$ has an infimum (denoted by $\wedge E$) and a supremum (denoted by $\vee E$). Furthermore, the countable chain condition implies that there is a countable subset D of E such that $\wedge E = \wedge D$ and $\vee E = \vee D$. For further reference we remark that in a complete Boolean algebra the following identities hold (Halmos, 1963):

De Morgan:

$$\left(\bigvee_i b_i \right)^c = \bigwedge_i b_i^c, \qquad \left(\bigwedge_i b_i \right)^c = \bigvee_i b_i^c. \tag{7.1}$$

Associative Laws: If I is the union of sets $I_j, j \in J$ then

$$\bigvee_{j \in J} \left(\bigvee_{i \in I_j} b_i \right) = \bigvee_{i \in I} b_i, \qquad \bigwedge_{j \in J} \left(\bigwedge_{i \in I_j} b_i \right) = \bigwedge_{i \in I} b_i. \tag{7.2}$$

Distributive laws:

$$\left(\bigvee_{i \in I} b_i \right) \wedge \left(\bigvee_{j \in J} c_j \right) = \bigvee_{i \in I, j \in J} (b_i \wedge c_j), \tag{7.3}$$

$$\left(\bigwedge_{i \in I} b_i \right) \vee \left(\bigwedge_{j \in J} c_j \right) = \bigwedge_{i \in I, j \in J} (b_i \vee c_j). \tag{7.4}$$

The following particular form of the associative law is often used:

$$\bigwedge_{i \in I_1, j \in I_2} (b_i \wedge c_j) = \left(\bigwedge_{i \in I_1} b_i \right) \wedge \left(\bigwedge_{j \in I_2} c_j \right). \tag{7.5}$$

A probability algebra arises from a Boolean σ-algebra, satisfying the countable chain condition, if a positive probability measure is added.

Definition 7.1 Probability Algebra. *Let $\mathcal{B}$ be a Boolean σ-algebra, satisfying the countable chain condition, with top element $\top$, bottom element $\bot$, and μ a normalized, positive measure on $\mathcal{B}$. That is μ is σ-additive on $\mathcal{B}$, $\mu(\top) = 1$, and $\mu(b) = 0$ implies $b = \bot$. Then $(\mathcal{B}, \mu)$ is called a probability algebra.*

Such probability algebras are studied in detail in (Kappos, 1969). We add an important property of probability algebras: clearly $\mu(\wedge b_i) \leq \inf_i \mu(b_i)$ and $\mu(\vee b_i) \geq \sup_i \mu(b_i)$ holds for any family of elements $\{b_i\}$. But there are important cases where equality hold. A subset D of $\mathcal{B}$ is called *downward (upward) directed*, if for every pair $b', b'' \in D$ there is an element $b \in D$ such that $b \leq b' \wedge b''(b \geq b' \vee b'')$.

Lemma 7.2 *If D is a downward (upward) directed subset of $\mathcal{B}$, then*

$$\mu(\wedge b_i) \;=\; \inf_i \mu(b_i), \qquad (\mu(\vee b_i) \;=\; \sup_i \mu(b_i)) \tag{7.6}$$

Proof. There is a countable subfamily of elements b_i which have the same meet. If $c_i, i = 1, 2, \ldots$ are these elements, then define $c_1' = c_1$ and select elements c_i' in the downward directed set such that $c_2' \leq c_1' \wedge c_2$, $c_3' \leq c_2' \wedge c_3, \ldots$. Then $c_1' \geq c_2' \geq c_3' \geq \ldots$ and this sequence has still the same infimum. However, by the continuity of probability we have now

$$\mu(\wedge b_i) \;=\; \mu(\wedge c_i') \;=\; \lim_{i \to \infty} \mu(c_i') \;\geq\; \inf_i \mu(b_i). \tag{7.7}$$

But this implies $\mu(\wedge b_i) = \inf_i \mu(b_i)$. The case of upwards directed sets is treated in the same way. $\square$

The elements b of a probability algebra can be considered as pieces of belief, which are measured by $\mu(b)$. A particular source of knowledge may yield uncertain information relative to a certain information algebra (Φ, D). In the simplest case, the uncertainty may be represented by a decomposition of the total belief $\top$ into a finite number of disjoint pieces of belief $b_1, \ldots, b_m$. This decomposition corresponds to a number of possible, but mutually incompatible interpretations of the knowledge. To each possible interpretation i corresponds a belief b_i and an associated element $\phi_i \in \Phi$, which represents the information asserted, if interpretation i happens to be the right one. We are going to formalize this informal idea as follows:

Definition 7.3 Simple Experiment. *A finite decomposition $\mathbf{b} = \{b_1, \ldots\ldots, b_m\}$ of the total belief $\top$ of a probability algebra $(\mathcal{B}, \mu)$ is called a* simple experiment.

Denote by $\mathcal{E}$ the set of all simple experiments in $(\mathcal{B}, \mu)$. We introduce a meet-operation between two experiments $\mathbf{b}_1$ and $\mathbf{b}_2$ of $\mathcal{E}$:

$$\mathbf{b}_1 \wedge \mathbf{b}_2 \;=\; \{b_1 \wedge b_2 : b_1 \in \mathbf{b}_1, b_2 \in \mathbf{b}_2, b_1 \wedge b_2 \neq \bot\}. \tag{7.8}$$

This operation is clearly commutative and associative, as well as idempotent. $\mathcal{E}$ is thus a semilattice, with a partial order $\mathbf{b}_1 \leq \mathbf{b}_2$ defined by $\mathbf{b}_1 = \mathbf{b}_1 \wedge \mathbf{b}_2$.

Definition 7.4 Simple Random Variable. *Let (Φ, D) be a domain-free information algebra and $\mathbf{b} \in \mathcal{E}$ a simple experiment. Let h be a mapping $h : \mathbf{b} \to \Phi$. Then h is called a* simple random variable *defined on $\mathbf{b}$.*

We denote by $\mathcal{H}_s$ the set of all simple random variables based on a probability algebra $(\mathcal{B}, \mu)$ and a domain-free information algebra (Φ, D).

There are a number of important special classes of random variables:

1. *Normalized Random Variables:* If $h(b) \neq z$ for all $b \in \mathbf{b}$, then the random variable h is called *normalized.*

2. *Canonical Random Variables:* If $h(b) \neq h(b')$ for $b \neq b'$, then the random variable h is called *canonical.*

3. *Vacuous Random Variable:* The random variable e, defined on $\{\top\}$ with $e(\top) = e$ is called *vacuous.*

Note that the vacuous random variable is both normalized and canonical. If h is a random variable defined on $\mathbf{b}$, then we can associate with it a corresponding unique canonical random variable $h^{\rightarrow}$. For any $b \in \mathbf{b}$ define $c = \vee\{b' : h(b') = h(b)\}$. That is we group all elements b in $\mathbf{b}$ with the same image $h(b)$ together. The elements c define a simple experiment $\mathbf{b}^{\rightarrow}$. On this simple experiment we define $h^{\rightarrow}(c) = h(b)$, if c groups the beliefs with value $h(b)$ together.

We can also associate a unique normalized random variable $h^{\downarrow}$ with any random variable h defined on $\mathbf{b}$, provided $h(b) \neq z$ for at least one element of $\mathbf{b}$. Define $c = \vee\{b : h(b) \neq z\}$. By $\mathcal{B} \wedge c$ we denote the Boolean algebra whose elements are of the form $b \wedge c$ for $b \in \mathcal{B}$. The operations of meet and join are inherited from $\mathcal{B}$: $(b \wedge c) \wedge (b' \wedge c) = (b \wedge b') \wedge c$, and similarly for the join. Complementation is relative to c: $(b \wedge c)^c = b^c \wedge c$. The top element of $\mathcal{B} \wedge c$ is c. It is clear that $\mathcal{B} \wedge c$ is still a σ-algebra satisfying the countable chain condition. By μ/c we denote the positive measure on $\mathcal{B} \wedge c$ defined by

$$\mu/c(b) \quad = \quad \frac{\mu(b)}{\mu(c)}. \tag{7.9}$$

This corresponds of course to a conditional probability, given c. $(\mathcal{B} \wedge c, \mu/c)$ is still a probability algebra. And $\mathbf{b} \wedge c = \{b \in \mathbf{b}, h(b) \neq z\}$ is a simple experiment in this probability algebra. Then $h^{\downarrow}$ is defined by $h^{\downarrow}(b) = h(b)$ for all $b \in \mathbf{b} \wedge c$.

Remark that $(h^{\downarrow})^{\rightarrow} = (h^{\rightarrow})^{\downarrow}$. So we may interchange the operations of normalizing and canonizing.

We are now going to define operations of combination and focusing in the set of simple random variables. Combination corresponds to the aggregation of the pieces of information represented by two or more random variables. Focusing on the other hand represents the extraction of the information encoded in the random variable relative to some domain x.

Combination: Let h_1 and h_2 be two simple random variables defined on $\mathbf{b}_1$ and $\mathbf{b}_2$. Combination means first the combination of the beliefs in the two sets $\mathbf{b}_1$ and $\mathbf{b}_2$ of possible interpretations. So we get the new set of possible interpretations $\mathbf{b}_1 \wedge \mathbf{b}_2$ consisting of the beliefs $b_1 \wedge b_2 \neq \bot$ of $b_1 \in \mathbf{b}_1$ and

$b_2 \in \mathbf{b}_2$. To the belief $b_1 \wedge b_2$ corresponds the combination $h_1(b_1) \otimes h_2(b_2)$ of the two individual pieces of information associated with b_1 and b_2. Thus, the combined random variable, denoted by $h_1 \otimes h_2$ is defined on $\mathbf{b}_1 \wedge \mathbf{b}_2$ by

$$(h_1 \otimes h_2)(b_1 \wedge b_2) \quad = \quad h_1(b_1) \otimes h_2(b_2). \tag{7.10}$$

Focusing: Let h be a simple random variable defined on $\mathbf{b}$ and $x \in D$. Then we define the simple random variable $h^{\Rightarrow x}$ on $\mathbf{b}$ by

$$(h^{\Rightarrow x})(b) \quad = \quad (h(b))^{\Rightarrow x}. \tag{7.11}$$

This is similar to the corresponding definition of operations between order-preserving mappings between information algebras. So it is no surprise that $\mathcal{H}_s$ with these operations forms an information algebra.

Theorem 7.5 ($\mathcal{H}_s, D$) *is a domain-free information algebra, if D has a top element.*

Proof. (1) *Semigroup.* Combination of simple random variables is clearly commutative. Associativity follows from the associativity of meets of simple experiments and of combination in Φ. The vacuous random variable is clearly the neutral element for combination.

(2) *Transitivity.* Suppose h is a simple random variable defined on $\mathbf{b}$ and $x, y \in D$. Then, for a $b \in \mathbf{b}$,

$$((h^{\Rightarrow x})^{\Rightarrow y})(b) \quad = \quad ((h(b))^{\Rightarrow x})^{\Rightarrow y} \quad = \quad (h(b))^{\Rightarrow x \wedge y} \quad = \quad (h^{\Rightarrow x \wedge y})(b).$$

(3) *Combination:* Let h_1 and h_2 be two simple random variables defined on $\mathbf{b}_1$ and $\mathbf{b}_2$. Then, both simple random variables $(h_1^{\Rightarrow x} \otimes h_2)^{\Rightarrow x}$ and $h_1^{\Rightarrow x} \otimes h_2^{\Rightarrow x}$ are defined on $\mathbf{b}_1 \wedge \mathbf{b}_2$. So, for $b = b_1 \wedge b_2 \in \mathbf{b}_1 \wedge \mathbf{b}_2$, we have

$$\begin{aligned} ((h_1^{\Rightarrow x} \otimes h_2)^{\Rightarrow x})(b) \quad &= \quad ((h_1(b_1))^{\Rightarrow x} \otimes h_2(b_2))^{\Rightarrow x} \\ &= \quad (h_1(b_1))^{\Rightarrow x} \otimes (h_2(b_2))^{\Rightarrow x} \quad = \quad (h_1^{\Rightarrow x} \otimes h_2^{\Rightarrow x})(b). \end{aligned}$$

(4) *Support:* D has a top element $\top$. Then for a simple random variable h defined on $\mathbf{b}$, and $b \in \mathbf{b}$, we obtain

$$(h^{\Rightarrow \top})(b) \quad = \quad (h(b))^{\Rightarrow \top} \quad = \quad h(b).$$

Thus we have $h^{\Rightarrow \top} = h$ and $\top$ is a support for h.

(5) *Idempotency:* Let h be a simple random variable defined on $\mathbf{b}$ and $x \in D$. Then both $h^{\Rightarrow x}$ and $h \otimes h^{\Rightarrow x}$ are defined on $\mathbf{b}$. Thus, for $b \in \mathbf{b}$,

$$(h \otimes h^{\Rightarrow x})(b) \quad = \quad h(b) \otimes (h^{\Rightarrow x})(b) \quad = \quad h(b) \otimes (h(b))^{\Rightarrow x} \quad = \quad h(b).$$

Hence we have $h \otimes h^{\Rightarrow x} = h$. $\qquad\qquad\square$

So we have a partial order in $\mathcal{H}_s$. Suppose h_1 defined on $\mathbf{b}_1$ and h_2 defined on $\mathbf{b}_2$. We have $h_1 \leq h_2$ if, and only if, $\mathbf{b}_1 \geq \mathbf{b}_2$, that is, $\mathbf{b}_1$ is coarser than $\mathbf{b}_2$, and $h_1(b_1) \leq h_2(b_2)$, whenever $b_1 \geq b_2$.

Given the information algebra $(\mathcal{H}_s, D)$, we consider its completion $\mathcal{H}$, which is a compact information algebra $(\mathcal{H}, \mathcal{H}_s, D)$ with the simple random variables as finite elements (see Theorem 6.36). We call the elements of $\mathcal{H}$ *(generalized) random variables*. Random variables are represented as ideals of simple random variables. Or, alternatively, for $h \in \mathcal{H}$ we have

$$h = \vee\{h_s : h_s \in \mathcal{H}_s, h_s \leq h\}. \tag{7.12}$$

We remind that $h_s \leq h$ is the same as $h_s \in h$, if we identify the embedding of $\mathcal{H}_s$ into $\mathcal{H}$ with $\mathcal{H}_s$. We recall also from (6.19) and (6.20) how combination and focusing is expressed in the completed algebra;

$$h_1 \otimes h_2 = \vee\{h' \in \mathcal{H}_s : h' \leq h_1' \otimes h_2', h_1' \in h_1, h_2' \in h_2\},$$
$$h^{\Rightarrow x} = \vee\{h' \in \mathcal{H}_s : h' \leq h''^{\Rightarrow x} \text{ for some } h'' \in h\}. \tag{7.13}$$

We have clearly

$$h_1 \otimes h_2 \leq \vee\{h_1' \otimes h_2' : h_1' \in h_1, h_2' \in h_2\}.$$

But, on the other hand,

$$\{h' = h_1' \otimes h_2' : h_1' \in h_1, h_2' \in h_2\}$$
$$\subseteq \{h' : h' \in \mathcal{H}_s, h' \leq h_1' \otimes h_2' \text{ for some } h_1' \in h_1, h_2' \in h_2\}.$$

Thus, we see that the following equality holds too,

$$h_1 \otimes h_2 = \vee\{h_1' \otimes h_2' : h_1' \in h_1, h_2' \in h_2\}. \tag{7.14}$$

Similarly, we find also that

$$h^{\Rightarrow x} = \vee\{h'^{\Rightarrow x} : h' \in h\}. \tag{7.15}$$

Note that combination and focusing of canonical random variables yields not necessarily canonical random variables. But we may define new operations for canonical random variables:

$$h_1 \otimes_c h_2 = (h_1 \otimes h_2)^{\rightarrow}, \tag{7.16}$$
$$h^{\Rightarrow_c x} = (h^{\Rightarrow x})^{\rightarrow}. \tag{7.17}$$

Similarly, although focusing of normalized random variables gives a normalized random variable, this is not necessarily the case for the combination of normalized random variables. Again we may define a new combination

$$h_1 \otimes_n h_2 = (h_1 \otimes h_2)^{\downarrow}. \tag{7.18}$$

Finally, for normalized, canonical random variables, we may define

$$h_1 \otimes_{n,c} h_2 = ((h_1 \otimes h_2)^{\downarrow})^{\rightarrow}, \tag{7.19}$$
$$h^{\Rightarrow_{n,c} x} = (h^{\Rightarrow x})^{\rightarrow}. \tag{7.20}$$

We leave it to the reader to verify that in each of these cases we obtain an information algebra. In each case we can, by completion, embed the information algebra into the completed algebra of generalized random variables.

7.1.2 Allocation of Probability

What is the belief allocated to an element $\phi \in \Phi$ by a random variable h? The answer to this question leads to a notion which corresponds to the notion of distribution functions for ordinary random variables. Let h be a simple random variable defined on an experiment $\mathbf{b}$. Any belief $b \in \mathbf{b}$ such that $\phi \leq h(b)$ can be thought to imply ϕ. Thus, we define the total belief allocated to ϕ by h as the join of all belief implying ϕ through h:

$$\rho_h(\phi) \;=\; \vee\{b \in \mathbf{b} : \phi \leq h(b)\}. \tag{7.21}$$

Note that $h \leq h'$ implies $\rho_h(\phi) \leq \rho_{h'}(\phi)$ for all $\phi \in \Phi$. We use this remark to extend the definition of the allocation of belief to a general random variable $h \in \mathcal{H}$: If h_s is a simple random variable such that $h_s \leq h$, then we expect $\rho_{h_s}(\phi) \leq \rho_h(\phi)$. Therefore, we define $\rho_h(\phi)$ as the supremum of the beliefs allocated to ϕ by all $h_s \leq h$:

$$\rho_h(\phi) \;=\; \vee\{\rho_{h_s}(\phi) : h_s \leq h\}. \tag{7.22}$$

Once the belief allocated to an information $\phi \in \Phi$ by a random variable h is defined, we are able measure it,

$$sp_h(\phi) \;=\; \mu(\rho_h(\phi)). \tag{7.23}$$

This is called the degree of support or of belief allocated to ϕ by the random variable h. It expresses the probability that ϕ is implied by h, or measures the amount of belief allocated to ϕ. The larger this probability the more credible becomes ϕ in the light of the uncertain information encoded by the random variable ϕ.

For a fixed random variable h, the allocation of belief ρ_h can be seen as a mapping from Φ into $\mathcal{B}$. Similarly, sp_h is a mapping from Φ into $[0,1]$. The following theorem states the basic properties of the mapping ρ_h.

Theorem 7.6 *Let $h \in \mathcal{H}$. Then the mapping ρ_h has the following properties:*

1. $\rho_h(e) = \top$.

2. $\rho_h(\phi_1 \otimes \phi_2) = \rho_h(\phi_1) \wedge \rho_h(\phi_2)$ for any elements $\phi_1, \phi_2 \in \Phi$.

3. $\rho_h(z) = \bot$, if h is normalized.

Proof. We prove the theorem first for simple random variables. So let $h \in \mathcal{H}_s$ defined on $\mathbf{b}$.

(1) follows since $e \leq h(b)$ for all $b \in \mathbf{b}$.

(2) Let $\phi_1, \phi_2 \in \Phi$, apply the definition of ρ_h and the distributive law of Boolean algebras:

$$\begin{aligned}
\rho_h(\phi_1) \wedge \rho_h(\phi_2) &= (\vee\{b_1 \in \mathbf{b} : \phi_1 \leq h(b_1)\}) \wedge (\vee\{b_2 \in \mathbf{b} : \phi_2 \leq h(b_2)\}) \\
&= \vee\{b_1 \wedge b_2 : b_1, b_2 \in \mathbf{b}, \phi_1 \leq h(b_1), \phi_2 \leq h(b_2)\}.
\end{aligned}$$

But we have $b_1 \wedge b_2 = \perp$, unless $b_1 = b_2$. Therefore we obtain

$$
\begin{aligned}
\rho_h(\phi_1) \wedge \rho_h(\phi_2) &= \vee\{b \in \mathbf{b} : \phi_1, \phi_2 \leq h(b)\} \\
&= \vee\{b \in \mathbf{b} : \phi_1 \otimes \phi_2 \leq h(b)\} \\
&= \rho_h(\phi_1 \otimes \phi_2).
\end{aligned} \tag{7.24}
$$

(3) follows from $h(b) < z$ if $h(b) \neq z$ for all $b \in \mathbf{b}$.

Now we turn to a general random variable $h \in \mathcal{H}$.

(1) Follows from (1) proved above, since

$$
\rho_h(e) = \vee\{\rho_{h_s}(e) : h_s \leq h\} = \top.
$$

(2) Let $\phi_1, \phi_2 \in \Phi$. We use the definition and the distributive law of Boolean algebras:

$$
\begin{aligned}
&\rho_h(\phi_1) \wedge \rho_h(\phi_2) \\
&= (\vee\{\rho_{h'}(\phi_1) : h' \in \mathcal{H}_s, h' \leq h\}) \wedge (\vee\{\rho_{h''}(\phi_2) : h'' \in \mathcal{H}_s, h'' \leq h\}) \\
&= \vee\{\rho_{h'}(\phi_1) \wedge \rho_{h''}(\phi_2) : h', h'' \in \mathcal{H}_s, h', h'' \leq h\}.
\end{aligned}
$$

But $h', h'' \leq h$ implies that $h' \otimes h'' = h_s \leq h$ and h_s is still a simple random variable. Therefore we conclude that

$$
\rho_{h'}(\phi_1) \leq \rho_{h_s}(\phi_1), \qquad \rho_{h''}(\phi_2) \leq \rho_{h_s}(\phi_2)
$$

such that

$$
\rho_{h'}(\phi_1) \wedge \rho_{h''}(\phi_2) \leq \rho_{h_s}(\phi_1) \wedge \rho_{h_s}(\phi_2) = \rho_{h_s}(\phi_1 \otimes \phi_2).
$$

This implies that

$$
\rho_h(\phi_1) \wedge \rho_h(\phi_2) \leq \vee\{\rho_{h_s}(\phi_1 \otimes \phi_2) : h_s \in \mathcal{H}_s, h_s \leq h\} = \rho_h(\phi_1 \otimes \phi_2).
$$

But $\rho_{h_s}(\phi_1) \geq \rho_{h_s}(\phi_1 \otimes \phi_2)$. Hence

$$
\begin{aligned}
\rho_h(\phi_1) &= \vee\{\rho_{h_s}(\phi_1) : h_s \leq h\} \geq \vee\{\rho_{h_s}(\phi_1 \otimes \phi_2) : h_s \leq h\} \\
&= \rho_h(\phi_1 \otimes \phi_2).
\end{aligned}
$$

Similarly, we obtain $\rho_h(\phi_2) \geq \rho_h(\phi_1 \otimes \phi_2)$, hence $\rho_h(\phi_1 \otimes \phi_2) \leq \rho_h(\phi_1) \wedge \rho_h(\phi_2)$ and therefore $\rho_h(\phi_1 \otimes \phi_2) = \rho_h(\phi_1) \wedge \rho_h(\phi_2)$.

(3) Assume h normalized, such that $\rho_{h_s}(z) = \perp$ for all simple random variables $h_s \leq h$. Hence

$$
\rho_h(z) = \vee\{\rho_{h_s}(z) : h_s \leq h\} = \perp.
$$

□

Mappings $\rho : \Phi \to \mathcal{B}$ from an information algebra Φ into a probability algebra $(\mathcal{B}, \mu)$ which satisfy properties (1) and (2) of the theorem above will be called *allocations of probability*. This concept has been introduced in (Shafer,

1973; Shafer, 1979). We shall study allocations of probability in detail in Section 7.3. Note that (2) means that ρ_h is a semigroup homomorphism.

As a consequence of this theorem we remark that $\phi_1 \leq \phi_2$ implies $\rho_h(\phi_1) \geq \rho_h(\phi_2)$.

The vacuous random variable induces the allocation defined by $\nu(\phi) = \bot$, if $\phi \neq e$, $\nu(e) = \top$. This is called the *vacuous allocation*.

Note also that $h^{\rightarrow}$ has the same allocation of probability as h itself, $\rho_{h^{\rightarrow}} = \rho_h$. This is not the case for the normalized random variable $h^{\downarrow}$. Here we clearly have

$$\rho_{h^{\downarrow}}(\phi) \quad = \quad \rho_h(\phi) \wedge c, \tag{7.25}$$

where c is the total belief allocated by h to non-zero elements.

7.1.3 Support Functions

We turn now to the mapping sp_h from Φ into the interval $[0, 1]$. The following theorem states fundamental properties of this function.

Theorem 7.7 *Let $h \in \mathcal{H}$. Then the mapping sp_h has the following properties:*

1. $sp_h(e) = 1$.

2. *Assume $\phi_1, \ldots \phi_m \geq \phi$. Then*

$$sp_h(\phi) \quad \geq \quad \sum_{\emptyset \neq I \subseteq \{1,\ldots,m\}} (-1)^{|I|+1} sp_h \left(\bigotimes_{i \in I} \phi_i \right). \tag{7.26}$$

3. $sp_h(z) = 0$, *if h is normalized.*

Proof. (1) follows from Theorem 7.6 (1), since

$$sp_h(e) \quad = \quad \mu(\rho_h(e)) \quad = \quad \mu(\top) \quad = \quad 1.$$

(2) Note that under the conditions of the theorem $\rho_h(\phi_i) \leq \rho_h(\phi)$ for all $i = 1, \ldots m$. This implies

$$\mu(\rho_h(\phi_1) \vee \cdots \vee \rho_h(\phi_m)) \quad \leq \quad \mu(\rho_h(\phi)) \quad = \quad sp_h(\phi).$$

If we apply the well known inclusion-exclusion formula from probability theory for the join on the left hand side above, we obtain the right hand side of (7.26).

(3) follows from Theorem 7.6 (3), since $sp_h(z) = \mu(\rho_h(z)) = \mu(\bot) = 0$.

$\square$

A function which satisfies (2) of the theorem above is called *monotone of order* ∞. Shafer called functions which satisfy (1) to (3) of this theorem *belief functions* and founded a mathematical theory of evidence, also called Dempster-Shafer theory, on them (Shafer, 1976; Shafer, 1979). Therefore we consider the developments in this chapter as an extension of Dempster-Shafer theory to information algebras. We claim that information algebras are the natural structures to develop Dempster-Shafer theory in its most general form.

We shall call functions, which satisfy (1) and (2) of the theorem *support functions* since they represent degrees of support induced on the elements of an information algebra by a random variable h.

Note that (2) of the theorem above shows in particular that $\phi_1 \leq \phi_2$ implies $sp_h(\phi_1) \geq sp_h(\phi_2)$. This is ordinary monotonicity. Less precise information has at least as much support as more precise information.

We remark, that $h^{\rightarrow}$ has the same support function as h. But this is not the case for $h^{\downarrow}$. Here we have for $\phi \neq z$,

$$sp_{h^{\downarrow}}(\phi) = \frac{sp_h(\phi)}{(1 - sp_h(z))}, \qquad (7.27)$$

and $sp_{h^{\downarrow}}(z) = 0$. Note that $1 - sp_h(z) = \mu(c)$ where c is total belief allocated to non-zero elements.

Above we have defined ρ_h in terms of allocations of probability of simple random variables. We remark that it is also possible to define sp_h in terms of support functions of simple random variables.

Theorem 7.8 *Let $h \in \mathcal{H}$ and $\phi \in \Phi$. Then*

$$sp_h(\phi) = \sup\{sp_{h'}(\phi) : h' \in h\}. \qquad (7.28)$$

Proof. By definition, we have

$$sp_h(\phi) = \mu(\rho_h(\phi)) = \mu(\vee\{\rho_{h'}(\phi) : h' \in h\}$$

If $h'_1, \ldots, h'_m \leq h$, then $h'_1, \ldots, h'_m \leq h' = h'_1 \otimes \cdots \otimes h'_m \leq h$, which implies $\rho_{h'_1}(\phi), \ldots, \rho_{h'_m}(\phi) \leq \rho_{h'}(\phi)$. This shows that $\{\rho_{h'}(\phi) : h' \in h\}$ is upwards directed. By Lemma 7.2 we obtain therefore

$$sp_h(\phi) = \sup\{\mu(\rho_{h'}(\phi)) : h' \in h\} = \sup\{sp_{h'}(\phi) : h' \in h\}. \quad (7.29)$$

$\square$

According to this result, support functions of generalized random variables may be approximated by support functions of simple random variables.

7.2 Probabilistic Argumentation Systems

7.2.1 Assumption-Based Information

In practice information is often encoded by sentences of a certain information system and uncertainty is represented through assumptions. This is an approach which parallels the one of the previous section to some degree. We show in this section how it works and that it leads like the former method to *allocations of probability* and to *support functions.*

The general approach is as follows. We assume that an information system $(\mathcal{L}, C, \mathcal{S})$ is used to express knowledge and information. Depending on unknown circumstances, the knowledge or information may take one of several possible forms. In order to formalize this idea, a set Ω of possible *assumptions* is introduced. One of the assumption $\omega \in \Omega$ corresponds to the actual, correct assumption, but it is unknown which one. Associated with each possible assumption $\omega \in \Omega$ is a statement $X(\omega) \subseteq \mathcal{L}$, which expresses what is known to be true, under the assumption ω. X can here be taken as a multivalued mapping from Ω into $\mathcal{L}$. The quadruple $(\Omega, X, \mathcal{L}, C)$ is called an *assumption-based information.*

As a first example we may consider *propositional logic.* Let $A = \{a_1, \dots \dots, a_m\}$ and $P = \{p_1, \dots, p_n\}$ be two disjoint sets of propositional symbols. A wff $\xi \in \mathcal{L}_{A \cup P}$ is called an *assumption-based knowledge.* This can be reduced to the general scheme described above: a literal l_i of a propositional symbol a_i is either a_i or $\neg a_i$. A set $\{l_1, \dots, l_m\}$ of m literals of propositional symbols $a_1, \dots, a_m$ is called a scenario. The set of all 2^m possible scenarios represents the set Ω of possible assumptions. If $\omega = \{l_1, \dots, l_m\}$ is a scenario, then we obtain $\xi_{A \leftarrow \omega}$ from ξ, if we replace in ξ every occurrence of a_i by $\top$, if $l_i = a_i$, and every a_i by $\bot$, if $l_i = \neg a_i$. The formula $\xi_{A \leftarrow \omega}$ represents $X(\omega)$. Such systems have been considered by (Laskey & Lehner, 1989) in relation to Dempster-Shafer theory of evidence. They are discussed in a systematic manner in (Kohlas & Monney, 1995; Haenni, Kohlas & Lehmann, 2000). We mention also that they are closely related to Assumption-Based Truth Maintenance (ATMS), a widely used technique in artificial intelligence (De Kleer, 1986).

As another example we look at linear equations. Let ω_i for $i = 1, \dots, m$ denote real-valued variables. Consider a system of linear equations

$$a_{0i} + \sum_{j=1}^{n} a_{ij} X_j + \omega_i = 0, \qquad i = 1, \dots, m. \tag{7.30}$$

Such systems are considered in regression analysis in mathematical statistics. In our point of view the vectors $(\omega_1, \dots, \omega_m)$ form the assumptions Ω. The ω_i are in this context also considered as disturbances. And the system of linear equations above, for a fixed set of values of ω_i, represents the set $X(\omega)$. Of course more general systems could be considered, mixing linear systems with both real-valued assumptions as well as discrete ones, including binary ones. In the last case we would mix propositional logic with linear or more general systems. We refer to (Anrig, et. al., 1997) for the description of a computer system treating such assumption-based information.

In the general case of an assumption-based information $(\Omega, X, \mathcal{L}, C)$ there may well be assumptions $\omega \in \Omega$ such that $C(X(\omega)) = \mathcal{L}$. Such assumptions are called *contradictory.* In applications they have to be considered as impossible, as excluded by the given information or knowledge. In propositional assumption-based systems these are inconsistent scenarios, that is, scenarios, which together with the knowledge base ξ are not satisfiable. In linear systems, these are disturbances which render the linear systems (7.30) contradictory, without solution.

Let $(\Omega, X, \mathcal{L}, C)$ represent an assumption-based information. If $\omega \in \Omega$ happens to be the correct assumption, then the statement $X(\omega)$ is valid. But then all equivalent statements, and in particular all consequences $C(X(\omega))$, are valid too. This allows to lift the assumption-based information to the elements of the information algebra associated with an information system $(\mathcal{L}, C)$.

In fact, let (Φ_C, S) be the information algebra associated with the information system $(\mathcal{L}, C, \mathcal{S})$. Then the multivalued mapping X of an assumption-based information induces a mapping $\Gamma : \Omega \to \Phi_C$ by

$$\Gamma(\omega) \;\; = \;\; C(X(\omega)). \tag{7.31}$$

Such a mapping Γ from a set of assumptions Ω can also be considered with respect to any arbitrary information algebra (Φ, D). $\Gamma(\omega)$ must be considered as the information which can be asserted, if the assumption ω holds. The $\Gamma(\omega)$ are called *focal information*. In fact, then all $\phi \leq \Gamma(\omega)$ can be deduced to hold. A quadruple $(\Omega, \Gamma, \Phi, D)$ is called a *hint*. Of course, a hint is quite similar to a random variable. But we make no measurability assumptions so far. That is why we prefer to speak of hints rather than of random variables. In accordance to the discussion above, we call assumptions ω for which $\Gamma(\omega) = z$ contradictory and for the same reasons as above we may want to eliminate them as impossible assumptions.

Consider now a piece of information $\phi \in \Phi$ which we may take as a hypothesis. What can be inferred from a hint $(\Omega, \Gamma, \Phi, D)$ with respect to this hypothesis? Can it be proved to hold necessarily in the light of the hint? This will in general not be possible as it is unknown, which of the assumptions ω holds actually. But at least, the set of assumptions ω which imply ϕ can be determined from the hint,

$$qs(\phi) \;\; = \;\; \{\omega \in \Omega : \phi \leq \Gamma(\omega)\}. \tag{7.32}$$

This is called the *quasi-support set* of ϕ. It gives a qualitative image of the support of a hypothesis ϕ by a hint. The larger this set, the more support is given to the hypothesis. Of course not all assumptions ω may be equally likely and this can influence the appreciation of the hypothesis. This aspect will be pursued in the following subsection.

Contradictory assumptions imply all possible hypotheses (because $\phi \leq z$ holds always). This is often not desirable, it does not make sense. We are only interested in assumptions which support a hypothesis and are not contradictory, that is, not excluded as impossible. Therefore, the support set defined by (7.32) is called the *quasi*-support set, whereas the proper *support set* is defined as

$$s(\phi) \;\; = \;\; \{\omega \in \Omega : \phi \leq \Gamma(\omega) \neq z\}. \tag{7.33}$$

$s(\phi)$ as well as $qs(\phi)$ can be looked at as mappings from Φ into the power set of Ω. The following theorem states a few fundamental properties of these mappings.

Theorem 7.9 *Let $s(\phi)$ and $qs(\phi)$ denote the support and quasi-support sets of a hint $(\Omega, \Gamma, \Phi, D)$. Then*

1. $s(e) = qs(e) = \Omega$.

2. $s(\phi \otimes \psi) = s(\phi) \cap s(\psi); \quad qs(\phi \otimes \psi) = qs(\phi) \cap qs(\psi)$.

3. $s(z) = \emptyset$.

Proof. (1) follows because $e \leq \phi$ for all $\phi \in \Phi$ and (3) since contradictory assumptions are excluded in support sets.

(2) Since $\phi \otimes \psi \leq \Gamma(\omega)$ implies $\phi, \psi \leq \Gamma(\omega)$, we have $s(\phi \otimes \psi) \subseteq s(\phi) \cap s(\psi)$. On the other hand, $\phi, \psi \leq \Gamma(\omega)$ implies $\phi \otimes \psi \leq \Gamma(\omega)$ and hence the other set inclusion $s(\phi \otimes \psi) \supseteq s(\phi) \cap s(\psi)$ holds also. The corresponding result for qs follows in the same way. □

Since $\psi \leq \phi$ signifies that $\psi \otimes \phi = \phi$ it follows from (3) of the theorem above that $s(\phi) \subseteq s(\psi)$. A mapping from an information algebra into a Boolean algebra, which satisfies (1) and (2) of Theorem 7.9 is called an *allocation of support*; if it satisfies also (3), then it is called *normalized*. It is evident that allocations of support resemble strongly to allocations of probability. The exact relation between the two notions will be worked out in the next subsection.

7.2.2 Probabilistic Argumentation Systems

The supports allocated to hypotheses ϕ in an information algebra by a hint or to statements Y in an information system by an assumption-based information can be considered as *arguments* in favor of the hypotheses: if one of the assumptions in the support happens to be the true one, then the hypothesis holds necessarily. Note however that, even if all assumptions in the support of a hypothesis turn out to be not valid, this does not yet prove the hypothesis to be wrong. It simply means that it can not be derived from the available knowledge encoded in the hint or the assumption-based information. So, the strength of the support of a hypothesis induced by a hint or an assumption-based information depends somehow on the size of the support set. Clearly, if $s(\psi) \subseteq s(\phi)$, then ψ is less supported than ϕ. This induces at least a partial order of strength of support.

This qualitative order of support can be improved, if a measure of likelihood of the assumptions is available. In fact in many cases it may be known that some assumptions are more probable than others to hold. We assume therefore that a *probability measure* is given over the assumptions. More precisely, we assume a Boolean σ-algebra $\mathcal{A}$ of subsets of Ω and a probability measure P on this algebra, such that $(\Omega, \mathcal{A}, P)$ becomes a probability space. As an example, in a propositional argumentation systems, each a_i may have a probability p_i. Each scenario $\omega = \{l_1, \ldots, l_m\}$ then gets the probability

$$p(\omega) \quad = \quad \prod_{l_i = a_i} p_i \prod_{l_i = \neg a_i} (1 - p_i),$$

if we assume the a_i stochastically independent. In the case of linear systems often a multidimensional Gaussian distribution is assumed for the disturbance vectors $(\omega_1, \ldots, \omega_m)$.

We extend therefore the concept of a hint by the introduction of a probability measure to become $(\Omega, \mathcal{A}, P, \Gamma, \Phi, D)$. In the following, when we speak of a hint, we mean such a probabilistic structure. This concept generalizes the idea of multivalued mappings introduced by Dempster (Dempster, 1967). Then not only support sets $s(\phi)$ for hypotheses $\phi \in \Phi$ are defined, but also the probability of such sets may be defined. Clearly, the larger the probability of a support set, the more support is given to the hypothesis. Therefore we define

$$qsp(\phi) \quad = \quad P(qs(\phi)) \tag{7.34}$$

to be the *degree of quasi-support* of the hypothesis ϕ induced by the hint $(\Omega, \mathcal{A}, P, \Gamma, \Phi, D)$. $qsp(\phi)$ is of course only defined if $qs(\phi)$ is measurable. If there are contradictory assumptions in Ω, then we argue as above that these assumptions are to be excluded as impossible, which implies that the probability measure P should be *conditioned* on the set $qs^c(z)$. That is, the probability space $(\Omega, \mathcal{A}, P)$ is to be replaced by the space $(\Omega', \mathcal{A}', P')$, where $\Omega' = qs^c(z), \mathcal{A}' = \mathcal{A} \cap qs^c(z)$ and $P'(A') = P(A')/P(qs^c(z))$ if $qs(z)$ is measurable and different from zero. If $qs(z)$ is not measurable, then the conditional probability is defined using the outer measure P^* of P by $P'(A') = P^*(A')/P^*(qs^c(z))$ (see (Neveu, 1964)). This new hint is called *normalized*. If $P(qs^c(z)) = 0$ or $P^*(qs^c(z)) = 0$, then the hint can not be normalized. Such hints do not represent proper information. The *degree of support* of a hypothesis ϕ can now be defined as

$$sp(\phi) \quad = \quad P'(s(\phi)) \tag{7.35}$$

provided the set $s(\phi)$ is measurable. It is here, however, for technical reasons convenient to continue to consider non-normalized hints; everything carries over at once to normalized ones.

What is unsatisfactory so far, is that in general there are hypotheses $\phi \in \Phi$ for which the degrees of quasi-support or of support are not defined because $qs(\phi)$ or $s(\phi)$ are not measurable sets. Note that we did not have this problem with random variables as defined in the previous Section 7.1. Of course, we could as usual in probability theory, require that the mapping $\Gamma(\omega)$ is measurable. But this restriction is neither natural nor necessary. We show now that even for non-measurable hypotheses the degree of support can be defined in a very natural way in the present framework. This will establish the link to Section 7.1.

Let $\mathcal{J}$ be the σ-ideal of P-null sets in $\mathcal{A}$ and let $\mathcal{B} = \mathcal{A}/\mathcal{J}$ be the quotient algebra of the equivalence classes modulo $\mathcal{J}$ (two sets $H', H'' \in \mathcal{A}$ are equivalent modulo $\mathcal{J}$ if $H' - H'' \in \mathcal{J}$ and $H'' - H' \in \mathcal{J}$). If $[H]$ denotes the equivalence class of the set H, then, for a countable family $H_i, i \in I$ of measurable sets,

$$[H^c] = [H]^c,$$

$$\bigvee_{i \in I} [H_i] = \left[\bigcup_{i \in I} H_i \right],$$

$$\bigwedge_{i \in I} [H_i] = \left[\bigcap_{i \in I} H_i \right]. \tag{7.36}$$

So $[H]$ defines a mapping from $\mathcal{A}$ into $\mathcal{B}$, called the projection. We denote $[\Phi]$ by $\top$ and $[\emptyset]$ by $\perp$. These are the top and bottom elements of $\mathcal{B}$.

It is well known (see for example (Halmos, 1963)), that $\mathcal{B}$ is a Boolean σ-algebra, satisfying the countable chain condition (see Section 7.1). By $\mu([H]) = P(H)$ a *normalized, positive* measure μ is defined on $\mathcal{B}$. So, $(\mathcal{B}, \mu)$ becomes a *probability algebra*.

We use now this construction of a probability algebra out of the probability space to extend the definition of the degrees of support qs and sp beyond elements ϕ for which $qs(\phi)$ or $s(\phi)$ are measurable. Even, if $qs(\phi)$ is not measurable, any $A \in \mathcal{A}$, such that $A \subseteq qs(\phi)$, represents an argument, which supports ϕ. To exploit this remark, define for every set $H \in \mathcal{P}(\Omega)$

$$\rho_0(H) = \vee \{ [H'] : H' \subseteq H, H' \in \mathcal{A} \}. \tag{7.37}$$

This mapping has properties similar to an allocation of support, it is in fact an allocation of probability, as the following theorem shows.

Theorem 7.10 *The application $\rho_0 : \mathcal{P}(\Omega) \to \mathcal{A}/\mathcal{J}$ as defined in (7.37) has the following properties:*

$$\rho_0(\Omega) = \top,$$
$$\rho_0(\emptyset) = \perp,$$

$$\rho_0 \left(\bigcap_{i \in I} H_i \right) = \bigwedge_{i \in I} \rho_0(H_i). \tag{7.38}$$

if $\{ H_i, i \in I \}$ is a countable family of subsets of Ω.

Proof. Clearly, $\rho_0(\Omega) = [\Omega] = \top \in \mathcal{A}/\mathcal{J}$. Similarly, $\rho_0(\emptyset) = [\Omega] = \perp \in \mathcal{A}/\mathcal{J}$.

In order to prove the remaining identity, let $H_i, i \in I$ be a countable family of subsets of Ω. For every index i, there is a countable family of sets $H'_j \in \mathcal{A}$ such that $H'_j \subseteq H_i$ and $\vee[H'_j] = [\cup H'_j] = \rho_0(H_i)$ since $\mathcal{B}$ satisfies the countable chain condition. Take $A_i = \cup H'_j$. Then $A_i \subseteq H_i$, $A_i \in \mathcal{A}$ and $P(A_i) =$

$\mu(\rho_0(H_i))$. Define $A = \cap_{i \in I} A_i \in \mathcal{A}$. It follows that $A \subseteq \cap_{i \in I} H_i$ and, because the projection is a σ-homomorphism, we obtain $[A] = \wedge_{i \in I}[A_i] = \wedge_{i \in I}\rho_0(H_i)$.

We are going to show now that $[A] = \rho_0(\cap_{i \in I} H_i)$ which proves then the theorem. For this it is sufficient to show that $P(A) = \mu(\rho_0(\cap_{i \in I} H_i))$. This is so, since $P(A) = \mu([A])$ and $A \subseteq \cap H_i$, hence $[A] \leq \rho_0(\cap H_i)$. Therefore, if $\mu([A]) = \mu(\rho_0(\cap H_i))$ we must well have $[A] = \rho_0(\cap H_i)$, since μ is positive.

Now, clearly $P(A) \leq \mu(\rho_0(\cap H_i))$. As above, we conclude that there is a $A' \in \mathcal{A}, A' \subseteq \cap H_i$ such that $P(A') = \mu(\rho_0(\cap H_i))$. $A' \cup (A - A') \subseteq \cap H_i$ implies that $P(A' \cup (A - A')) = P(A')$, hence $P(A - A') = 0$. Define $A_i' = A_i \cup (A - A') \subseteq H_i$. Then $A_i - A_i' = \emptyset$ and therefore,

$$\mu(\rho_0(H_i)) \;=\; P(A_i) \;\leq\; P(A_i') \;=\; P(A_i) + P(A_i' - A_i) \quad (7.39)$$
$$\leq\; \mu(\rho_0(H_i)).$$

This implies that $P(A_i' - A_i) = 0$, therefore we have $[A_i] = [A_i']$. From this we obtain that $\cap A_i' = \cap(A_i \cup (A' - A)) = (A' - A) \cup (\cap A_i) = (A' - A) \cup A = A \cup A' = A' \cup (A - A')$. But $\cap A_i'$ and $\cap A_i$ are equivalent, since $[\cap A_i'] = \wedge[A_i'] = \wedge[A_i] = [\cap A_i]$. This implies finally that $P(A) = P(\cap A_i) = P(\cap A_i') = P(A') + P(A - A') = P(A') = \mu(\rho_0(\cap H_i))$. This is what we wanted to prove. □

If we consider the power set of Ω as an information algebra with intersection as combination, then ρ_0 is an allocation of probability. In fact, it is not only a semigroup or intersection homomorphism, but a σ-homomorphism.

Take now $\mathcal{B} = \mathcal{A}/\mathcal{J}$ and consider the probability algebra $(\mathcal{B}, \mu)$. Then we can compose the allocation of support s from Φ into the power set $\mathcal{P}(\Omega)$ with the mapping ρ_0 from $\mathcal{P}(\Omega)$ into $\mathcal{B}$ to a mapping $\rho = \rho_0 \circ s : \Phi \to \mathcal{B}$. Since

$$\rho(e) \;=\; \rho_0(s(e)) \;=\; \rho_0(\Omega) \;=\; \top,$$

$$\rho(z) \;=\; \rho_0(s(z)) \;=\; \rho_0(\emptyset) \;=\; \bot,$$

$$\rho(\phi \otimes \psi) \;=\; \rho_0(s(\phi \otimes \psi)) \;=\; \rho_0(s(\phi) \cap s(\psi))$$
$$=\; \rho_0(s(\phi)) \wedge \rho_0(s(\psi)) \;=\; \rho(\phi) \wedge \rho(\psi),$$

it follows that ρ is a normalized allocation of probability on the information algebra Φ. In this way, a hint leads always to an allocation of probability, once a probability measure on the assumptions is introduced.

In particular, we may now define

$$sp(\phi) \;=\; \mu(\rho(\phi)).$$

This extends the function (7.35) to all elements ϕ of Φ. This function is a *support function* on Φ. The same holds for $\rho = \rho_0 \circ qs$, except that the allocation is not normalized.

Note furthermore that $\{[H'] : H' \subseteq H, H' \in \mathcal{A}\}$ is an upward directed family. Therefore, according to Lemma 7.2 we have

$$
\begin{aligned}
sp(\phi) \;=\; \mu(\rho(\phi)) \;&=\; \mu(\vee\{[A] : A \in \mathcal{A}, A \subseteq s(\phi)\}) \\
&=\; \sup\{\mu([A]) : A \in \mathcal{A}, A \subseteq s(\phi)\} \\
&=\; \sup\{P(A) : A \in \mathcal{A}, A \subseteq s(\phi)\} \\
&=\; P_*(s(\phi)) \qquad\qquad\qquad ,
\end{aligned}
$$

where P_* is the inner probability associated with P. This shows, that the degree of support of a piece of information ϕ is the inner probability of the allocated support $s(\phi)$. Support functions and inner probability measures are thus closely related. The result above is very appealing: any measurable set A, which is contained in $s(\phi)$ supports ϕ. So we expect $P(A) \leq sp(\phi)$. In the absence of further information, it is reasonable to take $sp(\phi)$ to be the least upper bound of the probabilities of A supporting ϕ.

As an example we may consider the Boolean algebra $\mathcal{P}(\Omega)$ of subsets of some set Ω as an information algebra, where combination is defined as set-intersection. Suppose that a probability space $(\Omega, \mathcal{A}, P)$ is defined. We may then consider the identity mapping of Ω into itself. This defines a hint $(\Omega, \mathcal{A}, P, id, \mathcal{P}(\Omega), \{\Omega\})$. It leads first to an allocation of support $s(H) = H$. The degree of support becomes then simply

$$
sp(H) \;=\; P_*(H)
$$

For measurable hypotheses H the inner probability measure equals the probability itself. This shows again the intimate relation between support functions and inner probability measures. Inner probability measures are in fact support functions.

7.2.3 Operations with Argumentation Systems

An assumption-based information $(\Omega_1, X_1, \mathcal{L}, C)$ relative to a certain information system may be complemented by a second assumption-based information $(\Omega_2, X_2, \mathcal{L}, C)$ coming from a different source. Two (or more) such assumption-based pieces of information should somehow be fused into one, combined assumption-based information. Indeed, it is clear how to do that: if in the first system $\omega_1 \in \Omega_1$ happens to be the correct assumption, and $\omega_2 \in \Omega_2$ is the correct assumption in the second system, then both the statements $X_1(\omega_1)$ and $X_2(\omega_2)$ are to be accepted. In other words under these circumstances the valid statement is $X_1(\omega_1) \cup X_2(\omega_2)$. And this statement can be associated with the combined assumption (ω_1, ω_2). Thus we may build the *combined* assumption-based system $(\Omega_1 \times \Omega_2, X, \mathcal{L}, C)$ where

$$
X(\omega_1, \omega_2) \;=\; X_1(\omega_1) \cup X_2(\omega_2).
$$

Here we assumed that the two systems are based on different sets of assumptions. This is in many cases a reasonable assumption. But it is not really necessary. Define $\Omega = \Omega_1 \times \Omega_2$ and $X_1'(\omega_1, \omega_2) = X_1(\omega_1), X_2'(\omega_1, \omega_2) = X_2(\omega_2)$, then

$(\Omega, X_1', \mathcal{L}, C)$ and $(\Omega, X_2', \mathcal{L}, C)$ are not essentially different from $(\Omega_1, X_1, \mathcal{L}, C)$ and $(\Omega_2, X_2', \mathcal{L}, C)$. So we may assume, without loss of generality, that the assumption-based pieces of information to be combined are defined relative to the same set of assumptions Ω. Then $(\Omega, X_1, \mathcal{L}, C)$ and $(\Omega, X_2, \mathcal{L}, C)$ are combined into $(\Omega, X, \mathcal{L}, C)$ where $X(\omega) = X_1(\omega) \cup X_2(\omega)$. Note that, as with allocations of probability, even if both initial systems contain no contradictory assumptions, this can no more be guaranteed for the combined assumption-based system, since $C(X_1(\omega) \cup X_2(\omega))$ may well equal $\mathcal{L}$. New information may thus prove certain assumptions to be impossible (because contradictory with the new knowledge).

Next we may consider the corresponding hints $(\Omega, \Gamma_i, \Phi_C, S), i = 1, 2$ associated with an information algebra (Φ_C, S) induced by the information system. We recall that $\Gamma_i = C(X_i(\omega))$. Then, corresponding to the combined assumption-based information, we have the hint $(\Omega, \Gamma, \Phi_C, S)$ where

$$
\begin{aligned}
\Gamma(\omega) \;&=\; C(X_1(\omega) \cup X_2(\omega)) \;=\; C(C(X_1(\omega)) \cup C(X_2(\omega))) \\
&=\; \Gamma_1(\omega) \otimes \Gamma_2(\omega).
\end{aligned}
$$

Thus combination of hints essentially amounts to combination of focal information. Again combination may prove certain assumptions ω to be contradictory or impossible, namely those for which $\Gamma_1(\omega) \otimes \Gamma_2(\omega) = \mathcal{L}$. We may sometimes want to modify the combination in such a way that contradictory assumptions are eliminated. The combined hint is then $(\Omega', \Gamma, \Phi_C, S)$ where

$$
\Omega' \;=\; \{\omega \in \Omega : \Gamma(\omega) = \Gamma_1(\omega) \otimes \Gamma_2(\omega) \neq \mathcal{L}\}.
$$

This we call *normalized* combination. It will only be applied to normalized hints.

How does combination of hints affect the support sets? For a $\phi \in \Phi$ we obtain for an unnormalized combination

$$
\begin{aligned}
qs(\phi) \;&=\; \{\omega \in \Omega : \phi \leq \Gamma_1(\omega) \otimes \Gamma_2(\omega)\} \\
&=\; \cup\{\omega \in \Omega : \phi_1 \leq \Gamma_1(\omega), \phi_2 \leq \Gamma_2(\omega), \phi \leq \phi_1 \otimes \phi_2\} \\
&=\; \cup\{qs_1(\phi_1) \cap qs_2(\phi_2) : \phi \leq \phi_1 \otimes \phi_2\}. \qquad (7.40)
\end{aligned}
$$

Here qs_1, qs_2 denote the allocations of (quasi-) supports of the two original hints. Based on this result, we define an operation of combination between allocations of support, by

$$
(qs_1 \otimes qs_2)(\phi) \;=\; \cup\{qs_1(\phi_1) \cap qs_2(\phi_2) : \phi \leq \phi_1 \otimes \phi_2\}. \qquad (7.41)
$$

Next, let $(\Omega, X, \mathcal{L}, C)$ be an assumption-based information and L a sublanguage of $\mathcal{L}$. A family of subsets $Y(\omega), \omega \in \Omega$ is called a *focused* version of the original assumption-based system, if

$$
C(Y(\omega)) \;=\; C(X(\omega))^{\Rightarrow L} \;=\; C(C(X(\omega)) \cap L) \qquad \text{for all } \omega \in \Omega.
$$

The sets $Y(\omega)$ contain, as far as consequences in the sublanguage L are concerned, exactly the same information as $X(\omega)$. In this sense they represent an extraction, a focus, of the information $X(\omega)$ with respect to the sublanguage L.

If we consider the associated hint $(\Omega, \Gamma, \Phi_C, S)$ with a lattice of sublanguages S, then for a $L \in S$ we may define, according to the discussion above

$$\Gamma^{\Rightarrow L}(\omega) \;=\; (\Gamma(\omega))^{\Rightarrow L} \;=\; C(\Gamma(\omega) \cap L).$$

Then, the new hint $(\Omega, \Gamma^{\Rightarrow L}, \Phi_C, S)$ is the original hint focused on the sublanguage L. It contains all the original information as far as L is concerned. This operation carries over to general hints $(\Omega, \Gamma, \Phi, D)$

Again we may ask how this focusing affects the support sets. Let for a $\phi \in \Phi$ and a $x \in D$

$$qs^{\Rightarrow x}(\phi) \;=\; \{\omega \in \Omega : \phi \leq \Gamma^{\Rightarrow x}(\omega)\}$$

denote the support sets for the focused hint. Then, if ω is such that $\phi \leq \Gamma^{\Rightarrow x}(\omega)$, take $\psi = \Gamma^{\Rightarrow x}(\omega) = \psi^{\Rightarrow x} \geq \phi$, hence $\omega \in qs(\psi)$. On the other hand, if ω is such that $\phi \leq \psi = \psi^{\Rightarrow x} \leq \Gamma(\omega)$ for some ψ, that is, $\omega \in qs(\psi)$, then $\phi \leq \Gamma^{\Rightarrow x}(\omega)$, since $\psi^{\Rightarrow x} \leq \Gamma(\omega)$ implies $\psi^{\Rightarrow x} \leq \Gamma^{\Rightarrow x}(\omega)$. So, if $\omega \in qs(\psi)$ for some $\psi = \psi^{\Rightarrow x} \geq \phi$, then $\omega \in qs^{\Rightarrow x}(\phi)$. From this it follows that

$$qs^{\Rightarrow x}(\phi) \;=\; \cup\{qs(\psi) : \phi \leq \psi = \psi^{\Rightarrow x}\}. \tag{7.42}$$

This shows, how the allocation of support for the focused hint can be obtained from the allocation of support of the original hint.

7.3 Allocations of Probability

7.3.1 Algebra of Allocations

Random variables or hints and probabilistic argumentation systems provide means to model explicitly the mechanisms which generate uncertain information. Alternatively, allocations of probability serve to directly assign beliefs to pieces of information. This is more in the spirit of a subjective description of belief, advocated especially by G. Shafer (Shafer, 1973; Shafer, 1976). In this view, allocations of probability are taken as the primitive elements, rather than random variables or hints. This is the point of view developed in this section (see also (Kohlas, 1995; Kohlas, 1997)).

Consider a domain-free information algebra (Φ, D) and a probability algebra $(\mathcal{B}, \mu)$. Allocations of probability are mapping $\rho : \Phi \to \mathcal{B}$ such that

(A1) $\rho(e) = \top$,

(A2) $\rho(\phi \otimes \psi) = \rho(\phi) \wedge \rho(\psi)$.

If furthermore $\rho(z) = \bot$ holds, then the allocation is called *normalized*. Note, that if $\phi \leq \psi$, that is, $\phi \otimes \psi = \psi$, then $\rho(\phi \otimes \psi) = \rho(\phi) \wedge \rho(\psi) = \rho(\psi)$, hence $\rho(\psi) \leq \rho(\phi)$.

We may think of an allocation of probability as the description of a body of belief obtained from a source of information. Two (or more) distinct sources of information will lead to the definition of two (or more) corresponding allocations of probability. Thus, in a general setting let A_Φ be the set of all allocations of probability on Φ in $(\mathcal{B}, \mu)$. Select two allocations $\rho_i, i = 1, 2$, from A_Φ. How can they be combined in order to synthesize the two bodies of information they represent into a single body?

The basic idea is as follows: consider a piece of information ϕ in Φ. If now ϕ_1 and ϕ_2 are two other pieces of information in Φ, such that $\phi \leq \phi_1 \otimes \phi_2$, then any belief allocated to ϕ_1 and to ϕ_2 by the two allocations ρ_1 and ρ_2 respectively, is a belief allocated to ϕ by the two allocations simultaneously. That is, the total belief $\rho(\phi)$ to be allocated to ϕ by the two allocations ρ_1 and ρ_2 together must be at least the combined, common support allocated to ϕ_1 and ϕ_2 by each of the two allocations respectively,

$$\rho(\phi) \;\geq\; \rho_1(\phi_1) \wedge \rho_2(\phi_2). \tag{7.43}$$

In the absence of other information, it seems then reasonable to define the combined belief in ϕ, as obtained from the two sources of information, as the least upper bound of all these implied beliefs,

$$\rho(\phi) \;=\; \vee\{\rho_1(\phi_1) \wedge \rho_2(\phi_2) : \phi \leq \phi_1 \otimes \phi_2\}. \tag{7.44}$$

This defines indeed a new allocation of probability.

Theorem 7.11 ρ *as defined by (7.44) is an allocation of probability.*

Proof. First, we have

$$\begin{aligned}
\rho(e) &= \vee\{\rho_1(\phi_1) \wedge \rho_2(\phi_2) : e \leq \phi_1 \otimes \phi_2\} \\
&= \rho_1(e) \wedge \rho_2(e) \;=\; \top.
\end{aligned}$$

So (A1) is satisfied.

Next, let $\psi_1, \psi_2 \in \Phi$. By definition we have

$$\rho(\psi_1 \otimes \psi_2) \;=\; \vee\{\rho_1(\phi_1) \wedge \rho_2(\phi_2) : \psi_1 \otimes \psi_2 \leq \phi_1 \otimes \phi_2\}.$$

Now, $\psi_1 \leq \psi_1 \otimes \psi_2$ implies that

$$\begin{aligned}
&\vee\{\rho_1(\phi_1) \wedge \rho_2(\phi_2) : \psi_1 \otimes \psi_2 \leq \phi_1 \otimes \phi_2\} \\
&\leq\; \vee\{\rho_1(\phi_1) \wedge \rho_2(\phi_2) : \psi_1 \leq \phi_1 \otimes \phi_2\}
\end{aligned}$$

and similarly for ψ_2. Thus, we have $\rho(\psi_1 \otimes \psi_2) \leq \rho(\psi_1), \rho(\psi_2)$, that is $\rho(\psi_1 \otimes \psi_2) \leq \rho(\psi_1) \wedge \rho(\psi_2)$.

On the other hand,

$$\begin{aligned}
&\{(\phi_1, \phi_2) : \psi_1 \otimes \psi_2 \leq \phi_1 \otimes \phi_2\} \\
&\supseteq\; \{(\phi_1, \phi_2) : \phi_1 = \phi_1' \otimes \phi_1'', \phi_2 = \phi_2' \otimes \phi_2'', \psi_1 \leq \phi_1' \otimes \phi_2', \psi_2 \leq \phi_1'' \otimes \phi_2''\}.
\end{aligned}$$

By the distributive law for complete Boolean algebras we obtain then

$$\rho(\psi_1 \otimes \psi_2)$$
$$\geq \quad \vee\{\rho_1(\phi_1' \otimes \phi_1'') \wedge \rho_2(\phi_2' \otimes \phi_2'') : \psi_1 \leq \phi_1' \otimes \phi_2', \psi_2 \leq \phi_1'' \otimes \phi_2''\}$$
$$= \quad \vee\{\rho_1(\phi_1') \wedge \rho_1(\phi_1'') \wedge \rho_2(\phi_2') \wedge \rho_2(\phi_2'') : \psi_1 \leq \phi_1' \otimes \phi_2', \psi_2 \leq \phi_1'' \otimes \phi_2''\}$$
$$= \quad (\vee\{\rho_1(\phi_1') \wedge \rho_2(\phi_2') : \psi_1 \leq \phi_1' \otimes \phi_2'\}) \wedge$$
$$\quad (\vee\{\rho_1(\phi_1'') \wedge \rho_2(\phi_2'') : \psi_2 \leq \phi_1'' \otimes \phi_2''\})$$
$$= \quad \rho(\psi_1) \wedge \rho(\psi_2). \tag{7.45}$$

This implies finally that $\rho(\psi_1 \otimes \psi_2) = \rho(\psi_1) \wedge \rho(\psi_2)$. Thus (A2) holds too and ρ is indeed an allocation of probability. □

In this way, in the set of allocations of probability A_Φ a binary combination operation is defined. We denote this operation by $\otimes$. Thus, ρ as defined by (7.44) is written as $\rho = \rho_1 \otimes \rho_2$. We note, that (7.44) is similar to (7.41). The following theorem gives us the elementary properties of this operation.

Theorem 7.12 *The combination operation, as defined by (7.44), is* commutative, associative, idempotent *and the vacuous allocation is the* neutral *element of this operation.*

Proof. The commutativity of (7.44) is evident. For the associativity note that for a $\phi \in \Phi$ we have, due to the associativity and distributivity of complete Boolean algebras

$$((\rho_1 \otimes \rho_2) \otimes \rho_3)(\phi)$$
$$= \quad \vee\{\rho_1 \otimes \rho_2(\phi_{12}) \wedge \rho_3(\phi_3) : \phi \leq \phi_{12} \otimes \phi_3\}$$
$$= \quad \vee\{\vee\{\rho_1(\phi_1) \wedge \rho_2(\phi_2) : \phi_{12} \leq \phi_1 \otimes \phi_2\} \wedge \rho_3(\phi_3) : \phi \leq \phi_{12} \otimes \phi_3\}$$
$$= \quad \vee\{\rho_1(\phi_1) \wedge \rho_2(\phi_2) \wedge \rho_3(\phi_3) : \phi \leq \phi_1 \otimes \phi_2 \otimes \phi_3\}.$$

For $(\rho_1 \otimes (\rho_2 \otimes \rho_3))(\phi)$ we obtain exactly the same result in the same way. This proves associativity.

To show idempotency consider

$$\rho \otimes \rho(\phi) \quad = \quad \vee\{\rho(\phi_1) \wedge \rho(\phi_2) : \phi \leq \phi_1 \otimes \phi_2\}$$
$$= \quad \vee\{\rho(\phi_1 \otimes \phi_2) : \phi \leq \phi_1 \otimes \phi_2\} \quad = \quad \rho(\phi)$$

since the last supremum is attained for $\phi_1 = \phi_2 = \phi$.

Finally let ν denote the vacuous allocation. Then, for any allocation ρ and any $\phi \in \Phi$ we have, noting that $\nu(\psi) = \bot$, unless $\psi = e$, in which case $\nu(e) = \top$,

$$\rho \otimes \nu(\phi) \quad = \quad \vee\{\rho(\phi_1) \wedge \nu(\phi_2) : \phi \leq \phi_1 \otimes \phi_2\} \quad = \quad \rho(\phi).$$

This shows that ν is the neutral element for combination. □

This theorem shows that A_Φ is a commutative and idempotent *semigroup*, that is, a *semilattice*. Indeed, a partial order between allocations can be introduced as usual by defining $\rho_1 \leq \rho_2$ if $\rho_1 \otimes \rho_2 = \rho_2$. This means that for all $\phi \in \Phi$,

$$\rho_1 \otimes \rho_2(\phi) \;=\; \vee\{\rho_1(\phi_1) \wedge \rho_2(\phi_2) : \phi \leq \phi_1 \otimes \phi_2\} \;=\; \rho_2(\phi).$$

We have therefore always $\rho_1(\phi_1) \wedge \rho_2(\phi_2) \leq \rho_2(\phi)$ if $\phi \leq \phi_1 \otimes \phi_2$. Take now $\phi_1 = \phi$ and $\phi_2 = e$, such that $\phi \leq \phi \otimes e = \phi$, to obtain $\rho_1(\phi) \wedge \rho_2(e) = \rho_1(\phi) \leq \rho_2(\phi)$. Thus we have $\rho_1 \leq \rho_2$ if, and only if, $\rho_1(\phi) \leq \rho_2(\phi)$ for all $\phi \in \Phi$.

As an example consider allocations ρ of the form

$$\rho_\phi(\psi) \;=\; \begin{cases} \top & \text{if } \psi \leq \phi, \\ \bot & \text{otherwise,} \end{cases} \tag{7.46}$$

for some $\phi \in \Phi$. Such allocations are called *deterministic*. They express the fact that full belief is allocated to the information ϕ. Now, for $\phi_1, \phi_2 \in \Phi$ we have

$$\begin{aligned} \rho_{\phi_1} \otimes \rho_{\phi_2}(\psi) &= \vee\{\rho_{\phi_1}(\psi_1) \wedge \rho_{\phi_2}(\psi_2) : \psi \leq \psi_1 \otimes \psi_2\} \\ &= \left\{ \begin{array}{ll} \top & \text{if } \psi \leq \phi_1 \otimes \phi_2, \\ \bot & \text{otherwise} \end{array} \right\} = \rho_{\phi_1 \otimes \phi_2}(\psi). \quad (7.47) \end{aligned}$$

So, the combination of allocations produces the expected result.

Next we turn to the operation of focusing of an allocation of probability. Let ρ be an allocation of probability on an information algebra (Φ, D) and x a domain in D. Just as it is possible to focus an information ϕ from Φ to the domain x, it should also be possible to focus the belief in the allocation ρ to the information supported by the domain x. This means to extract the information related to x from ρ. Thus, for a $\phi \in \Phi$ consider the beliefs allocated to pieces of information ψ which are supported by x and which entail ϕ, i.e. $\phi \leq \psi = \psi^{\Rightarrow x}$. The part of the belief allocated to ϕ and relating to the domain x, $\rho^{\Rightarrow x}(\phi)$ is then at least $\rho(\psi)$,

$$\rho^{\Rightarrow x}(\phi) \;\geq\; \rho(\psi) \qquad \text{for any } \psi = \psi^{\Rightarrow x} \geq \phi. \tag{7.48}$$

In the absence of other information, it seems again reasonable to define $\rho^{\Rightarrow x}(\phi)$ to be the least upper bound of all these implied supports,

$$\rho^{\Rightarrow x}(\phi) \;=\; \vee\{\rho(\psi) : \psi = \psi^{\Rightarrow x} \geq \phi\}. \tag{7.49}$$

This is similar to (7.42) and it defines indeed an allocation of probability.

Theorem 7.13 $\rho^{\Rightarrow x}(\phi)$ *as defined by (7.49) is an allocation of probability.*

Proof. We have by definition

$$\rho^{\Rightarrow x}(e) \;=\; \vee\{\rho(\psi) : \psi = \psi^{\Rightarrow x} \geq e\} \;=\; \rho(e) \;=\; \top,$$

since $e = e^{\Rightarrow x}$. Thus (A1) is verified.

Again by definition,

$$\rho^{\Rightarrow x}(\phi_1 \otimes \phi_2) = \vee\{\rho(\psi) : \psi = \psi^{\Rightarrow x} \geq \phi_1 \otimes \phi_2\}.$$

From $\phi_1, \phi_2 \leq \phi_1 \otimes \phi_2$ it follows that $\rho^{\Rightarrow x}(\phi_1 \otimes \phi_2) \leq \rho^{\Rightarrow x}(\phi_1), \rho^{\Rightarrow x}(\phi_2)$ and thus $\rho^{\Rightarrow x}(\phi_1 \otimes \phi_2) \leq \rho^{\Rightarrow x}(\phi_1) \wedge \rho^{\Rightarrow x}(\phi_2)$.

On the other hand,

$$\{\psi : \psi = \psi^{\Rightarrow x} \geq \phi_1 \otimes \phi_2\}$$
$$\supseteq \{\psi = \psi_1 \otimes \psi_2 : \psi_1 = \psi_1^{\Rightarrow x} \geq \phi_1, \psi_2 = \psi_2^{\Rightarrow x} \geq \phi_2\}.$$

From this we obtain, using the distributive law for complete Boolean algebras,

$$\rho^{\Rightarrow x}(\phi_1 \otimes \phi_2)$$
$$\geq \vee\{\rho(\psi_1 \otimes \psi_2) : \psi_1 = \psi_1^{\Rightarrow x} \geq \phi_1, \psi_2 = \psi_2^{\Rightarrow x} \geq \phi_2\}$$
$$= \vee\{\rho(\psi_1) \wedge \rho(\psi_2) : \psi_1 = \psi_1^{\Rightarrow x} \geq \phi_1, \psi_2 = \psi_2^{\Rightarrow x} \geq \phi_2\}$$
$$= (\vee\{\rho(\psi_1) : \psi_1 = \psi_1^{\Rightarrow x} \geq \phi_1\}) \wedge (\vee\{\rho(\psi_2) : \psi_2 = \psi_2^{\Rightarrow x} \geq \phi_2\})$$
$$= \rho(\psi_1) \wedge \rho(\psi_2).$$

This proves property (A2) for an allocation of support. □

The following remarks are easily verified:

$$\rho^{\Rightarrow x}(\phi) \leq \rho(\phi) \quad \text{for all } \phi \in \Phi,$$
$$\phi = \phi^{\Rightarrow x} \quad \text{implies} \quad \rho^{\Rightarrow x}(\phi) = \rho(\phi). \qquad (7.50)$$

Thus we see that $\rho^{\Rightarrow x} \leq \rho$. Furthermore, if $\rho^{\Rightarrow x} = \rho$, then the allocation of probability ρ is said to be supported by the domain x. The following theorem shows that allocations of probability, furnished with the operations of combination and focusing as defined so far, satisfy the axioms of domain-free information algebras.

Theorem 7.14 1. *Transitivity. For $\rho \in A_\Phi$ and $x, y \in D$,*

$$(\rho^{\Rightarrow x})^{\Rightarrow y} = \rho^{\Rightarrow x \wedge y}. \qquad (7.51)$$

2. *Combination. For $\rho_1, \rho_2 \in A_\Phi$ and $x \in D$*

$$(\rho_1^{\Rightarrow x} \otimes \rho_2)^{\Rightarrow x} = \rho_1^{\Rightarrow x} \otimes \rho_2^{\Rightarrow x}. \qquad (7.52)$$

3. *Idempotency. For $\rho \in A_\Phi$ and $x \in D$*

$$\rho \otimes \rho^{\Rightarrow x} = \rho. \qquad (7.53)$$

Proof. (1) By definition and the associative law for complete Boolean algebras

$$(\rho^{\Rightarrow x})^{\Rightarrow y}(\phi) = \vee\{\rho^{\Rightarrow x}(\psi) : \psi = \psi^{\Rightarrow y} \geq \phi\}$$
$$= \vee\{\vee\{\rho(\psi') : \psi' = \psi'^{\Rightarrow x} \geq \psi\} : \psi = \psi^{\Rightarrow y} \geq \phi\}$$
$$= \vee\{\rho(\psi') : \psi' = \psi'^{\Rightarrow x} \geq \psi = \psi^{\Rightarrow y} \geq \phi\}.$$

Also by definition, we have

$$\rho^{\Rightarrow x \wedge y}(\phi) \;=\; \vee\{\rho(\psi) : \psi = \psi^{\Rightarrow x \wedge y} \geq \phi\}$$
$$=\; \vee\{\rho(\psi) : \psi = (\psi^{\Rightarrow x})^{\Rightarrow y} \geq \phi\}.$$

If ψ is supported by $x \wedge y$, then it is supported both by x and by y (Lemma 3.6 (6)). Take here now $\psi' = \psi$. Then it follows that $\psi' = \psi'^{\Rightarrow x} \geq \psi = \psi^{\Rightarrow y} \geq \phi$. This implies that

$$(\rho^{\Rightarrow x})^{\Rightarrow y}(\phi) \;\geq\; \rho^{\Rightarrow x \wedge y}(\phi).$$

On the other hand, $\psi' = \psi'^{\Rightarrow x} \geq \psi = \psi^{\Rightarrow y} \geq \phi$ implies that $\psi'^{\Rightarrow y} = (\psi'^{\Rightarrow x})^{\Rightarrow y} \geq \psi^{\Rightarrow y} \geq \phi$. Furthermore, $\psi'^{\Rightarrow y} \leq \psi'$ implies $\rho(\psi') \leq \rho(\psi'^{\Rightarrow y})$. Thus we obtain

$$(\rho^{\Rightarrow x})^{\Rightarrow y}(\phi) \;\leq\; \vee\{\rho(\psi'^{\Rightarrow y}) : \psi'^{\Rightarrow y} = (\psi'^{\Rightarrow y})^{\Rightarrow x} \geq \phi\}.$$

Now set in this formula $\psi'^{\Rightarrow y} = \psi$, such that $\psi^{\Rightarrow y} = \psi$. Then we obtain

$$(\rho^{\Rightarrow x})^{\Rightarrow y}(\phi) \;\leq\; \vee\{\rho(\psi) : \psi = \psi^{\Rightarrow y} = (\psi^{\Rightarrow y})^{\Rightarrow x} \geq \phi\}$$
$$=\; \vee\{\rho(\psi) : \psi = \psi^{\Rightarrow y} = \psi^{\Rightarrow x \wedge y} \geq \phi\}$$
$$=\; \rho^{\Rightarrow x \wedge y}(\phi).$$

This proves transitivity of focusing of allocations of probability.

(2) Again, by definition and the distributive and associative laws of complete Boolean algebras, for any $\phi \in \Phi$

$$\rho_1^{\Rightarrow x} \otimes \rho_2^{\Rightarrow x}(\phi) = \vee\{\rho_1^{\Rightarrow x}(\phi_1) \wedge \rho_2^{\Rightarrow x}(\phi_2) : \phi \leq \phi_1 \otimes \phi_2\}$$
$$=\; \vee\{(\vee\{\rho_1(\psi_1) : \psi_1 = \psi_1^{\Rightarrow x} \geq \phi_1\})$$
$$\wedge (\vee\{\rho_2(\psi_2) : \psi_2 = \psi_2^{\Rightarrow x} \geq \phi_2\}) : \phi \leq \phi_1 \otimes \phi_2\}$$
$$=\; \vee\{\rho_1(\psi_1) \wedge \rho_2(\psi_2) : \psi_1 = \psi_1^{\Rightarrow x} \geq \phi_1, \psi_2 = \psi_2^{\Rightarrow x} \geq \phi_2, \phi \leq \phi_1 \otimes \phi_2\}$$
$$=\; \vee\{\rho_1(\psi_1) \wedge \rho_2(\psi_2) : \psi_1 = \psi_1^{\Rightarrow x}, \psi_2 = \psi_2^{\Rightarrow x}, \phi \leq \psi_1 \otimes \psi_2\}.$$

Also by definition,

$$\rho_1^{\Rightarrow x} \otimes \rho_2(\phi) \;=\; \vee\{\rho_1^{\Rightarrow x}(\phi_1) \wedge \rho_2(\phi_2) : \phi \leq \phi_1 \otimes \phi_2\}$$

and therefore, again by definition, and the associative and distributive laws of complete Boolean algebras

$$(\rho_1^{\Rightarrow x} \otimes \rho_2)^{\Rightarrow x}(\phi)$$
$$=\; \vee\{(\vee\{\rho_1^{\Rightarrow x}(\phi_1) \wedge \rho_2(\phi_2) : \psi \leq \phi_1 \otimes \phi_2\}) : \phi \leq \psi = \psi^{\Rightarrow x}\}$$
$$=\; \vee\{\rho_1^{\Rightarrow x}(\phi_1) \wedge \rho_2(\phi_2) : \phi \leq \psi = \psi^{\Rightarrow x} \leq \phi_1 \otimes \phi_2\}$$
$$=\; \vee\{(\vee\{\rho_1(\psi_1) : \phi_1 \leq \psi_1 = \psi_1^{\Rightarrow x}\}) \wedge \rho_2(\phi_2) : \phi \leq \psi = \psi^{\Rightarrow x} \leq \phi_1 \otimes \phi_2\}$$
$$=\; \vee\{\rho_1(\psi_1) \wedge \rho_2(\phi_2) : \phi_1 \leq \psi_1 = \psi_1^{\Rightarrow x}, \phi \leq \psi = \psi^{\Rightarrow x} \leq \phi_1 \otimes \phi_2\}$$
$$=\; \vee\{\rho_1(\psi_1) \wedge \rho_2(\psi_2) : \psi_1 = \psi_1^{\Rightarrow x}, \phi \leq \psi = \psi^{\Rightarrow x} \leq \psi_1 \otimes \psi_2\}.$$

Now, whenever $\psi_1 = \psi_1^{\Rightarrow x}, \psi_2 = \psi_2^{\Rightarrow x}, \phi \leq \psi_1 \otimes \psi_2$, take $\psi = \psi_1 \otimes \psi_2$. Then $\psi^{\Rightarrow x} = \psi$ and $\phi \leq \psi = \psi^{\Rightarrow x} \leq \psi_1 \otimes \psi_2, \psi_1 = \psi_1^{\Rightarrow x}$. This shows that

$$\rho_1^{\Rightarrow x} \otimes \rho_2^{\Rightarrow x}(\phi) \leq (\rho_1^{\Rightarrow x} \otimes \rho_2)^{\Rightarrow x}(\phi).$$

On the other hand, whenever $\psi_1 = \psi_1^{\Rightarrow x}, \phi \leq \psi = \psi^{\Rightarrow x} \leq \psi_1 \otimes \psi_2$, then $\phi \leq \psi = \psi^{\Rightarrow x} \leq (\psi_1 \otimes \psi_2)^{\Rightarrow x} = \psi_1 \otimes \psi_2^{\Rightarrow x}$ and $\psi_2 \geq \psi_2^{\Rightarrow x}$. Further, $\psi_2 \geq \psi_2^{\Rightarrow x}$ implies that $\rho_2(\psi_2) \leq \rho_2(\psi_2^{\Rightarrow x})$. Therefore (renaming $\psi_2^{\Rightarrow x}$ by ψ_2), we obtain

$$
\begin{aligned}
&(\rho_1^{\Rightarrow x} \otimes \rho_2)^{\Rightarrow x}(\phi) \\
&\leq \quad \vee\{\rho_1(\psi_1) \wedge \rho_2(\psi_2) : \psi_1 = \psi_1^{\Rightarrow x}, \psi_2 = \psi_2^{\Rightarrow x}, \phi \leq \psi = \psi^{\Rightarrow x} \leq \psi_1 \otimes \psi_2\} \\
&= \quad \vee\{\rho_1(\psi_1) \wedge \rho_2(\psi_2) : \psi_1 = \psi_1^{\Rightarrow x}, \psi_2 = \psi_2^{\Rightarrow x}, \phi \leq \psi_1 \otimes \psi_2\},
\end{aligned}
$$

since we may always take $\psi = \psi_1 \otimes \psi_2$ such that $\psi^{\Rightarrow x} = (\psi_1 \otimes \psi_2)^{\Rightarrow x} = \psi_1 \otimes \psi_2$. This shows that

$$\rho_1^{\Rightarrow x} \otimes \rho_2^{\Rightarrow x}(\phi) \geq (\rho_1^{\Rightarrow x} \otimes \rho_2)^{\Rightarrow x}(\phi).$$

And this proves (2).

(3) We have $\rho \otimes \rho^{\Rightarrow x} \geq \rho$. On the other hand, $\rho^{\Rightarrow x} \leq \rho$, hence $\rho \otimes \rho^{\Rightarrow x} \leq \rho \otimes \rho = \rho$ by the idempotency of combination. This proves the idempotency. □

As an example consider for a $\phi \in \Phi$ the deterministic allocation,

$$\rho_\phi^{\Rightarrow x}(\psi) = \vee\{\rho_\phi(\psi') : \psi' = \psi'^{\Rightarrow x} \geq \psi\}.$$

This equals $\top$, if there is a $\psi' = \psi'^{\Rightarrow x} > \psi$ such that $\psi' \leq \phi$, and $\bot$ otherwise. But, we have $\psi' = \psi'^{\Rightarrow x} \leq \phi$ if, and only if, $\psi' = \psi'^{\Rightarrow x} \leq \phi^{\Rightarrow x}$. This shows that $\rho_\phi^{\Rightarrow x}(\psi) = \rho_{\phi \Rightarrow x}(\psi)$ or $\rho_\phi^{\Rightarrow x} = \rho_{\phi \Rightarrow x}$.

The last theorem, together with Theorem 7.12 shows that (A_Φ, D) is an *information algebra*, except that the existence of a support is not necessarily guaranteed. The existence of a support is guaranteed, if D has a top element. So, the algebraic structure of (Φ, D) carries over to A_Φ. In fact, this algebra can be seen as an extension of Φ since the elements of Φ can be regarded as deterministic allocations. The mapping $\phi \mapsto \rho_\phi$ is an embedding of (Φ, D) in (A_Φ, D). The allocation $\zeta(\phi) = \bot$ for all ϕ is a null element in (A_Φ, D), since $\rho \otimes \zeta = \zeta$ for all allocations ρ. Also clearly $\zeta^{\Rightarrow x} = \zeta$. So the algebra satisfies the nullity axiom.

7.3.2 Normalized Allocations

Even if both allocations of probability ρ_1, ρ_2 are normalized, it is not excluded that the combination $\rho_1 \otimes \rho_2$ is no more normalized. Combining information from two distinct source may introduce contradictions. In this section we discuss normalization in relation to combination and focusing of allocation of probability. For that purpose let us introduce a further operation on allocations, the normalization. For an allocation of probability $\rho \in A_\Phi$ define $\rho^\downarrow$ as follows

$$\rho^\downarrow(\phi) \;=\; \rho(\phi) \wedge \rho^c(z), \tag{7.54}$$

similar as in (7.25). That is, we remove from $\rho(\phi)$ the part of the allocated probability, that goes to the null element z. This is the symbolic part of conditioning. For the null allocation ζ we define $\zeta^\downarrow = \zeta$. $\rho^\downarrow$ is again an allocation of probability. The following theorem shows that we may as well normalize a combined allocation of probability *after* combination and the same holds also for focusing.

Theorem 7.15   *1. For $\rho_1, \rho_2 \in A_\Phi$*

$$(\rho_1^\downarrow \otimes \rho_2^\downarrow)^\downarrow \;=\; (\rho_1 \otimes \rho_2)^\downarrow. \tag{7.55}$$

2. For any $\rho \in A_\Phi$

$$(\rho^\downarrow)^{\Rightarrow x} \;=\; (\rho^{\Rightarrow x})^\downarrow. \tag{7.56}$$

Proof. (1) We have

$$
\begin{aligned}
\rho_1^\downarrow \otimes \rho_2^\downarrow(\phi) \;&=\; \vee\{(\rho_1(\phi_1) \wedge \rho_1^c(z)) \wedge (\rho_2(\phi_2) \wedge \rho_2^c(z)) : \phi \le \phi_1 \otimes \phi_2\} \\
&=\; \rho_1 \otimes \rho_2(\phi) \wedge (\rho_1^c(z) \wedge \rho_2^c(z)).
\end{aligned}
$$

From this we obtain

$$
\begin{aligned}
(\rho_1^\downarrow &\otimes \rho_2^\downarrow)^\downarrow(\phi) \\
&=\; \rho_1 \otimes \rho_2(\phi) \wedge (\rho_1^c(z) \wedge \rho_2^c(z)) \wedge ((\rho_1 \otimes \rho_2(z)) \wedge (\rho_1^c(z) \wedge \rho_2^c(z)))^c \\
&=\; \rho_1 \otimes \rho_2(\phi) \wedge (\rho_1 \otimes \rho_2(z))^c \wedge (\rho_1^c(z) \wedge \rho_2^c(z)).
\end{aligned}
$$

Note that

$$
\begin{aligned}
\rho_1 \otimes \rho_2(z) \;&=\; \vee\{\rho_1(\phi_1) \otimes \rho_2(\phi_2) : z \le \phi_1 \otimes \phi_2\} \\
&\ge\; \rho_1(z), \rho_2(z).
\end{aligned}
$$

To see this, take in the join above $\phi_1 = z$ and $\phi_2 = e$. Thus we have

$$(\rho_1 \otimes \rho_2(z))^c \;\le\; \rho_1^c(z) \wedge \rho_2^c(z).$$

This shows finally that

$$(\rho_1^\downarrow \otimes \rho_2^\downarrow)^\downarrow(\phi) \;=\; \rho_1 \otimes \rho_2(\phi) \wedge (\rho_1 \otimes \rho_2(z))^c \;=\; (\rho_1 \otimes \rho_2)^\downarrow(\phi).$$

This proves the first part of the theorem.

(2) For the second part note that $\rho^{\Rightarrow x}(z) = \rho(z)$. Then we have

$$
\begin{aligned}
(\rho^{\downarrow})^{\Rightarrow x}(\phi) &= \vee\{\rho^{\downarrow}(\psi) : \phi \leq \psi = \psi^{\Rightarrow x}\} \\
&= \vee\{\rho(\phi) \wedge \rho^c(z) : \phi \leq \psi = \psi^{\Rightarrow x}\} \\
&= \vee\{\rho(\phi) : \phi \leq \psi = \psi^{\Rightarrow x}\} \wedge \rho^c(z) \\
&= \rho^{\Rightarrow x}(\phi) \wedge (\rho^{\Rightarrow x}(z))^c = (\rho^{\Rightarrow x})^{\downarrow}(\phi). \qquad (7.57)
\end{aligned}
$$

This proves the second part of the theorem. □

Clearly, if ρ is a normalized allocation of probability, then $\rho = \rho^{\downarrow}$, hence $(\rho^{\downarrow})^{\downarrow} = \rho^{\downarrow}$. Let now $A_\Phi^{\downarrow}$ denote the subset of all normalized allocation of probability in A_Φ. Note that according to Theorem 7.15, (2), $A_\Phi^{\downarrow}$ is closed under the operation of focusing. But it is not closed under combination. However, we can define a new operation of combination within $A_\Phi^{\downarrow}$. Namely, for $\rho_1, \rho_2 \in A_\Phi^{\downarrow}$ define $\rho_1 \oplus \rho_2 = (\rho_1 \otimes \rho_2)^{\downarrow}$. It follows from Theorem 7.15, that this operation is commutative and associative. It is also clear that it is idempotent and that the vacuous allocation of probability is still its neutral element. Theorem 7.14 carries also over to this new system.

Theorem 7.16 *1. Transitivity. For $\rho \in A_\Phi^{\downarrow}$ and $x, y \in D$,*

$$
(\rho^{\Rightarrow x})^{\Rightarrow y} = \rho^{\Rightarrow x \wedge y}. \qquad (7.58)
$$

2. Combination. For $\rho_1, \rho_2 \in A_\Phi^{\downarrow}$ and $x \in D$

$$
(\rho_1^{\Rightarrow x} \oplus \rho_2)^{\Rightarrow x} = \rho_1^{\Rightarrow x} \oplus \rho_2^{\Rightarrow x}. \qquad (7.59)
$$

3. Idempotency. For $\rho \in A_\Phi^{\downarrow}$ and $x \in D$

$$
\rho \oplus \rho^{\Rightarrow x} = \rho. \qquad (7.60)
$$

Proof. Transitivity is inherited from A_Φ, since $A_\Phi^{\downarrow}$ is closed under focusing. Then we have further

$$
\begin{aligned}
(\rho_1^{\Rightarrow x} \oplus \rho_2)^{\Rightarrow x} &= ((\rho_1^{\Rightarrow x} \otimes \rho_2)^{\downarrow})^{\Rightarrow x} = ((\rho_1^{\Rightarrow x} \otimes \rho_2)^{\Rightarrow x})^{\downarrow} \\
&= (\rho_1^{\Rightarrow x} \otimes \rho_2^{\Rightarrow x})^{\downarrow} = \rho_1^{\Rightarrow x} \oplus \rho_2^{\Rightarrow x}
\end{aligned}
$$

This proves (2).

Finally we obtain

$$
\rho \oplus \rho^{\Rightarrow x} = (\rho \otimes \rho^{\Rightarrow x})^{\downarrow} = \rho^{\downarrow} = \rho. \qquad (7.61)
$$

And this proves idempotency (3). □

According to this theorem $(A_\Phi^{\downarrow}, D)$ becomes itself an information algebra, which we call the *normalized* algebra associated to (A_Φ, D).

The mapping $\rho \mapsto \rho^{\downarrow}$ is a homomorphism by Theorem 7.15 and also because the vacuous allocation is normalized. This underlines a close similarity between normalization of allocations and scaling of valuations (Section 3.7), especially set potentials.

7.3.3   Random Variables and Allocations

As we can also combine and focus random variables and hints, or perform similar operations on argumentation systems, the question arises, whether these operations lead to the operations of combination and focusing of the corresponding allocations of probability as introduced above. In other words, we ask for example, whether the mapping $h \mapsto \rho_h$ is a homomorphism (or even an isomorphism) with respect to the operations of combination and focusing for allocations of probability as introduced above. The answer is negative in the general case. It is however affirmative in most cases of practical relevance. We remark first, that the answer is affirmative in the restricted case of simple random variables.

Theorem 7.17 *Let h_1, h_2, h be simple random variables, defined on simple experiments in a probability algebra $(\mathcal{B}, \mu)$ with values in a domain-free information algebra (Φ, D). Then, for all $\phi \in \Phi$ and $x \in D$,*

$$\rho_{h_1 \otimes h_2}(\phi) \;=\; \vee\{\rho_{h_1}(\phi_1) \wedge \rho_{h_2}(\phi_2) : \phi \leq \phi_1 \otimes \phi_2\}, \tag{7.62}$$

$$\rho_{h^{\Rightarrow x}}(\phi) \;=\; \vee\{\rho_h(\psi) : \phi \leq \psi = \psi^{\Rightarrow x}\}. \tag{7.63}$$

Proof. (1) Assume that h_1 is defined on the simple experiment $\mathbf{b}_1$ and h_2 on the simple experiment $\mathbf{b}_2$. From the definition of an allocation of probability, and the distributive and associative laws for complete Boolean algebra, we obtain

$$\begin{aligned}
&\vee\{\rho_{h_1}(\phi_1) \wedge \rho_{h_2}(\phi_2) : \phi \leq \phi_1 \otimes \phi_2\} \\
&\quad = \quad \vee\{(\vee\{b_1 \in \mathbf{b}_1 : \phi_1 \leq h_1(b_1)\})) \\
&\qquad\qquad \wedge (\vee\{b_2 \in \mathbf{b}_2 : \phi_2 \leq h_2(b_2)\})) : \phi \leq \phi_1 \otimes \phi_2\} \\
&\quad = \quad \vee\{\vee\{b_1 \wedge b_2 \in \mathbf{b}_1 \wedge \mathbf{b}_2 : \phi_1 \leq h_1(b_1), \phi_2 \leq h_2(b_2)\} : \phi \leq \phi_1 \otimes \phi_2\} \\
&\quad = \quad \vee\{b_1 \wedge b_2 \in \mathbf{b}_1 \wedge \mathbf{b}_2 : \phi_1 \leq h_1(b_1), \phi_2 \leq h_2(b_2), \phi \leq \phi_1 \otimes \phi_2\}.
\end{aligned}$$

But $\phi \leq \phi_1 \otimes \phi_2$, $\phi_1 \leq h_1(b_1)$ and $\phi_2 \leq h_2(b_2)$ if, and only if, $\phi \leq h_1(b_1) \otimes h_2(b_2)$. So we conclude that

$$\begin{aligned}
&\vee\{\rho_{h_1}(\phi_1) \wedge \rho_{h_2}(\phi_2) : \phi \leq \phi_1 \otimes \phi_2\} \\
&\quad = \quad \vee\{b_1 \wedge b_2 \in \mathbf{b}_1 \wedge \mathbf{b}_2 : \phi \leq h_1(b_1) \otimes h_2(b_2)\} \\
&\quad = \quad \vee\{b_1 \wedge b_2 \in \mathbf{b}_1 \wedge \mathbf{b}_2 : \phi \leq h_1 \otimes h_2(b_1 \wedge b_2)\} \tag{7.64} \\
&\quad = \quad \rho_{h_1 \otimes h_2}(\phi).
\end{aligned}$$

(2) Assume that h is defined on the simple experiment $\mathbf{b}$. Then $h^{\Rightarrow x}$ is also defined on $\mathbf{b}$. The associative law of complete Boolean algebra gives us then,

$$\begin{aligned}
\vee\{\rho_h(\psi) : \phi \leq \psi = \psi^{\Rightarrow x}\} &= \vee\{\vee\{b \in \mathbf{b} : \psi \leq h(b)\} : \phi \leq \psi = \psi^{\Rightarrow x}\} \\
&= \vee\{b \in \mathbf{b} : \phi \leq \psi = \psi^{\Rightarrow x} \leq h(b)\}.
\end{aligned}$$

But, $\phi \leq \psi = \psi^{\Rightarrow x} \leq h(b)$ holds if, and only if, $\phi \leq (h(b))^{\Rightarrow x} = h^{\Rightarrow x}(b)$. Hence we see that

$$\vee\{\rho_h(\psi) : \phi \leq \psi = \psi^{\Rightarrow x}\} = \vee\{b \in \mathbf{b} : \phi \leq h^{\Rightarrow x}(b)\} \qquad (7.65)$$

$$= \rho_{h^{\Rightarrow x}}(\phi).$$

$\square$

As far as allocations of probability, which are induced by simple random variables, are concerned, this theorem shows that the combination and focusing of allocations as defined in (7.44) and (7.49) reflects correctly the corresponding operations on the underlying random variables. Let $\mathcal{R}_s$ be the image of $\mathcal{H}_s$ under the mapping $h \mapsto \rho_h$. That is $\mathcal{R}_s$ is the set of all allocations of probability which are induced by simple random variables in $(\mathcal{B}, \mu)$. The mapping $h \mapsto \rho_h$ is a *homomorphism*, since

$$\rho_{h_1 \otimes h_2} = \rho_{h_1} \otimes \rho_{h_2},$$
$$\rho_{h^{\Rightarrow x}} = \rho_h^{\Rightarrow x}.$$

On the right hand side we understand the operations of combination and focusing as those defined above by (7.44) and (7.49). Also the vacuous random variable maps to the vacuous allocation. Thus we conclude that $\mathcal{R}_s$ is a subalgebra of the information algebra A_Φ. We remark that, if we restrict the mapping $h \mapsto \rho_h$ to *canonical* random variables, then the mapping becomes an isomorphism.

In any case the mapping is order-preserving. This implies that ideals $h \in \mathcal{H}$, that is, generalized random variables, are mapped into ideals $\{\rho_{h'} : h' \in h\}$. Thus, we can extend the mapping from $\mathcal{H}_s$ to $\mathcal{R}_s$ to a mapping from the completion $\mathcal{H}$ of $\mathcal{H}_s$ to the completion $\mathcal{R}$ of $\mathcal{R}_s$.

$$\rho_h = \vee\{\rho_{h'} : h' \leq h\}$$

and also (compare (7.22))

$$\rho_h(\phi) = \vee\{\rho_{h'}(\phi) : h' \leq h\} \qquad (7.66)$$

It is an open question whether $\mathcal{R}$ equals A_Φ, that is, whether any allocation of probability is induced by a (generalized) random variable, see however (Shafer, 1979; Kohlas, 1993)). Another question is, whether the extended mapping is still a homomorphism (relative to the operations of combination and focusing defined by (7.44) and (7.49))? The answer is still positive.

Theorem 7.18 *Let $h_1, h_2 \in \mathcal{H}$ be (generalized) random variables, and $x \in D$. Then*

$$\rho_{h_1 \otimes h_2} = \rho_{h_1} \otimes \rho_{h_2},$$
$$\rho_{h^{\Rightarrow x}} = \rho_h^{\Rightarrow x}. \qquad (7.67)$$

Proof. (1) The mapping $h \mapsto \rho_h$ from the compact information algebra of random variables $(\mathcal{H}, \mathcal{H}_s, D)$ into the algebra of allocations of probability is continuous. Therefore, by Lemma 6.43 (2) and (7.14),

$$\rho_{h_1 \otimes h_2} \;=\; \rho_{\vee\{h_1' \otimes h_2' : h_1' \in h_1, h_2' \in h_2\}} \;=\; \vee\{\rho_{h_1' \otimes h_2'} : h_1' \in h_1, h_2' \in h_2\}.$$

On the other hand, for every $\phi \in \Phi$, we obtain, using the associative and distributive laws of Boolean algebra,

$$
\begin{aligned}
\rho_{h_1} &\otimes \rho_{h_2}(\phi) \\
&= \vee\{\rho_{h_1}(\phi_1) \wedge \rho_{h_2}(\phi_1) : \phi \le \phi_1 \otimes \phi_2\} \\
&= \vee\{(\vee\{\rho_{h_1'}(\phi_1) : h_1' \in h_1\}) \\
&\qquad \wedge(\vee\{\rho_{h_2'}(\phi_2) : h_2' \in h_2\}) : \phi \le \phi_1 \otimes \phi_2\} \\
&= \vee\{\rho_{h_1'}(\phi_1) \wedge \rho_{h_2'}(\phi_2) : h_1' \in h_1, h_2' \in h_2, \phi \le \phi_1 \otimes \phi_2\} \\
&= \vee\{\vee\{\rho_{h_1'}(\phi_1) \wedge \rho_{h_2'}(\phi_2) : \phi \le \phi_1 \otimes \phi_2\} : h_1' \in h_1, h_2' \in h_2\} \\
&= \vee\{\rho_{h_1' \otimes h_2'}(\phi) : h_1' \in h_1, h_2' \in h_2\}.
\end{aligned}
$$

The last equality follows from Theorem 7.17 (7.62). This proves that $\rho_{h_1 \otimes h_2} = \rho_{h_1} \otimes \rho_{h_2}$.

(2) Again by continuity, we obtain from (7.15)

$$\rho_{h \Rightarrow x} \;=\; \rho_{\vee\{h' \Rightarrow x : h' \in h\}} \;=\; \vee\{\rho_{h' \Rightarrow x} : h' \in h\}.$$

But, we have also, by Theorem 7.17 (7.63),

$$
\begin{aligned}
\rho_h^{\Rightarrow x}(\phi) &= \vee\{\rho_h(\psi) : \phi \le \psi = \psi^{\Rightarrow x}\} \\
&= \vee\{\vee\{\rho_{h'}(\psi) : h' \in h\} : \phi \le \psi = \psi^{\Rightarrow x}\} \\
&= \vee\{\vee\{\rho_{h'}(\psi) : \phi \le \psi = \psi^{\Rightarrow x}\} : h' \in h\} \\
&= \vee\{\rho_{h' \Rightarrow x}(\phi) : h' \in h\}.
\end{aligned}
$$

This proves that $\rho_{h \Rightarrow x} = \rho_h^{\Rightarrow x}$. $\square$

As a consequence of this theorem, we note that formulas (7.14) and (7.15) for random variables, translate into similar formulas for allocations of probability. Indeed,

$$
\begin{aligned}
\rho_{h_1 \otimes h_2} &= \vee\{\rho_h : h \in h_1 \otimes h_2\} \\
&= \vee\{\rho_h : h \le h_1' \otimes h_2' \text{ for some } h_1' \in h_1, h_2' \in h_2\} \\
&= \vee\{\rho : \rho \le \rho_1' \otimes \rho_2' \text{ for some } \rho_1' \in \rho_{h_1}, \rho_2' \in \rho_{h_2}\}.
\end{aligned}
$$

Thus, for $\rho_1, \rho_2 \in \mathcal{R}$, we may define

$$\rho_1 \otimes \rho_2 \;=\; \vee\{\rho \in \mathcal{R}_s : \rho \le \rho_1' \otimes \rho_2' \text{ for some } \rho_1' \in \rho_1, \rho_2' \in \rho_2\}.$$

From this, we deduce in the same way as (7.14) that

$$\rho_1 \otimes \rho_2 \;=\; \vee\{\rho_1' \otimes \rho_2' : \rho_1' \in \rho_1, \rho_2' \in \rho_2\}. \tag{7.68}$$

By the same argument, we obtain

$$\rho^{\Rightarrow x} \;=\; \vee\{\rho'^{\Rightarrow x} : \rho' \in \rho\}.$$

We may also ask, how operations with argumentation systems (see Section 7.2.3) extend to related operations with the corresponding allocations of probability? As shown before, probabilistic argumentation systems induce allocations of probability (see Section 7.2.2). Thus, if $(\Omega, \mathcal{A}, P, \Gamma, \Phi, D)$ is a hint with a probability measure on the assumptions and ρ_0 the application from Φ into the probability algebra $\mathcal{B} = \mathcal{A}/\mathcal{J}$ defined by (7.37), then $\rho = \rho_0 \circ qs$ is an allocation of probability. Now, consider two generalized hints $(\Omega, \mathcal{A}, P, \Gamma_1, \Phi, D)$ and $(\Omega, \mathcal{A}, P, \Gamma_2, \Phi, D)$ together with combined hint $(\Omega, \mathcal{A}, P, \Gamma, \Phi, D)$, where $\Gamma(\omega) = \Gamma_1(\omega) \otimes \Gamma_1(\omega)$. We ask, how the allocation of probability of the two combined hint is related to the allocation of probability of the two original hints. Also, if $x \in D$, then how is the allocation of probability induced by the focused hint $(\Omega, \mathcal{A}, P, \Gamma^{\Rightarrow x}, \Phi, D)$ related to the allocation of probability of the hint $(\Omega, \mathcal{A}, P, \Gamma, \Phi, D)$? This question is answered by the following theorem.

Theorem 7.19

1. *Let* $qs_1, qs_2 : \Phi \rightarrow \Omega$ *be two allocations of support in the probability space* $(\Omega, \mathcal{A}, P)$ *and* $\rho_0 : \mathcal{P}(\Omega) \rightarrow \mathcal{A}/\mathcal{J}$ *as defined by (7.37). Then*

$$(\rho_0 \circ qs_1) \otimes (\rho_0 \circ qs_2) \;\leq\; \rho_0 \circ (qs_1 \otimes qs_2). \tag{7.69}$$

2. *Let* qs *be an allocation of support,* ρ_0 *as above and* $x \in D$. *Then*

$$(\rho_0 \circ qs)^{\Rightarrow x} \;\leq\; \rho_0 \circ qs^{\Rightarrow x}. \tag{7.70}$$

The operations on the left hand sides are assumed to be defined by (7.44) and (7.49), whereas those on the right hand side are defined by (7.41) and (7.42) respectively.

Proof. (1) Since

$$\rho_0 \left(\bigcup_{i \in I} H_i \right) \;\geq\; \bigvee_{i \in I} \rho_0(H_i). \tag{7.71}$$

we obtain, using (7.41) that

$$
\begin{aligned}
\rho_0 \circ (qs_1 \otimes qs_2)(\phi) \;&=\; \rho_0(\cup\{qs_1(\phi_1) \cap qs_2(\phi_2) : \phi \leq \phi_1 \otimes \phi_2\}) \\
&\geq\; \vee\{\rho_0(qs_1(\phi_1) \cap qs_1(\phi_1)) : \phi \leq \phi_1 \otimes \phi_2\} \\
&=\; \vee\{\rho_0(qs_1(\phi_1)) \wedge \rho_0(qs_2(\phi_2)) : \phi \leq \phi_1 \otimes \phi_2\} \\
&=\; (\rho_0 \circ qs_1) \otimes (\rho_0 \circ qs_2)(\phi). \tag{7.72}
\end{aligned}
$$

(2) Similarly, we obtain also

$$\rho_0 \circ qs^{\Rightarrow x}(\phi) = \rho_0(\cup\{qs(\psi) : \psi = \psi^{\Rightarrow x} \geq \phi\}) \qquad (7.73)$$
$$\geq \vee\{\rho_0(qs(\psi)) : \psi = \psi^{\Rightarrow x} \geq \phi\} = (\rho_0 \circ qs)^{\Rightarrow x}(\phi).$$

$\square$

In general, there is not equality between the two ways to obtain the allocation of probability of the combined argumentation systems. In fact, this theorem shows, that in passing to the allocations of probability some information is lost. That is why argumentation systems should first be combined and only then the allocation of probability taken.

Of course there are important cases where equality holds. For example, this is the case, if Φ is countable and all $qs(\phi)$ are measurable. Then, for countable families H_i of measurable sets we have

$$\rho_0 \left(\bigcup_{i \in I} H_i\right) = \left[\bigcup_{i \in I} H_i\right] = \bigvee_{i \in I} [H_i] = \bigvee_{i \in I} \rho_0(H_i). \qquad (7.74)$$

More generally this holds, if the family of sets $qs(\phi)$ is countable and all $qs(\phi)$ are measurable.

7.3.4   Labeled Allocations and their Algebra

To any domain-free information algebra (Φ, D) a corresponding labeled algebra can be associated by considering labeled pieces of information (ϕ, x), where x is a support of ϕ (Chapter 3). This can also be done with respect to the information algebra (A_Φ, D) of allocations. Any allocation ρ can be labeled by a support x of it, provided it has a support. Alternatively, the restriction ρ_x of ρ to the domain x can be considered. It will be shown in this section that this latter approach leads in a natural way to an algebra *isomorphic* to the labeled information algebra associated with (A_Φ, D).

To start with we consider a *labeled* information algebra (Φ, D). As usual, Φ_x denotes the set of valuations with domain x. Consider now an allocation $\rho : \Phi_x \to \mathcal{B}$, such that

(A1) $\rho(e_x) = \top$,

(A2) for $\phi, \psi \in \Phi_x, \rho(\phi \otimes \psi) = \rho(\phi) \wedge \rho(\psi)$.

Here $\mathcal{B}$ is as usual a complete Boolean algebra. Such a mapping will be called a *labeled* allocation with domain x. We define as its label $d(\rho)$ its domain x. We note that, as before, for $\phi, \psi \in \Phi_x$, $\phi \leq \psi$ implies $\rho(\phi) \geq \rho(\psi)$.

Like with the former domain-free allocations we define some operations with labeled allocations. First, the labeled allocation ν_x defined by

$$\nu_x(\phi) = \bot \qquad \text{for all } \phi \in \Phi_x, \phi \neq e_x,$$
$$\nu_x(e_x) = \top \qquad\qquad\qquad\qquad\qquad (7.75)$$

is called the *vacuous* allocation on domain x. Next, if ρ_x and ρ_y are two labeled allocations on domains x and y respectively, then we define their combination ρ in analogy to (7.44) for $\phi \in \Phi_{x \vee y}$

$$\rho(\phi) \;=\; \vee\{\rho_x(\phi_x) \wedge \rho_y(\phi_y) : d(\phi_x) = x, d(\phi_y) = y, \phi \leq \phi_x \otimes \phi_y\}. \tag{7.76}$$

Note that here, as well as in the sequel, the partial order of labeled information within a domain is used (see Section 6.2). We denote this mapping also as a product, $\rho = \rho_x \otimes \rho_y$ and write $\rho(\phi) = \rho_x \otimes \rho_y(\phi)$.

Furthermore, if ρ is a labeled allocation with domain x and y a domain with $y \leq x$, then a mapping $\rho^{\downarrow}$ is defined by

$$\rho^{\downarrow y}(\phi) \;=\; \rho(\phi^{\uparrow x}) \qquad \text{for all } \phi \in \Phi_y. \tag{7.77}$$

Here $\phi^{\uparrow x} = \phi \otimes e_x$ is the vacuous extension of ϕ to domain x. $\rho^{\downarrow y}$ is called the *marginal* of ρ with respect to y.

The following theorem states that $\rho_x \otimes \rho_y$ and $\rho^{\downarrow y}$ are indeed again labeled allocations.

Theorem 7.20 *$\rho_x \otimes \rho_y$ is a labeled allocation with domain $x \vee y$ and $\rho^{\downarrow y}$ is a labeled allocation with domain y.*

Proof. The first statement is proved exactly as Theorem 7.11 noting that $d(\phi_x \otimes \phi_y) = x \vee y$.

As for the second statement, note first that $\rho^{\downarrow y}(e_y) = \rho(e_y^{\uparrow x}) = \rho(e_x) = \top$. This proves property (A1) of a labeled allocation. Furthermore, for $\phi, \psi \in \Phi_y$, we obtain

$$\rho^{\downarrow y}(\phi \otimes \psi) \;=\; \rho((\phi \otimes \psi)^{\uparrow x}) \;=\; \rho(\phi^{\uparrow x} \otimes \psi^{\uparrow x})$$
$$=\; \rho(\phi^{\uparrow x}) \wedge \rho(\psi^{\uparrow x}) \;=\; \rho^{\downarrow y}(\phi) \wedge \rho^{\downarrow y}(\psi).$$

This proves property (A2) for labeled allocations. $\qquad\square$

As an inverse to the operation of marginalization, which restricts a labeled allocation to a smaller domain, we may also define the operation of extension of a labeled allocation to a finer domain. Formally, if ρ is a labeled allocation with domain x and y a domain such that $x \leq y$, then we define

$$\rho^{\uparrow y} \;=\; \rho \otimes \nu_y. \tag{7.78}$$

This gives then, according to (7.76) for a $\phi \in \Phi_y$

$$\rho^{\uparrow y}(\phi) \;=\; \vee\{\rho(\psi) : \psi \in \Phi_x, \phi \leq \psi^{\uparrow y}\}. \tag{7.79}$$

This makes sense: the support allocated to a ϕ in Φ_y by a labeled allocation with domain $x \leq y$ is the supremum of all supports allocated to pieces of information ψ in domain x which imply ϕ. By Theorem 7.20 above, $\rho^{\uparrow y}$ is a labeled allocation with domain y; it is called the *vacuous extension* of ρ to y.

The next theorems show that the labeled allocations form themselves a labeled information algebra.

Theorem 7.21 *Let (Φ, D) be a labeled information algebra. Then the following hold*

1. Semigroup: *Combination of labeled allocations of probability is associative and commutative. For all $x \in D$ there exists an element ν_x such that $\rho \otimes \nu_x = \rho$ if $d(\rho) = x$.*

2. Labeling: *For two labeled allocations ρ_1, ρ_2 on (Φ, D) we have $d(\rho_1 \otimes \rho_2) = d(\rho_1) \vee d(\rho_2)$.*

3. Marginalization: *If for a labeled allocation $x \leq d(\rho)$ then $d(\rho^{\downarrow x}) = x$.*

4. Transitivity: *For a labeled allocation ρ and domains $x, y \in D$ such that $x \leq y \leq d(\rho)$ we have $(\rho^{\downarrow y})^{\downarrow x} = \rho^{\downarrow x}$.*

5. Combination: *For two labeled allocations ρ_x, ρ_y with $d(\rho_x) = x, d(\rho_y) = y$ we have $(\rho_x \otimes \rho_y)^{\downarrow x} = \rho_x \otimes \rho_y^{\downarrow x \wedge y}$.*

6. Stability: *For $x, y \in D$ with $x \leq y$ we have $\nu_y^{\downarrow x} = \nu_x$.*

7. Idempotency: *For a labeled allocation ρ and $x \in D$ with $x \leq d(\rho)$ we have $\rho \otimes \rho^{\downarrow x} = \rho$.*

Proof. (1) Commutativity of combination is evident. Associativity is proved as in Theorem 7.12. If ν_x is the vacuous allocation on domain x and ρ a labeled allocation with domain x, then for any $\phi \in \phi_x$

$$\rho \otimes \nu_x(\phi) \quad = \quad \vee\{\rho(\psi) : d(\psi) = x, \phi \leq \psi\} \quad = \quad \rho(\phi).$$

since $\nu_x(\psi) = \perp$ unless $\psi = e_x$, in which case $\nu_x(e_x) = \top$.

(2) follows from the definition of combination.

(3) is already proved in Theorem 7.20

(4) For a $\phi \in \Phi_x$ we have, by definition, if $d(\rho) = z$,

$$(\rho^{\downarrow y})^{\downarrow x}(\phi) \quad = \quad \rho^{\downarrow y}(\phi^{\uparrow y}) \quad = \quad \rho((\phi^{\uparrow y})^{\uparrow z}) \quad = \quad \rho(\phi^{\uparrow z}) \quad = \quad \rho^{\downarrow x}(\phi).$$

(5) For a $\phi \in \Phi_x$ we have

$$(\rho_x \otimes \rho_y)^{\downarrow x}(\phi) = (\rho_x \otimes \rho_y)(\phi^{\uparrow x \vee y}$$
$$= \quad \vee\{\rho_x(\phi_x) \wedge \rho_y(\phi_y) : d(\phi_x) = x, d(\phi_y) = y, \phi^{\uparrow x \vee y} \leq \phi_x \otimes \phi_y\}.$$

On the other hand,

$$(\rho_x \otimes \rho_y^{\downarrow x \wedge y})(\phi)$$
$$= \quad \vee\{\rho_x(\phi_x) \wedge \rho_y^{\downarrow x \wedge y}(\phi_{x \wedge y}) : d(\phi_x) = x, d(\phi_{x \wedge y}) = x \wedge y, \phi \leq \phi_x \otimes \phi_{x \wedge y}\}$$
$$= \quad \vee\{\rho_x(\phi_x) \wedge \rho_y(\phi_{x \wedge y}^{\uparrow y}) : d(\phi_x) = x, d(\phi_{x \wedge y}) = x \wedge y, \phi \leq \phi_x \otimes \phi_{x \wedge y}\}.$$

Now, for $\phi \in \Phi_x$, $\phi^{\uparrow x \vee y} \leq \phi_x \otimes \phi_y$ implies $\phi \leq (\phi_x \otimes \phi_y)^{\downarrow x} = \phi_x \otimes \phi_y^{\downarrow x \wedge y}$ and, for $\phi \in \Phi_y$, $(\phi^{\downarrow x \wedge y})^{\uparrow y} \leq \phi$, hence $\rho_y((\phi^{\downarrow x \wedge y})^{\uparrow y}) \geq \rho_y(\phi)$. It follows that $(\rho_x \otimes \rho_y)^{\downarrow x}(\phi) \leq (\rho_x \otimes \rho_y^{\downarrow x \wedge y})(\phi)$. But $\phi \leq \phi_x \otimes \phi_{x \wedge y}$ implies that $\phi^{\uparrow x \vee y} = \phi \otimes e_{x \vee y} \leq \phi_x \otimes \phi_{x \wedge y} \otimes e_{x \vee y} = \phi_x \otimes \phi_{x \wedge y}^{\uparrow y}$. This shows then that $(\rho_x \otimes \rho_y)^{\downarrow x}(\phi) \geq (\rho_x \otimes \rho_y^{\downarrow x \wedge y})(\phi)$ and thus finally $(\rho_x \otimes \rho_y)^{\downarrow x}(\phi) = (\rho_x \otimes \rho_y^{\downarrow x \wedge y})(\phi)$.

(6) For $\phi \in \Phi_x, \phi \neq e_x$ it follows that $\phi^{\uparrow y} \neq e_y$. We have therefore by definition

$$\nu_y^{\downarrow x}(\phi) = \nu_y(\phi^{\uparrow y}) = \perp.$$

But $e_x^{\uparrow y} = e_y$, hence

$$\nu_y^{\downarrow x}(e_x) = \nu_y(e_x^{\uparrow y}) = \nu_y(e_y) = \top.$$

This shows that $\nu_y^{\downarrow x}$ is the vacuous allocation on domain x and thus equals ν_x.

(7) Let $d(\rho) = y$ and $\phi \in \Phi_y$. Then we have

$$\begin{aligned}
\rho \otimes \rho^{\downarrow x}(\phi) &= \vee\{\rho(\phi_y) \wedge \rho^{\downarrow x}(\phi_x) : d(\phi_y) = y, d(\phi_x) = x, \phi \leq \phi_y \otimes \phi_x\} \\
&= \vee\{\rho(\phi_y) \wedge \rho(\phi_x^{\uparrow y}) : d(\phi_y) = y, d(\phi_x) = x, \phi \leq \phi_y \otimes \phi_x\} \\
&= \vee\{\rho(\phi_y \otimes \phi_x^{\uparrow y}) : d(\phi_y) = y, d(\phi_x) = x, \phi \leq \phi_y \otimes \phi_x\} \\
&= \vee\{\rho(\phi_y \otimes \phi_x^{\uparrow y}) : d(\phi_y) = y, d(\phi_x) = x, \phi \leq \phi_y \otimes \phi_x^{\uparrow y}\} \\
&\leq \rho(\phi).
\end{aligned}$$

But we have also $\phi = \phi \otimes \phi^{\downarrow x} = \phi \otimes (\phi^{\downarrow x})^{\uparrow y}$ and therefore $\rho(\phi) \leq \rho \otimes \rho^{\downarrow x}(\phi)$ which proves that $\rho(\phi) = \rho \otimes \rho^{\downarrow x}(\phi)$. □

It has been noted in Section 3.2 that to any domain-free information algebra (Φ, D) a labeled information algebra (Φ^*, D) with

$$\Phi^* = \{(\phi, x) : \phi \in \Phi, x \in D \text{ support of } \phi\}.$$

can be associated, such that $d(\phi') = x$ if $\phi' = (\phi, x)$ and combination and marginalization are defined by

$$\begin{aligned}
(\phi_1, x) \otimes (\phi_2, y) &= (\phi_1 \otimes \phi_2, x \vee y), \\
(\phi, x)^{\downarrow y} &= (\phi^{\Rightarrow y}, y) \qquad \text{for } y \leq x.
\end{aligned} \tag{7.80}$$

The same construction applies in particular to the information algebra (A_Φ, D) of allocations. Let thus

$$A_\Phi^* = \{(\rho, x) : \rho \in A_\Phi, x \in D \text{ support of } \rho\}.$$

Then combination and marginalization can be defined as in (7.80) and (A_Φ^*, D) becomes in this way a labeled information algebra. Only allocations ρ which have a support are in A_Φ^*.

Note however that the elements (ρ, x) are not labeled allocations in the sense defined before. But a labeled allocation ρ_x defined by $\rho_x(\phi') = \rho(\phi)$ if $\phi' = (\phi, x) \in \Phi_x^*$ can be associated with it. This defines in fact an *isomorphism* between the two labeled information algebras. That is what the following theorem says.

Theorem 7.22   *1. If $(\rho^{(1)}, x)$ and $(\rho^{(2)}, y)$ belong to (A_Φ^*, D), then*

$$\rho_x^{(1)} \otimes \rho_y^{(2)} = (\rho^{(1)} \otimes \rho^{(2)})_{x \vee y}. \tag{7.81}$$

2. If ν is the vacuous allocation in (A_Φ, D), then ν_x is the vacuous allocation on Φ_x for all $x \in D$.

3. If (ρ, x) belongs to (A_Φ^*, D), and $y \le x, y \in D$, then

$$\rho_x^{\downarrow y} = (\rho^{\Rightarrow y})_y. \tag{7.82}$$

Proof. (1) Let $\phi' = (\phi, x \vee y)$ belong to $\Phi_{x \vee y}^*$ and consider

$$\rho_x^{(1)} \otimes \rho_y^{(2)}(\phi')$$
$$= \vee\{\rho_x^{(1)}(\phi_x') \wedge \rho_y^{(2)}(\phi_y') : \phi_x' = (\phi_x, x), \phi_y' = (\phi_y, y), \phi' \le \phi_x' \otimes \phi_y'\}$$
$$= \vee\{\rho^{(1)}(\phi_x) \wedge \rho^{(2)}(\phi_y) : \phi_x = \phi^{\Rightarrow x}, \phi_y = \phi^{\Rightarrow y}, \phi \le \phi_x \otimes \phi_y\}.$$

By assumption, x is a support of $\rho^{(1)}$ that is

$$\rho_1(\phi_x) = \rho_1^{\Rightarrow x}(\phi_x) = \vee\{\rho_1(\psi_x) : \phi_x \le \psi_x = \psi_x^{\Rightarrow x}\}$$

and similarly for $\rho^{(2)}$ which has y as a support. This implies then,

$$(\rho^{(1)} \otimes \rho^{(2)})_{x \vee y}(\phi') = \rho^{(1)} \otimes \rho^{(2)}(\phi)$$
$$= \vee\{\rho^{(1)}(\phi_x) \wedge \rho^{(2)}(\phi_y) : \phi \le \phi_x \otimes \phi_y\}$$
$$= \vee\{\left(\vee\{\rho^{(1)}(\psi_x) : \phi_x \le \psi_x = \psi_x^{\Rightarrow x}\}\right)$$
$$\wedge \left(\vee\{\rho^{(2)}(\psi_y) : \phi_y \le \psi_y = \psi_y^{\Rightarrow y}\}\right) : \phi \le \phi_x \otimes \phi_y\}$$
$$= \vee\{\rho^{(1)}(\psi_x) \wedge \rho^{(2)}(\psi_y) : \phi \le \psi_x \otimes \psi_y, \psi_x = \psi_x^{\Rightarrow x}, \psi_y = \psi_y^{\Rightarrow y}\}$$
$$= \rho_x^{(1)} \otimes \rho_y^{(2)}(\phi').$$

(2) We have, for $\phi' = (\phi, x)$, that $\nu_x(\phi') = \nu(\phi) = \perp$, if $\phi \ne e$. And we have $\nu_x(e, x) = \nu(e) = \top$.

(3) Let $\rho = \rho^{\Rightarrow x}, y \le x$ and $\phi' = (\phi, y)$. We have to prove that $\rho_x^{\downarrow y}(\phi) = \rho^{\Rightarrow y}(\phi)$. Then

$$(\rho^{\Rightarrow y})(\phi) = \rho(\phi)$$

since $\phi = \phi^{\Rightarrow y}$. On the other hand, for $\psi' = (\psi, x)$, we have $\rho_x(\psi') = \rho(\psi)$. We obtain therefore

$$\rho_x^{\downarrow y}(\phi') = \rho_x(\phi'^{\uparrow x}) = \rho(\phi) = \rho^{\Rightarrow y}(\phi) \tag{7.83}$$

since $y \le x$ implies that $\phi = \phi^{\Rightarrow x}$. Thus we have in fact $(\rho^{\Rightarrow y})_y = \rho_x^{\downarrow y}$. $\square$

This theorem shows that we have two essentially equivalent ways to construct labeled information algebras of allocations from algebras of allocations.

We note that, as with algebras of allocations, we can consider *normalized* algebras of labeled allocations, which correspond to the labeled versions of normalized algebras of allocations.

7.4   Independent Sources

7.4.1   Independent Random Variables

In many cases sources of information, represented by random variables are independent. What does this mean? Let (Φ, D) be a domain-free information algebra and $(\mathcal{B}, \mu)$ a probability algebra.

Definition 7.23 Independent Experiments and Simple Random Variables. *A finite set of experiments* $\mathbf{b}_1, \ldots, \mathbf{b}_n$ *is called* independent, *if* $b_1 \wedge \cdots \wedge b_n \neq \bot$ *and*

$$\mu(b_1 \wedge \cdots \wedge b_n) \;=\; \mu(b_1) \cdots \mu(b_n) \tag{7.84}$$

for all $b_1 \in \mathbf{b}_1, \ldots, b_n \in \mathbf{b}_n$
   A finite set of simple random variables $h_1, \ldots, h_n$, *defined on* $\mathbf{b}_1, \ldots, \mathbf{b}_n$, *and with values in an information algebra* (Φ, D), *is called* independent, *if the experiments* $\mathbf{b}_1, \ldots, \mathbf{b}_n$ *are independent.*

If $h_1, \ldots, h_n$ are independent simple random variables, defined on experiments $\mathbf{b}_1, \ldots, \mathbf{b}_n$, then $h_1 \otimes \cdots \otimes h_n$ is a simple random variable defined on the experiment $\mathbf{b}_1 \wedge \cdots \wedge \mathbf{b}_n$. When considering independent simple random variables h, then it is convenient to define the associated *basic probability assignment (bpa)*. To define the bpa, form the canonical random variable $h^{\to}$ associated to h. This random variable is defined on some experiment $\mathbf{b} = \{b_1, \ldots, b_m\}$, and we have $h^{\to}(b_i) \neq h^{\to}(b_j)$ for any pair $b_i \neq b_j$ of elements of $\mathbf{b}$. The elements $h^{\to}(b_1), \ldots, h^{\to}(b_m)$ are called *focal elements* and the probabilities $\mu(b_1), \ldots, \mu(b_m)$ the associated *basic probabilities*. These two elements form the basic probability assignment associated with the simple random variable h. In fact, we may say that the probability $\mu(b_i)$ is assigned to the focal element $h^{\to}(b_i)$.
   Note that the support function sp_h of a simple random variable is fully determined by the basic probability assignment associated with h. In fact, for a $\phi \in \Phi$, we have

$$sp_h(\phi) \;=\; \sum_{h^{\to}(b_i) \geq \phi} \mu(b_i). \tag{7.85}$$

We may thus consider the basic probability assignment as a possible, and often very convenient, representation of the belief function sp_h. This has been demonstrated in (Shafer, 1976). Indeed, the support function of the combination of a number of independent simple random variables can be expressed very simply in terms of basic probability assignments.
   Let h_1 and h_2 be two independent simple random variables with focal element $\phi_{1,1}, \ldots, \phi_{1,m}$ and $\phi_{2,1}, \ldots, \phi_{2,n}$ respectively, whereas the corresponding basic probabilities are $\mu_{1,1}, \ldots, \mu_{1,m}$ and $\mu_{2,1}, \ldots, \mu_{2,n}$. Then it can easily be verified, that the combined random variable has focal elements $\phi_{1,i} \otimes \phi_{2,j}$.

Note that some of these elements may be identical. The corresponding basic
probabilities of these focal elements are

$$\mu(\phi) \quad = \quad \sum_{\phi_{1,i} \otimes \phi_{2,j} = \phi} \mu_{1,i} \mu_{2,j}. \tag{7.86}$$

If the normalized combination is considered, then all elements $\phi_{1,i} \otimes \phi_{2,j} = z$
are eliminated and the basic probabilities are renormalized,

$$\mu(\phi) \quad = \quad \frac{\sum_{\phi = \phi_{1,i} \otimes \phi_{2,j}} \mu_{1,i} \mu_{2,j}}{\sum_{\phi_{1,i} \otimes \phi_{2,j} = z} \mu_{1,i} \mu_{2,j}}. \tag{7.87}$$

This is called *Dempster's rule* (Shafer, 1976). So, in the case of independent simple random variables, the corresponding support functions are best
combined in terms of their bpa.

In a similar way, the bpa of a the focus $h^{\Rightarrow x}$ of a simple random variable
can also be simply expressed by the bpa of h. Indeed, it is easy to verify that
$h^{\Rightarrow x}$ has the following basic probabilities:

$$\mu(\phi) \quad = \quad \sum_{\phi_i^{\Rightarrow x} = \phi} \mu_i, $$

if h has the focal elements $\phi_1, \ldots, \phi_n$ with the probabilities $\mu_1, \ldots, \mu_n$.

The definition of independence can be extended to generalized random
variables.

Definition 7.24 Independence of Random Variables. *A finite set of
random variables* $h_1, \ldots, h_n$ *is called* independent, *if every set of simple random variables* $h'_1 \in h_1, \ldots, h'_n \in h_n$ *is independent.*

Since support functions of independent simple random variables can be
combined in a simple way, we may ask, whether this is also the case for independent generalized random variables. This is true in the following way:
each generalized random variable can be approximated by simple ones. And
its support function can similarly be approximated by the support functions
of the approximating simple random variables, see (7.28). The following theorem tells us, that the support function of the combined random variables can
be approximated in the same way by the support function of the combined
approximating simple random variables. This theorem holds in general, not
only for independent random variables. But it becomes especially interesting
in the latter case, since for independent simple random variables we know how
to combine their support functions by Dempster's rule.

Theorem 7.25 *Let* h_1, h_2 *and* h *be random variables in a domain-free information algebra* (Φ, D). *Then, for any* $\phi \in \Phi$, *and* $x \in D$,

$$sp_{h_1 \otimes h_2}(\phi) \quad = \quad \sup\{sp_{h'_1 \otimes h'_2} : h'_1 \in h_1, h'_2 \in h_2\}, \tag{7.88}$$

$$sp_{h^{\Rightarrow x}}(\phi) \quad = \quad \sup\{sp_{h' \Rightarrow x}(\phi) : h' \in h\}. \tag{7.89}$$

Proof. (1) Since the mapping $h \mapsto \rho_h$ is continuous we obtain

$$\rho_{h_1 \otimes h_2}(\phi) = \rho_{\vee\{h_1' \otimes h_2' : h_1' \in h_1, h_2' \in h_2\}}(\phi)$$
$$= \vee\{\rho_{h_1' \otimes h_2'}(\phi) : h_1' \in h_1, h_2' \in h_2\}.$$

The set $\{\rho_{h_1' \otimes h_2'}(\phi) : h_1' \in h_1, h_2' \in h_2\}$ is upwards directed. Therefore, we conclude that (Lemma 7.2)

$$sp_{h_1 \otimes h_2}(\phi) = \mu(\rho_{h_1 \otimes h_2}(\phi))$$
$$= \mu(\vee\{\rho_{h_1' \otimes h_2'}(\phi) : h_1' \in h_1, h_2' \in h_2\})$$
$$= \sup\{\mu(\rho_{h_1' \otimes h_2'}(\phi)) : h_1' \in h_1, h_2' \in h_2\}$$
$$= \sup\{sp_{h_1' \otimes h_2'}(\phi) : h_1' \in h_1, h_2' \in h_2\}.$$

(2) is proved similarly. First, we have by continuity of the mapping $h \mapsto \rho_h$

$$\rho_{h \Rightarrow x}(\phi) = \rho_{\vee\{h' \Rightarrow x : h' \in h\}}(\phi) = \vee\{\rho_{h' \Rightarrow x}(\phi) : h' \in h\}.$$

The set on the right is upwards directed. Hence we obtain (Lemma 7.2)

$$sp_{h \Rightarrow x}(\phi) = \mu(\rho_{h \Rightarrow x}(\phi))$$
$$= \mu(\vee\{\rho_{h' \Rightarrow x}(\phi) : h' \in h\})$$
$$= \sup\{\mu(\rho_{h' \Rightarrow x}(\phi)) : h' \in h\}$$
$$= \sup\{sp_{h' \Rightarrow x}(\phi) : h' \in h\}.$$

$\square$

7.4.2 Algebra of Bpas

In practice we often have random variables defined on different probability algebras with values in a fixed information algebra. Each random variable corresponds to a certain source of information. We want then to assume that these sources are "independent", and combine the random variables under this assumption. Or, alternatively, we may as well work with the support functions they induce.

If h is some simple random variable defined on some experiment **b**, then we have seen that we may switch to the associated bpa. Therefore, we take these elements in this section as the primitives. Formally, a bpa on a domain-free information algebra (Φ, D) can be described by a mapping $m : \Phi \rightarrow [0, 1]$, which has the following properties:

1. The set $F_m = \{\phi : m(\phi) > 0\}$ is *finite*.

2. $\sum_{\phi \in F_m} m(\phi) = 1$.

The elements of F_m are called *focal elements* of the bpa m. If m is a bpa, then we shall write sums like

$$\sum_{\phi \in S} m(\phi),$$

where S is a subset of Φ. This sum is always understood to be the finite sum over the elements $\phi \in S \cap F_m$. We denote the set of all bpas in Φ by $P(\Phi)$.

As we have seen before (Section 7.4.1), any simple random variable with values in Φ determines a bpa. Inversely, for any bpa on Φ there is a simple random variable which induces the bpa. In fact, for a bpa m, take $\mathcal{B}$ to be the power set of F_m. This is a finite, hence complete Boolean algebra. Define a probability measure μ on this algebra by putting $\mu(\psi) = m(\psi)$ for $\psi \in F_m$. This defines a probability algebra $(\mathcal{B}, \mu)$. The set F_m determines then a simple experiment in this probability algebra. And the mapping $h(\psi) = \psi$ for $\psi \in F_m$ is a simple random variable, which induces the bpa m on Φ. Therefore,

$$sp_m(\phi) \;\;=\;\; \sum_{\psi \geq \phi} m(\psi)$$

defines a support function, the support function associated with the bpa m.

We define an operation of *combination* between bpas, motivated by (7.86) as follows: if m_1 and m_2 are two bpas on a domain-free information algebra (Φ, D), then the combined bpa, denoted by $m_1 \oplus m_2$ is defined through

$$m_1 \oplus m_2(\phi) \;\;=\;\; \sum_{\psi_1 \otimes \psi_2 = \phi} m_1(\psi_1) m_2(\psi_2).$$

Clearly this defines a bpa. If m_1 and m_2 are the bpas of two *independent* random variables h_1 and h_2, then $m_1 \oplus m_2$ is the bpa of the simple random variable $h_1 \oplus h_2$.

Furthermore, we introduce also an operation of focusing of bpas, which is also inspired by focusing of random variables. If m is a bpa on (Φ, D), and x a domain in D, then the bpa m, focused on x, is denoted by $m^{\Rightarrow x}$. It is defined through

$$m^{\Rightarrow x}(\phi) \;\;=\;\; \sum_{\psi^{\Rightarrow x} = \phi} m(\psi). \tag{7.90}$$

Again, this is a bpa. In fact, if m is the bpa of a simple random variables h, then $m^{\Rightarrow x}$ is the bpa of the variable $h^{\Rightarrow x}$.

Thus, we have defined an operation of combination and one of focusing in the set $P(\Phi)$ of bpas. Clearly, combination is no more *idempotent*. So this is not an information algebra. But it is a valuation algebra. This algebra is similar to the valuation algebra of set potentials, see Section 2.3.6. It represents in fact a generalization of set potentials to general information algebras. The m-functions correspond to bpas on the information algebra of finite sets.

Theorem 7.26 *If* (Φ, D) *is a domain-free information algebra, then* $(P(\Phi), D)$ *with combination and focusing as defined in (7.90) and (7.90) is a domain-free valuation algebra.*

Proof. (1) *Semigroup.* Clearly, the operation of combination is commutative and associative. The bpa m_e with $F_{m_e} = \{e\}$ and $m_e(e) = 1$ is the neutral element of combination.

(2) *Transitivity.* Let m be a bpa, and $x, y \in D$, then by definition (7.90)

$$
\begin{aligned}
(m^{\Rightarrow x})^{\Rightarrow y}(\phi) &= \sum_{\psi^{\Rightarrow y} = \phi} m^{\Rightarrow x}(\psi) = \sum_{\psi^{\Rightarrow y} = \phi} \sum_{\eta^{\Rightarrow x} = \psi} m(\eta) \\
&= \sum_{(\eta^{\Rightarrow x})^{\Rightarrow y} = \phi} m(\eta) = \sum_{\eta^{\Rightarrow x \wedge y} = \phi} m(\eta) \\
&= m^{\Rightarrow x \wedge y}(\phi).
\end{aligned}
$$

So, we have indeed $(m^{\Rightarrow x})^{\Rightarrow y} = m^{\Rightarrow x \wedge y}$.

(3) *Combination.* Suppose m_1 and m_2 two bpas, and $x \in D$. Let $m' = (m_1^{\Rightarrow x} \oplus m_2)^{\Rightarrow x}$ and $m'' = m_1^{\Rightarrow x} \oplus m_2^{\Rightarrow x}$. We compute

$$
\begin{aligned}
m'(\phi) &= \sum_{\epsilon^{\Rightarrow x} = \phi} (m_1^{\Rightarrow x} \oplus m_2)(\epsilon) \\
&= \sum_{\epsilon^{\Rightarrow x} = \phi} \sum_{\epsilon_1 \otimes \psi_2 = \epsilon} m_1^{\Rightarrow x}(\epsilon_1) m_2(\psi_2) \\
&= \sum_{\epsilon^{\Rightarrow x} = \phi} \sum_{\epsilon_1 \otimes \psi_2 = \epsilon} \Big(\sum_{\psi_1^{\Rightarrow x} = \epsilon_1} m_1(\psi_1) \Big) m_2(\psi_2) \\
&= \sum_{(\psi_1^{\Rightarrow x} \otimes \psi_2)^{\Rightarrow x} = \phi} m_1(\psi_1) m_2(\psi_2)
\end{aligned}
$$

In the same way, we obtain that

$$
m''(\phi) = \sum_{\psi_1^{\Rightarrow x} \otimes \psi_2^{\Rightarrow x} = \phi} m_1(\psi_1) m_2(\psi_2).
$$

Since $(\psi_1^{\Rightarrow x} \otimes \psi_2)^{\Rightarrow x} = \psi_1^{\Rightarrow x} \otimes \psi_2^{\Rightarrow x}$, we conclude that $F_{m'} = F_{m''}$ and therefore $m' = m''$.

(4) *Neutrality.* It is evident that $m_e^{\Rightarrow x} = m_e$, for all domains $x \in D$.

(5) *Support.* Suppose m a bpa with focal elements F_m. Each focal element $\psi \in F_m$ has a support, and the supremum of the supports of all these supports is a support for every focal element (see Lemma 3.6 (6)). Let this domain be x. Then we have

$$
m^{\Rightarrow x}(\phi) = \sum_{\psi^{\Rightarrow x} = \phi} m(\psi) = \sum_{\psi = \phi} m(\psi) = m(\phi).
$$

So we see that $m^{\Rightarrow x} = m$ and x is a support for m. Note that we have proved here a stronger version of the support axiom than required in Section 3.2, for the case that D has no top element. $\square$

So this illustrates the possibility of the structure of a valuation algebra of support functions on an information algebra. However, support functions induced by a bpa represent only a subset of all support functions. It remains an open question whether and how general support functions on an information algebra can be given the structure of a valuation algebra and how this relates to the underlying random variables (see (Shafer, 1979; Kohlas, 1993)).

We remark further, that random variables with values in a Boolean information algebra are of particular interest. The duality of these algebras induces a corresponding duality of allocations of probability. This is well-known. We refer to (Shafer, 1976; Shafer, 1979; Kohlas & Monney, 1995).

References

Anrig, B., Haenni, R., Kohlas J., & Lehmann, N., 1997. Assumption-Based Modeling Using ABEL. *In:* Gabbay, D., Kruse, R., Nonnengart, A. & Ohlbach, H.J. (eds), *First Int. Joint Conf. on Qualitative and Quantitative Practical Reasoning; ECSQUARU-FAPR'97* , Lecture Notes Artif. Intell. Springer.

Arnborg, S., Corneil, D., & Proskurowski, A., 1987. Complexity of Finding Embeddings in a k-Tree. *SIAM J. of Algebraic and Discrete Methods* , **38**, 277–284.

Barwise J., & Seligman, J., 1997. *The Logic of Distributed Systems.* Cambridge University Press, Cambridge U.K.

Bergstra, J.A., Heering, J., & Klint, P., 1990. Module Algebra. *J. of the Association for Computing Machinery* , **37**, (2), 335–372.

Besnard, P., & Kohlas, J., 1995. Evidence Theory Based on General Consequence Relations. *Int. J. of Found. of Comp. Science*, **6**, 119–135.

Clifford, A.H., & Preston, G.B., 1967. *The Algebraic Theory of Semigroups.* Amer. Math. Soc.; Providence, Rhode Island.

Cowell, R.G., Dawid, A.P., Lauritzen, S.L., & Spiegelhalter, D.J., 1999. *Probabilistic Networks and Expert Systems.* Springer, New York.

Croisot, R. 1953. Demi-groupes inversifs et demi-groupes réunions de demi-groupes simples. *Ann. Sci. Ecole norm. Sup.*, **79** (3), 361–379.

Davey, B.A., & Priestley, H.A., 1990. *Introduction to Lattices and Order.* Cambridge University Text, Cambridge.

Dawid, A.P., 1979. Conditional Independence in Statistical Theory. *Journal of the Royal Statistical Soc.*, **B 41**, 1–31.

Dawid, A.P., 1992. Applications of a General Propagation Algorithm for Probabilistic Expert Systems. *Statistics and Computing* , **2**, 25–36.

Dawid, A.P., 1998. Conditional Independence. *Encyclopedia of Statistical Sciences: Update*, **2**, 146–155.

Dechter,R., 1999. Bucket Elimination: A Unifying Framework for Reasoning. *Artificial Intelligence*, **113** , 41–85.

De Kleer, J., 1986. An Assumption-Based TMS. *Artificial Intelligence* , **28**, 127–162.

Dempster, A.P., 1967. Upper and Lower Probabilities Induced by a Multivalued Mapping. *Annals of Math. Stat.*, **113**, 41–85.

Gabbay, D.M., & Smets, Ph. (eds), 2000. *Handbook of Defeasible Reasoning and Uncertainty Management Systems, Vol. 5.* Kluwer, Dordrecht.

Haenni, R., Kohlas J., & Lehmann, N., 2000. Probabilistic Argumentation Systems. *In:* Kohlas, J., & Moral, S. (eds), *Handbook of Defeasible Reasoning and Uncertainty Management Systems*, vol. 5: Algorithms for Uncertainty and Defeasible Reasoning. Kluwer, Dordrecht.

Halmos, P., 1950. *Measure Theory.* Van Nostrand-Reinhold, London.

Halmos, P., 1963. *Lectures on Boolean Algebras.* Van Nostrand Company, Princeton, NJ.

Henkin, L., Monk, J.D., & Tarski, A., 1971. *Cylindric Algebras.* Studies in Logic and the Foundations of Mathematics, North Holland.

Hewitt, E., & Zuckerman, H.S., 1956. The l_1 algebra of a commutative semigroup. *Amer. Math. Soc.*, **83**, 70–97.

Jensen F.V., Lauritzen, S.L., & Olesen K.G., 1990. Bayesian Updating in Causal Probabilistic Networks by Local Computation. *Computational Statistics Quarterly*, **4**, 269–282.

Kappos, D.A., 1969. *Probability Algebras and Stochastic Spaces.* Academic Press, New York and London.

Kohlas, J., 1993. Support- and Plausibility Functions Induced by Filter-Valued Maps. *Int. J. of General Systems*, **21**, 343–363.

Kohlas, J., 1995. Mathematical Foundations of Evidence Theory. *In:* Coletti., J., Dubois, D. & Scozzafova, R. (eds), *Mathematical Models for Handling Partial Knowledge in Artificial Intelligence.* Plenum Press, New York, 31–64.

Kohlas, J., 1997. Allocation of Arguments and Evidence Theory. *Theor. Comp. Science*, **171**, 221–246.

Kohlas, J., Haenni, R., & Moral, S., 1999. Propositional Information Systems. *J. Logic Computat.*, **9**, 651–681.

Kohlas, J., & Monney, P.-A., 1995. *A Mathematical Theory of Hints*. Springer, Lecture Notes in Economics and Mathematical Systems **425**, Berlin.

Kohlas J., & Shenoy, P., 2000. Computation in Valuation Algebras. *In:* Kohlas, J., & Moral, S. (eds), *Handbook of Defeasible Reasoning and Uncertainty Management Systems*, vol. 5: Algorithms for Uncertainty and Defeasible Reasoning, 5–39. Kluwer, Dordrecht.

Kohlas, J., & Staerk,R., 1996. *Information Algebras and Information Systems*. Tech. Rept. 96-14, University of Fribourg, Institute for Informatics.

Laskey, K.B., & Lehner, P.E., 1989. Assumptions, Beliefs and Probabilities. *Artificial Intelligence*, **41**, 65–77.

Lauritzen, S.L., 1996. *Graphical Models*. Clarendon Press, Oxford.

Lauritzen, S.L., & Jensen F.V., 1997. Local Computations With Valuations From a Commutative Semigroup. *Ann. of Math. and Art. Int.*, **21**, 51–69.

Lauritzen, S.L., & Spiegelhalter, D.J., 1988. Local Computations With Probabilities on Graphical Structures and their Application to Expert Systems. *J. of Royal Stat. Soc.*, **50**(2), 157–224.

Liu, L., 1996. A Theory of Gaussian Belief Functions. *Int. J. of Approximate Reasoning*, **14**, 95–126.

Maier, D., 1983. *The Theory of Relational Databases*. Computer Science Press, Rockville, Md.

Marquis, P., 2000. Consequence Finding Algorithms. *In:* Kohlas, J., & Moral, S. (eds), *Handbook of Defeasible Reasoning and Uncertainty Management Systems*, vol. 5: Algorithms for Uncertainty and Defeasible Reasoning. Kluwer, Dordrecht.

Mengin, J., & Wilson, N., 1999. Logical Deduction Using the Local Computation Framework. *In:* Hunter, A. & Parson, S. (eds), *European Conf. on Symbolic and Quantitative Approaches to Reasoning and Uncertainty; ECSQUARU'99* , Lecture Notes Artif. Intell. Springer.

Monney, P.-A., 2000. *Assumption-Based Reasoning with Functional Models*. Habilitation Thesis, University of Fribourg (Switzerland), Seminar of Statistics.

Neveu, J., 1964. *Bases mathématiques du calcul des probabilités*. Masson, Paris.

Pearl, J., 1988. *Probabilistic Reasoning in Intelligent Systems*. Morgan Kaufmann Publ. Inc.

Renardel de Lavalette G.R., 1992. Logical Semantics of Modularization. *Pages 306–315 of:* Börger, E., Jäger, G., Kleine Büning, H. & Richter, M.M. (eds), *Computer Science Logic, Selected Papers from CSL '91.* Springer Lecture Notes in Computer Science 626.

Shafer, G., 1973. *Allocation of Probability: A Theory of Partial Belief.* Dissertation, Princeton University.

Shafer, G., 1976. *A Mathematical Theory of Evidence.* Princeton University Press.

Shafer, G., 1979. Allocations of Probability. *The Annals of Probability,* **7**, 827–839.

Shafer, G., 1991. *An Axiomatic Study of Computation in Hypertrees.* Working Paper 232. School of Business, University of Kansas.

Shafer, G., 1996. *Probabilistic Expert Systems.* Society for Industrial and Applied Mathematics, Philadelphia.

Shenoy, P.P., 1989. A Valuation-Based Language for Expert Systems. *Int. J. of Approximate Reasoning,* **3**, 383–411.

Shenoy, P.P., 1992. Valuation-Based Systems: A Framework for Managing Uncertainty in Expert Systems. *Pages 83–104 of:* Zadeh, L.A., & Kacprzyk, (eds), *Fuzzy Logic for the Management of Uncertainty.* John Wiley & Sons.

Shenoy, P.P., 1994. Using Dempster-Shafer's Belief Function Theory in Expert Systems. *Pages 395–414 of:* Yager, R.R., Kacprzyk, J., & Fedrizzi, M. (eds), *Advances The Dempster-Shafer Theory of Evidence.* John Wiley & Sons.

Shenoy, P.P., 1997 a). Conditional Independence in Valuation-Based Systems. *Int. J. of Approximate Reasoning,* **10**, 203–234.

Shenoy, P.P., 1997 b). Binary Join Trees for Computing Marginals in the Shenoy Shafer Architecture. *Int. J. of Approximate Reasoning,* **17**, 239–263.

Shenoy, P.P., & Shafer, G., 1990. Axioms for Probability and Belief Function Propagation. *Pages 169–198 of:* Shachter, R.D., Levitt, T.S., Lemmer, J.F. & Kanal, L.N. (eds), *Uncertainty in Artificial Intelligence, 4.* North Holland.

Smets, P., 1998. The Transferable Belief Model for Quantified Belief Representation. *In:* Gabbay, D.M., & Smets, Ph. (eds), 2000. *Handbook of Defeasible Reasoning and Uncertainty Management Systems, Vol. 1,* 267–301, Kluwer, Dordrecht.

Spohn, W., 1988. Ordinal Conditional Functions: A Dynamic Theory of Epistemic States. *Pages 105–134 of:* Harper, W.L. & Skyrms, B. (eds), *Causation in Decision, Belief Change, and Statistics, II.* Kluwer Academic Publishers.

Studeny, M., 1993. Formal Properties of Conditional Independence in Different Calculi of AI. *Pages 341–348 of:* Clarke, M., & Kruse, R. & Moral, S. (eds), *Symbolic and Quantitative Approaches to Reasoning and Uncertainty.* Springer, Berlin.

Studeny, M., 1995. Conditional Independence and Natural Conditional Functions. *Int. J. of Approximate Reasoning* , **12**, 43–68.

Tamura, T., & Kimura, N., 1954. On Decompositions of a Commutative Semigroup. *Kodai Math. Sem. Rep.*, 109–112.

Wojcicki, R., 1988. *Theory of Logical Calculi. Basic Theory of Consequence Operations.* Kluwer Academic Publishers, Dordrecht, Boston, London.

Zadeh, L.A., 1978. Fuzzy Sets as a Basis for a Theory of Possibility. *Fuzzy Sets and Systems*, **1**, 3–28.

Zadeh, L.A., 1979. A Theory of Approximate Reasoning. *Pages 149–194 of:* Ayes, J.E., & Michie, D. & Mikulich, L.I. (eds), *Machine Intelligence.* vol **9**, Ellis Horwood, Chichester, UK.

Index

additive potential
    conditional independence, 147
    maximization, 147
    regular valuation algebra, 147
algebraic lattice, 196
allocation
    of belief, 215
    of probability, 209, 216, 227
      deterministic, 230
      labeled, 240
      normalized, 227, 233
      vacuous, 217
    of support, 221
      normalized, 221
architecture for local computation
    HUGIN, 96, 114, **114**
    Lauritzen-Spiegelhalter, 96, **112**
    Shenoy-Shafer, 96, **109**
    with division, 111
assumption, 219
assumption-based
    information, 219
      linear equations, 219
      propositional logic, 219
      truth maintenance, 219
atom, 171
atomic
    algebra, 173
    closed algebra, 173
    composed algebra, 173

automorphism, 57
axioms
    of domain-free information algebra, 162
    of domain-free valuation algebra, 49
    of domain-free valuation algebra with null valuations, 66
    of domain-free valuation algebra with variable elimination, 54
    of labeled information algebra, 160
    of labeled valuation algebra, 11
    of labeled valuation algebra with null elements, 65
    of valuation algebra with partial marginalization, 37
    of valuation algebra with variable elimination, 13

b-function, 29
basic probability assignment, 25, 245, 247
    valuation algebra, 248
belief function, 24, 29, 217
Boolean
    algebra, 50
      complete, 210
      with variable elimination, 50

Other titles in the DMTCS series:

Combinatorics, Complexity, Logic:
Proceedings of DMTCS '96
D. S. Bridges, C. S. Calude, J. Gibbons,
S. Reeves, I. Witten (Eds)
981-3083-14-X

Formal Methods Pacific '97: Proceedings
of FMP '97
L. Groves and S. Reeves (Eds)
981-3083-31-X

The Limits of Mathematics: A Course on
Information Theory and the Limits of
Formal Reasoning
Gregory J. Chaitin
981-3083-59-X

Unconventional Models of Computation
C. S. Calude, J. Casti and M. J. Dinneen
(Eds)
981-3083-69-7

Quantum Logic
K. Svozil
981-4021-07-5

International Refinement Workshop and
Formal Methods Pacific '98
J. Grundy, M. Schwenke and T. Vickers
(Eds)
981-4021-16-4

Computing with Biomolecules: Theory
and Experiments
Gheorghe Paun (Ed)
981-4021-05-9

People and Ideas in Theoretical
Computer Science
C. S. Calude (Ed)
981-4021-13-X

Combinatorics, Computation and Logic:
Proceedings of DMTCS'99 and CATS'99
C. S. Calude and M. J. Dinneen (Eds)
981-4021-56-3

Polynomials: An Algorithmic Approach
M. Mignotte and D. Stefanescu
981-4021-51-2

The Unknowable
Gregory J. Chaitin
981-4021-72-5

Sequences and Their Applications:
Proceedings of SETA '98
C. Ding, T. Helleseth and H. Niederreiter
(Eds)
1-85233-196-8

Finite versus Infinite: Contributions to
an Eternal Dilemma
Cristian S. Calude and Gheorghe Paun
(Eds)
1-85233-251-4

Network Algebra
Gheorge Stefanescu
1-85233-195-X

Exploring Randomness
Gregory J. Chaitin
1-85233-417-7

Unconventional Models of Computation
(UMC2K)
I. Antoniou, C.S. Calude and M.J. Dineen
(Eds)
1-85233-415-0

Lattice Functions and Equations
S. Rudeanu
1-85233-266-2

Sequences and their Applications –
SETA '01
T. Helleseth, P.V. Kumar and K. Yang (Eds)
1-85233-529-7

Turing's Connectionism
C. Teuscher
1-85233-475-4